TROPHIC
ECOLOGY

TROPHIC ECOLOGY

James E. Garvey
Southern Illinois University
Carbondale, Illinois, USA

Matt R. Whiles
Southern Illinois University
Carbondale, Illinois, USA

CRC Press
Taylor & Francis Group
Boca Raton London New York

CRC Press is an imprint of the
Taylor & Francis Group, an **informa** business

CRC Press
Taylor & Francis Group
6000 Broken Sound Parkway NW, Suite 300
Boca Raton, FL 33487-2742

First issued in paperback 2018

ISBN-13: 978-1-4987-5846-8 (hbk)
ISBN-13: 978-0-367-13865-3 (pbk)

Library of Congress Cataloging-in-Publication Data

Names: Garvey, James E., author. | Whiles, Matt R., author.
Title: Trophic ecology / James E. Garvey and Matt R. Whiles.
Description: Boca Raton : Taylor & Francis, 2016. | Includes bibliographical references and index.
Identifiers: LCCN 2016011136 | ISBN 9781498758468
Subjects: LCSH: Food chains (Ecology) | Animals--Food. | Bioenergetics.
Classification: LCC QH541.15.F66 G37 2016 | DDC 577/.16--dc23
LC record available at https://lccn.loc.gov/2016011136

Visit the Taylor & Francis Web site at
http://www.taylorandfrancis.com

and the CRC Press Web site at
http://www.crcpress.com

Contents

SECTION I   Concepts and Patterns

SECTION II Mechanisms at the Organismal Scale

SECTION III Diet Data, Modeling, and Energetics Approaches

SECTION IV   Community and Ecosystem Concepts

SECTION V   Quantifying Material Flux and Synthesis

Preface

This book is the result of a collaboration between Jim and Matt, who come from different backgrounds and ecological paradigms. Jim Garvey grew up academically exploring organismal approaches to food web dynamics and effects of predation on prey populations, and Matt Whiles comes from a group of ecologists who study ecosystem-level processes, with a strong focus on patterns of production among ecosystems. What we discovered over the years is that the Rosetta stone for communicating between organismal and ecosystem ecologists is through mutual interests in natural history and trophic ecology. The edges of organismal, community, and ecosystem approaches in ecology are blurring, especially with the growing availability of new techniques for assessing trophic interactions and their implications for ecosystems. Given this realization, we think it is time to provide a formal text for both newcomers to the discipline as well as seasoned professionals looking for new ideas and refreshers on old topics.

We thank the many talented individuals who have reviewed many iterations of this book. Dr. Jim Lovvorn, the Aquatic Ecology Lab at The Ohio State University, and Dr. Alan Covich, Steve Borrego, Stacy Gucciardo, James Hankins, Jennifer Ireland, Jordan Keifer, Neil Rude, Jessica Sandstrom, Tyler Schartel, Jodi Vandermyde, Ashley Astroth, Krystin Calhoun, Kurt Campbell, Bradley Cox, Samantha Emberton, Brian Jones, Jeremy Lackowski, David Richards, Haley Rinella, Justin Rosenquist, Jatin Soneji, Abby Thomas, Kathryn Weatherwax, Sydney Youngs, Allison Asher, John Bowzer, Jeffrey Gersch, Jeffrey Hillis, Kenson Kanczuzewski, Elliott Kittel, Nicolle Macvey, Brittany Szynkowski, Matthew Young, and Andrew Young are among the many professionals and students who reviewed portions of this book and provided helpful feedback.

This book benefitted especially from conversations with Dr. Alison Coulter and Dr. David Coulter. Long hours were spent discussing exactly what trophic ecology should be. A portion of this book reflects those views. Dr. David Glover was instrumental in providing both conceptual and quantitative feedback about the topics covered in this book. Many hours were spent chatting with Dr. Bill Hintz about food web interactions, which was influential on this book. Dr. Marybeth Brey also reviewed portions of this book, especially the diet analysis chapter, finding at least some of the errors. For this, we owe her much gratitude. Also, Anthony Porreca served as a guinea pig, using some of these chapters to prepare for his PhD preliminary exam. For this, we apologize.

Several topics in this book were presented at the 2015 National American Fisheries Society Conference in Portland, Oregon as a workshop. Attendance was full with a waitlist, suggesting that there is much interest in trophic ecology and its applications. We would like to thank the participants for asking great questions and providing useful feedback. It was clear from this class that the level of knowledge about applications, especially stable isotopes, fatty acids, and the many statistical tools available, varies highly among practicing ecologists. This book should level the playing field and serve as both a guide for newcomers and a reference for those looking for new ideas or trying to recall old techniques.

We do not intend for this book to be the definitive text for ecology. However, to our knowledge, no such book as this exists, perhaps for good reason or perhaps because it just was not the right time until now. There are many detailed books covering all the topics herein. However, this text serves as a one-stop shop for familiarizing ecologists with a broad range of sophisticated topics in trophic ecology. As with all books, researching, learning, and writing were time-consuming, taking much time away from family and friends. We thank them all for their patience and support.

Authors

Jim Garvey is a professor of ecology and the director of the Center for Fisheries, Aquaculture, and Aquatic Sciences at Southern Illinois University. He received his BA degree in zoology from Miami University, Oxford, Ohio, in 1990, and his MS and PhD degrees in zoology from The Ohio State University Aquatic Ecology Laboratory in 1992 and 1997, respectively. He worked as a postdoc in biology at Queen's University, Kingston, Ontario, doing his research at the Queen's University Biological Station. He has published more than 100 journal articles and book chapters on bioenergetics, trophic interactions, and diet analysis, primarily in aquatic ecosystems with invasive species. Much of this book was envisioned while Jim taught an upper-level undergraduate course in trophic ecology and organized several gut analysis workshops for professional development. He sits on numerous science panels for research and environmental policy, has served as an elected officer and associate editor for the American Fisheries Society, and is interested in developing novel technological and statistical tools for unraveling food web interactions.

Matt Whiles is a professor of zoology, an interim director of the Cooperative Wildlife Research Laboratory, and the director of the Center for Ecology at Southern Illinois University. He earned his BS degree in biology at Kansas State University in 1988, his MS degree in entomology at the University of Georgia in 1991, and his PhD degree in ecology at the University of Georgia in 1995. Matt has published over 100 journal articles and book chapters on freshwater ecosystem ecology, invertebrate ecology, and secondary production, and is a coauthor of the popular textbook, *Freshwater Ecology, Second Edition*. He is an avid natural historian with strong interests in how animals influence ecosystem structure and function, particularly in streams and wetlands. He is also the past president of the Society for Freshwater Science.

1 Introduction and History of Trophic Ecology

1.1 APPROACH

This is not a traditional ecology textbook. It shows that *trophic ecology* is a means to bridge the gap between general ecology texts, including those focusing on community ecology, and the dense, theoretically rich books on food web theory, stoichiometry, and ecosystem processes. We first make a case for why this book is needed for upper-level undergraduate students, graduate students, and practicing ecologists looking for ideas and refreshers on concepts and techniques used in trophic research. The history of trophic ecology is old and varied. We will provide a brief overview of how trophic relationships have been considered for millennia and still are relevant to research today. The organization of the book will be described, giving a road map for how the information in the book should be used. Each chapter is intended to build on the next, reinforcing complementary concepts and approaches in trophic ecology.

1.2 NEED FOR TROPHIC ECOLOGY

Ecology is a broad discipline, which some scientists argue is devoid of specific laws and theories (Peters 1980), although many testable generalities do exist (Dodds 2009). Ecology literally means the study of home and is broadly defined as the study of the distribution and abundance of organisms on Earth. Anyone walking through a forest or snorkeling in a coral reef realizes that this is a terribly complex job for scientists. Given humanity's dependence on the function of the living environment, it behooves us to not walk away from this complexity but to try to unravel and understand it (Ulanowicz 2004).

Ecology is divided into several subdisciplines, based on their scale of organization (Figure 1.1). Physiological ecologists study ecological interactions at the level of the individual, incorporating natural selection, energetics, physiology, and behavior into their research. Population ecologists determine how groups of individuals of the same species interact, often using molecular genetics to study population structure. Populations have many characteristics that are unique relative to individuals, such as density, behavioral interactions, death rates, birth rates, and age structure. These characteristics cause populations to behave in complex and often surprising ways (Figure 1.1). Community ecologists are interested in the interactions among species. Consumption, cooperation, and competition are all processes that structure the relative abundance of species within communities, with networks of food webs being an emergent character (Figure 1.1). Ecosystem ecology is the study of the flow of energy and matter within and among communities and their physical environment. The ecosystem concept, being rooted in holism, may discount organismal interactions at

Individual	Population	Community	Ecosystem	Landscape

Genes	Gene frequency	Species richness	Nutrient dynamics	Biogeochemistry
Cells	Density	Predation	Stoichiometry	Erosion
Growth	Growth rate	Competition	Subsidies	Water retention
Fecundity	Birth rate	Mutualism	Trophic levels	Light levels
Energetics	Death rate	Size structure	Trophic production	Soil retention
	Production	Food webs	Temperature	Flow rate
	Size structure	*Energetics*	Sequestration	*Energetics*
	Age structure		*Energetics*	
	Energetics			

FIGURE 1.1 Levels of organization in ecology and some of their emergent properties (i.e., characteristics that make them unique and often are difficult to predict from the sum of their components at lower scales of organization). One commonality among all of them is that they must follow the basic laws of energy conservation or energetics.

finer resolutions (Box 1.1). Ecosystems, like all other organizational levels of ecology, have complex *emergent behaviors* that are difficult to characterize as the sum of their component parts (Figure 1.1).

Ecosystem ecology is not the only level of organization concerned with the transfer of energy and matter in an environmental context. A common thread connects all scales of ecological organization, because interacting cells, organisms, communities,

BOX 1.1

Ecosystem ecology may be the most difficult of the subdisciplines to grasp. In the purest sense, ecosystems have structure and function that may differ from the organisms within them. A prairie ecosystem is defined by its physical location and by the types of plants that dominate the community. The ecosystem is defined not necessarily by the plant species present but the broad interactions among the soil, the plant biomass, and the atmosphere surrounding it. Ecosystem ecologists quantify integrated ecosystem inputs and outputs like oxygen, carbon dioxide, water, and biomass production, which can be converted to energy uptake and potential transfer. A lake ecosystem is composed of a host of organisms including viruses, bacteria, phytoplankton, rooted plants, invertebrates, and fish, all in a deceptively simple bowl of water. Rather than studying how these organisms interact, ecosystem ecologists measure many of the same parameters as the prairie ecologists. In fact, ecosystem ecologists often categorize and compare ecosystems based on emergent properties such as nutrient uptake and primary production. Ecosystem ecologists realize that these inputs and outputs are the results of potentially billions of organismal interactions. Determining these mechanisms is less important than capturing the integrated function of ecosystems.

and ecosystems are like conduits transmitting power, materials, and even information from lower levels of organization to higher ones—all ecologists need to appreciate these consumptive, or trophic, relationships (Peden et al. 1974). Trophic ecology is the study of how energy and nutrients are exchanged at all ecological scales, accounting for mechanistic, evolutionarily relevant processes. We believe that authors like Jones and Lawton (1995) were circling around this idea and Schmitz et al. (2008) provided a beginning framework. This book appears to be the first time trophic ecology has gotten the formal definition it deserves, focusing on conceptual, theoretical, and quantitative approaches for linking energy fixation, consumption, and energy flow in ecosystems.

1.3 DEFINING TROPHIC ECOLOGY

The term *trophic* derives from the Greek word *trophe* or *trophus*, meaning nourishment. The root is found in the English word *trophy*, which means a reward usually taken through hunting or fighting. Thus, trophic relates to taking food forcefully from another organism. Early natural historians considered this an ideal way to describe the typically violent transaction of predation. In ecology today, the term trophic has evolved into an adjective used to describe a group of organisms with a similar set of energetic and nutritional needs and perhaps equivalent modes of energy capture through consumption. Popularized by ecologists like Charles Sutherland Elton (1927) in the development of pyramids of trophic production, adopting the suffix *troph* is a convenient method for categorizing these groups. To illustrate, autotroph means an organism that derives its own (auto) food reward (troph). This may seem elementary, and stopping at this point, trophic ecology should be simply defined as the ecology of deriving nourishment, which is equivalent to the subdiscipline of *foraging ecology* or the assessment of food habits. This is a deceptively simple concept as we will see because trophic relationships are relevant to all levels of ecological organization.

1.4 HISTORY OF TROPHIC ECOLOGY

A search of the popular Web of Science database reveals that the term *trophic ecology* was used in 1712 publications in the past 30 years, with its citation rate increasing geometrically. The term is used most often by ecologists interested in organismal or physiological approaches, with an interest in placing the organism into an ecosystem context (Table 1.1). Why is this term used so often but no formal description exists? In this section, we suggest that trophic ecology, being older than the discipline of ecology itself, has existed as a concept for centuries and is gaining popularity among ecologists as organismal ecologists frame individual, discrete consumptive interactions into the flows of energy and materials described by ecosystem ecologists. Many ecology historians write about how early ecologists discovered ecosystem ecology in the nineteenth century. In our view, most of these early natural historians and ecologists were exploring trophic ecology, not ecosystem ecology, which is a much more recent subdiscipline. In fact, most self-described ecosystem ecologists are trophic ecologists with the intent of informing how ecosystems work by isolating the important nodes and flows within networks (Box 1.1).

TABLE 1.1

Number of Citations from 1961 through 2014 for Articles on Trophic Ecology Using Various Terms

Category	Term	Number of Citations
Biome	Ocean or marine	764
	Freshwater	456
	Terrestrial	306
	Microbial	101
	Soil	71
	Fish	832
Organism	Arthropods	527
	Plants	312
	Mammals	238
	Other inverts	218
	Birds	200
	Reptiles	73
	Amphibians	46
	Physiology	690
Organizational	Population	609
	Community	487
	Ecosystem	399
	Food web	323
	Diet	1081
Term	Nitrogen or phosphorus or carbon	462
	Stable isotope	371
	Niche	292
	Predation	214
	Gut	167
	Trophic level	156
	Competition	133
	Nutrients	105
	Biodiversity	99
	Metabolic theory	98
	Fatty acid	77
	Efficiency	44
	DNA	39
	Bioenergetics	20
	Network	18
	Photosynthesis	16
	Decomposition	15
	Connectivity or connectance	14
	Stoichiometry	8
	Trophic pyramid	0

FIGURE 1.2 Painting by Pieter Brueghel the Elder, created in 1557, clearly shows the relationships among fish predator feeding and diets.

Humans have undoubtedly intuitively understood the trophic connections among autotrophy, herbivory, and carnivory as least since human groups became agrarian and needed to find large tracts of land on which to feed domesticated livestock or captured fishes from the oceans (Figure 1.2). Productive autotrophy was necessary for higher trophic-level ecosystem services such as producing plentiful milk and meat from goats and cattle. The idea that trophic relationships among organisms were interrelated transactions across multiple trophic levels was noted by Arabic philosopher Al Jahiz (or Al Jahith) in his *Book of Animals* written about 1300 years ago in what is now Iraq (Peters 1968). It is not surprising that as the laws of thermodynamics were being worked out by physicists in the nineteenth century to describe how energy (e.g., heat) was transferred among materials, western natural historians began imagining that trophic relationships were governed by similar relationships. Lorenzo Camerano, an Italian biologist, generated the first known diagram of trophic relationships in 1880 (Cohen 1994).

Stephen Forbes, an Illinois biologist, wrote the "Lake as a Microcosm" in 1887. In this paper, he described the lake biota from phytoplankton to fish as being interdependent. Although the term *food web* or *trophic level* did not emerge, he was most assuredly thinking about consumptive effects and the transfer of energy and matter through the lake. There is no evidence in the paper that he was thinking of processes that emerged from the lake such as secondary production, oxygen consumption, and carbon sequestration, which are largely ecosystem constructs.

Charles Elton was a British ecologist who was credited for formalizing trophic constructs such as the trophic pyramid (Chapter 2) and food webs (Chapter 10) in his seminal book *Animal Ecology* first published in 1927. Elton was thinking about organismal interactions and beginning to formalize what we define herein as trophic ecology. Elton's concepts are seemingly simple, but are, in reality, complex. As we will see, trophic interactions are complicated by factors such as competition within trophic levels, cannibalism, indirect feedback loops, induced defenses, omnivory, and trait-mediated interactions. The concept of trophic levels captures broad categorizations of interacting food webs in an environmental template, which provide an ecosystem service such as food production or consumption. Thus, Elton was generating an ecosystem-level construct, although there is no evidence that he realized this at the time (Box 1.1).

In 1942, Raymond Lindeman's paper "The Trophic-Dynamic Aspect of Ecology" was published posthumously. It is considered by many as the birth of the concept of ecosystem ecology, because he included energetic transactions and the concept of flux into the ecological literature. The lore is that his ideas were considered too radical by reviewers, and his adviser, G. Evelyn Hutchinson, a well-regarded limnologist, had to fight to get the paper accepted for publication in *Ecology*. In our view, this paper harkened the birth of formal trophic ecology and hinted at ecosystem ecology, which was to be developed thereafter when ecosystems were defined and described a couple of decades later.

In the 1960s, ecology seemed to split into two approaches, *holistic* and *reductionistic*, which led to confusion about trophic ecology's role in all subdisciplines that persist to this day. A similar argument over ecological succession of communities occurred between the reductionist plant ecologist Henry Gleason and holistic plant ecologist Fredric Clements decades earlier. Nelson Hairston and colleagues published a paper (Hairston et al. 1960) suggesting that biomass and production in trophic levels could be predicted by the relative role of the mechanisms of competition and predation. Although trophic ecology was a familiar term to ecologists in the mid-twentieth century, it was cemented as a stand-alone concept in Soviet Russian ecologist's V.S. Ivlev's 1961 book *Experimental Ecology of the Feeding of Fishes*. Interestingly, Ivlev's work in the 1930s on energy balance in fishes (Ivlev 1939a,b) was cited by Raymond Lindeman and clearly influenced the ecosystem-level ideas Lindeman developed in his 1942 paper. Whereas Lindeman was considering energetic transactions among multiple trophic levels within organisms and the environment, Ivlev was more concerned about developing mechanistic ways to quantify the diets of fishes and trace the relative sources of energy fueling their populations. So, in this context, the term trophic ecology was an individualistic, organismal, and perhaps *autecological* concept, with energy transfer leading to growth and reproduction. At the time, Ivlev (1961) tried to define exactly what the discipline of trophic ecology encompassed, saying, "The question of the true content of trophic ecology is a debatable one. The feeding of an animal is such a complete phenomenon that it would be possible without a great deal of distortion to include in trophic ecology a considerable proportion of all biological problems" (p. 3). He realized that there was holistic potential but never recognized (or did not intend to recognize) this as ecosystem ecology as Lindeman had done nearly two decades earlier. The 1960s harkened a reductionistic search for interactions shaping food webs in communities that continues to the present. Robert Paine conducted experiments in the rocky intertidal of the Pacific Ocean that showed how species interactions shape the species distributions of mussels and seastars, through trophic (i.e., consumptive) effects (Paine 1966). Large-scale, mechanistic, experimental approaches by Gene Likens and colleagues were occurring in the Hubbard Brook Experimental Forest in New Hampshire. During that decade, Eugene and Howard Odum pursued a holistic approach to trophic ecology, where the term *ecosystem* came to fore (Odum 1969). Meant to avoid getting lost in the complex weeds of organismal interactions, their approach was to treat ecological systems as *black boxes* through which energy and matter were transformed. These black boxes could be trophic levels, lakes, forests, oceans, and the entire Earth. Mechanism, in their view, was less important than the function or emergent properties of the relevant units within ecosystems (Figure 1.1). Similar searches for generalities scaling all levels of ecology are still being conducted (e.g., macroecology; Brown et al. 2004; Chapter 8). Ecological network analysis of Ulanowicz et al. (2014), in which complex ecosystems can be described in simple terms, is another example. These ecologists believe that somewhere deep within the complex webs of feeding relationships lie general patterns.

In the 1980s and 1990s, the trophic cascade hypothesis arose, popularized by Stephen Carpenter et al. (1987). This was an attempt to bridge the holistic and reductionistic approaches in ecology that began to split in the 1960s. These investigators began long-term experiments in which they manipulated the trophic structure of whole lakes and searched for mechanistic (e.g., population and community) and emergent changes (e.g., nutrient dynamics, stability; Chapter 10). Their intent was to unify consumptive, top-down mechanistic effects in ecosystems with bottom-up influences of energy availability. This was the beginning of a formal framework for trophic ecology, although the term was never explicitly defined in the literature. Rather, a heated debate occurred among ecologists about whether the relative roles of energy-driven and consumptive processes were generalizable across ecosystems, leading to more confusion than enlightenment for most ecologists.

The last decade has seen a precipitous rise in the use of the term trophic ecology in the literature, because many researchers are interested in energy and matter transfer from a variety of perspectives, including continuing to assess the validity of concepts such as the mechanistic trophic cascade hypothesis; the holistic, Odum-style approach to ecosystems; and in between, engineering-like approaches such as ecological network analyses. We conclude this book by hypothesizing that trophic ecology embraces all of these approaches, finding a common thread among them (Chapter 15).

1.5 BOOK APPROACH

We attempt to find general patterns in trophic ecology (see Chapter 15). We harken back to the concepts developed by Elton, Ivlev, and Lindeman and tie natural history, evolution, consumer interactions, energetics, and biogeochemistry into a mechanistic approach to understand how ecosystems behave (Allen and Gillooly 2009).

In Chapter 2, we revisit the classic approach to trophic ecology, following the concepts developed by Elton (1927) and taught in introductory ecology classes around the world. Trophic levels within trophic pyramids are useful conceptual starting points and foreshadowed the holistic, ecosystem-level views of the Odums. We will find that emergent properties within ecosystems provide support for grouping organisms into discrete functional units. Trophic levels indeed exist in nature and have unique properties that make them important for research and conservation management.

Organisms can be roughly categorized into trophic levels that might be defined as *autotrophs, consumers, scavengers, decomposers,* and *omnivores* (Chapter 3). As we will see, autotrophs and consumers (especially apex carnivores) tend to be overemphasized in ecology, whereas the scavengers and so-called decomposers in the microbial loop are deemphasized. At the level of ecosystem function and biogeochemical cycling, we ignore the interactions between scavengers and decomposers at our peril. Omnivory, which means feeding across trophic levels, occurs in many

ecosystems. However, the process is still in need of better understanding from ecological and evolutionary standpoints.

In the realm of individual, population, and community ecology, there exist interactions that are shaped by natural selection and lead to unique properties of trophic
levels and ultimately ecosystems. Many organisms, from single cells to complex
vertebrates, react to prey and the environment, leading to trade-offs between risk and
consumptive income (Chapter 4). Population and community interactions are driven
by selective pressures, which influence the rate by which energy moves between
trophic levels. In the past 500 million years, *extreme heterotrophs* called predators arose (Chapter 5), possessing all kinds of interesting behaviors and anatomical
characteristics that allow them to procure prey. As predators evolved, prey evolved
unique ways to avoid being consumed (Chapter 6). The selective arms race that
occurs between predators and prey regulates trophic interactions and thereby the
efficiency of energy transfer between trophic levels.

Many methods have arisen to quantify consumptive patterns in ecology. In
Chapter 7, we address ways to quantify and analyze dietary data from consumers.
These data provide a snapshot of the consumptive income of predators. Chapter 8
focuses on an alternative way to assess dietary income by quantifying the energetic
intake of organisms by knowing their metabolic, growth, and reproductive needs.
Energy flow among organisms is important as well and can be scaled to ecosystems
through network analysis. Alternately, this chapter briefly shows that energetic processes within organisms may scale up to entire ecosystems by identifying large-scale
commonalities in energetics across all organisms (i.e., *macroecology*). Collecting
individual diets and developing models of energetics for organisms are but two
ways of determining patterns of consumption. In Chapter 9, we review models of
consumption and show that sometimes the sheer intake of energy does not explain
patterns of growth and trophic transfer. Organisms have nutritional needs that are
missed in traditional ecological studies of trophic ecology, but may be the linchpin
underlying the growth, survival, and reproduction of organisms and the transfer of
energy through ecosystems.

At this juncture of the book, we begin to move away from organismal-based
categorizations in trophic ecology and more toward the foundations of flux. Food
webs were formally developed by Elton (1927) and put into an energetic context by
Lindeman (1942). Community ecologists adopted them because they consider species as their basic units. But they have ecosystem characteristics as well (Figure 1.1).
As we will see, food webs are terribly complex, even in communities with few species.
Chapter 10 shows techniques for categorizing food webs and searching for important
connections within them. Each node within a food web (or trophic level) has a rate
of growth, reproduction, and death called *secondary production* (Chapter 11). We
determine models for calculating the rate of turnover of biomass (i.e., production),
in these important groupings in ecosystems, which provides insight into the rate of
energy transfer within and among ecosystems.

The nutrients available within ecosystems drive trophic interactions (Chapter 12).
Trophic ecologists recognize that the relative concentrations of nutrients may be as
important as the total concentration in the ecosystem. If one nutrient becomes unlimited, then another nutrient may become limited. This is called *ecological stoichiometry*

and may provide the key to understanding many trophic interactions, biodiversity, and ultimately the efficiency of energy transfer in ecosystems. The ratio of two commonly limiting nutrients, phosphorus and nitrogen, in ecosystems is related to processes occurring within the cells of organisms that may scale up to the level of ecosystems as large as the open ocean (Chapter 12).

The past decade has given rise to the widespread use of tracers in ecosystems. Elements that occur rarely in the environment and vary spatially can provide insight into the past foraging behavior of organisms (Chapter 13). Common elements vary in mass, creating *stable isotopes*. These isotopes accumulate in unique ways depending on patterns of autotrophic and heterotrophic production in ecosystems. Isotopes of carbon often provide clues about the location of primary producers, whereas nitrogen isotopes become fractionated (i.e., enriched) in consumers. Thus, combining isotopes in analyses of prey and consumers can be used to identify trophic relationships and the most likely pathways of energy through ecosystems.

Most biologically produced molecules are rapidly catabolized by consumers or degraded in the environment. Molecules are often not useful for tracing trophic interactions. Fatty acids, the components of phospholipids, triacylglycerols, and wax esters, are conserved in many organisms when consumed (Chapter 14). Many long-chained, unsaturated fatty acids are not synthesized by heterotrophs and must be obtained from autotrophs. Plants and algae differ in their ability to synthesize and accumulate fatty acids, providing information about the geographic source of energy in the environment.

1.6 SYNTHESIS

In the final chapter (Chapter 15), we formally identify the common thread that is trophic ecology. Trophic interactions have been and will be critical in the present and future. As with the high biodiversity of the planet, the diversity of trophic interactions is also high. As human activities simplify Earth's ecosystems through extinctions, fragmentation, and climate change, the biological diversity that leads to complex trophic interactions will decline and be lost for a long time until they can be restored by evolution. Generalities do arise from trophic ecology. Statistical tools, simulation modeling, monitoring, and experimentation will need to be developed to make sense of the exponentially rising amount of information available to ecologists. Ecology is indeed dividing into more specific subdisciplines. Trophic ecology and evolutionary biology are two commonalities providing a common language for ecologists, allowing for syntheses across the entire discipline.

QUESTIONS AND ASSIGNMENTS

1. What are the levels of organization of ecology and how does trophic ecology contribute to these levels?
2. What is the definition of trophic ecology?
3. From where does the term *trophic* derive?
4. How is the term *trophic ecology* used by ecologists in the literature?
5. What is the significance of *The Book of Animals* written in Arabia?
6. Who created the first known trophic diagram?

7. What was Charles Elton's contribution to trophic ecology? How does this compare to Raymond Lindeman's contribution?
8. Describe the differences between holistic and reductionistic approaches in trophic ecology, explaining how ecologists like Ivlev, Hairston, Paine, and Odum contributed to these different pathways.
9. Is trophic ecology used frequently in the ecological literature presently?
10. How is the organization of this book used to teach us about trophic ecology and prepare us for complex topics incorporating both evolutionary and ecological approaches in the transfer of energy and materials?
11. Why are complex trophic interactions important to understand?

Section I

Concepts and Patterns

2 Trophic Pyramids and Trophic Levels

2.1 APPROACH

Trophic pyramids and trophic levels are a familiar topic for most students of ecology. The intent of this chapter is to provide a foundation for the sophisticated concepts and techniques to follow in this book. We will review trophic levels and provide many examples of the diversity within them, showing that many of the trophic interactions occurring in ecosystems today were likely occurring up to 400 million years ago. The trophic pyramid concept and trophic levels may be somewhat misleading and oversimplistic, but processes of natural selection and self-organization do result in constructs that lead to meaningful groupings within ecosystems. It is in this context that important exceptions such as omnivory and ecosystem subsidies can be better understood and explored.

2.2 INTRODUCTION

One of the simplest ways to convey trophic relationships in ecology is through the use of trophic pyramids (Elton 1927; Figure 2.1a), which are taught in grade schools, high schools, and introductory college courses, but rarely used in research (Chapter 1). The central question for this chapter is whether trophic pyramids are useful for driving research or are creating misleading oversimplifications. Traditionally, a trophic pyramid's base is composed of the autotrophs, which support the primary consumers, secondary consumers, and additional trophic levels if they are present (Figure 2.1a). The beauty of the pyramid as a conceptual construct is that it adheres to the basic laws of physics, in which energy transfer from one trophic level of the pyramid to the next is limited by the constraints of thermodynamics (Figure 2.1a). Energy is eventually lost as heat in any biological process and is unavailable for use as work. Thus, in the simplest sense, the autotroph base has to be the largest and trophic levels shrink dramatically as useful (non-heat) energy is transferred up the pyramid and heat is lost. A pyramid should look like a series of stacked stepping-stones, with each subsequent stone becoming smaller, because of the net loss of energy that occurs between each trophic level.

As with any scientific conceptual model, trophic pyramids are simplistic; they miss many important trophic interactions such as detritivory, scavenging, and omnivory. However, they do provide a useful starting organizational structure for exploring trophic interactions. Pyramids are best characterized as the amount of energy captured by the autotrophs and then transferred to each subsequent trophic level. The energy can be measured in many ways, including calories of heat, the

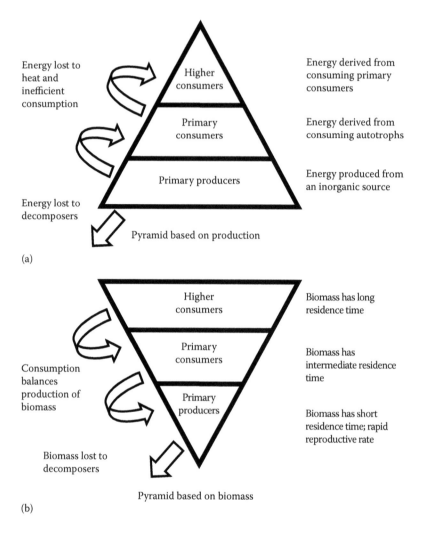

FIGURE 2.1 (a) Trophic pyramid based on production of biomass or energy in an ecosystem. Arrows depict energy transferred from one trophic level to another or lost to decomposition. (b) An inverted pyramid of standing biomass in an ecosystem. Arrows represent consumption of production within each trophic level or lost to decomposers. Not all biomass pyramids are inverted in ecosystems. Organic matter may accumulate. These pyramids do not include the processes of omnivory or the impact of energy subsidies from other ecosystems.

number of carbon bonds, ash-free dry mass, or the production of total biomass. The amount of standing biomass in a trophic level, especially at the base, may actually be quite low depending on the rate that it is being consumed by the next trophic level. Thus, if a single snapshot in time of all the organic matter in the ecosystem was taken, a pyramid composed of the actual biomass present may appear to be inverted (Figure 2.1b). For example, the rate by which cows graze grass in a pasture may meet the rate at which the grass replaces its biomass. Thus, it would appear that the grass is not growing because the net production is immediately being removed. There is more to this than differences in production among trophic levels. Huryn (1996) describes Allen's paradox as an intriguing example, in which K. Radway Allen could not use the production of lower trophic levels to explain the standing biomass of trout in streams. Only when production subsidies from adjacent ecosystems were considered (e.g., subsidies; Chapter 10) was the paradox explained. Trophic pyramids rely on income from sources other than local autotrophy to maintain their foundations.

We will begin our exploration of trophic relationships by briefly visiting the autotrophs in an ecosystem. Autotrophs, also called the primary producers, were likely the first organisms on Earth and form the basis and ultimately the limitations for complex trophic relationships on the planet. The wealth of this chapter will focus on the primary consumers, namely the herbivores, which consume autotrophic production. *Herbivores* have evolved complex ways to exploit the energy stored by plants. We will then introduce the concept of *carnivory*, where organisms, including plants, have evolved ways to consume herbivores and other animals. We end the chapter by assessing just how common the generalist feeders known as omnivores are in ecosystems and whether the presence of this group and other problems threaten the validity of the trophic pyramid concept.

2.3   AUTOTROPHS

Primary-producing autotrophs such as algae and vascular plants (i.e., the photoautotrophs) have a heavy burden on Earth. They are tasked with capturing energy from an external inorganic source and storing it for all biological processes occurring on Earth. Photosynthesis is a highly conserved chemical reaction that occurs within almost all autotrophs whereby energy from sunlight is used to fix inorganic carbon into carbohydrates (Figure 2.2). The net energy sources are photons (usually from sunlight) and electron donors such as hydrogen sulfide, acids, some organic molecules, or more commonly water. Electrons cascading through biologically mediated chemical transport chain reactions in chloroplasts generate energy that is harvested for carbon storage. This process is typically quantified by ecologists as the net carbon captured in ecosystems and varies dramatically across the surface of the planet (Table 2.1). It is important to note that autotrophs do not completely generate food resources on their own. They rely on a host of factors such as nutrients, appropriate temperatures, substrates, and perhaps even heterotrophic hosts to be able to manufacture food energy from an inorganic source.

In an abstract sense, algae and plants can be considered *light consumers*. In fact, not all light is the same. Many species of plants and algae can coexist because they rely on different light wavelengths to fuel their photosynthetic machinery. There are multiple molecular forms of chlorophyll (e.g., a, b, d, f) plus accessory pigments such as carotenoids. These pigments work together to maximize energy absorption

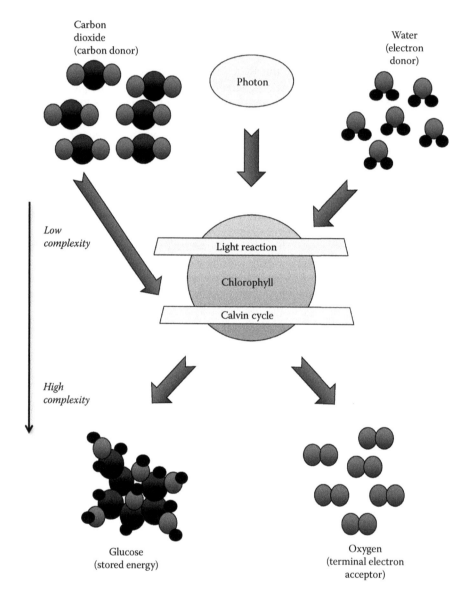

FIGURE 2.2 Autotrophic process of photosynthesis. Carbon dioxide gas combines with water in the chloroplasts of algae and plants. Photons excite chlorophyll molecules in the chloroplast, initiating the light reaction where electrons are used to synthesize ATP (the primary energy source in cells). The electrons are then used to initiate the production of NADPH, an electron recipient. Electrons within chlorophyll molecules are replenished by catalyzing water and releasing free oxygen. In the Calvin cycle (i.e., dark reaction), ATP and NADPH are used by the enzyme RuBisCo to convert carbon dioxide molecules to more complex carbohydrates, which contain stored energy. Reduced oxygen molecules are the waste product of photosynthesis.

TABLE 2.1

Heterogeneity in Autotrophic Production in Major Ecosystems of the Earth Wetlands Are the Most Productive Ecosystems on the Planet. Deserts and the Open Ocean Have Very Low Productivity

Ecosystem	Net Primary Production (g m^{-2} year^{-1})	Biomass (kg m^{-2})	Ratio of Biomass Accumulation
Terrestrial			
Tropical forest	1800	42	23
Temperate forest	1250	32	26
Boreal forest	800	20	25
Shrubland	600	6	10
Temperate grassland	700	4	6
Tundra and alpine	500	1.5	3
Desert	140	0.6	4
Wetland	2500	15	6
Aquatic			
Open ocean	125	0.003	0.02
Continental shelf	360	0.01	0.03
Algal beds and reefs	2000	2	1.00
Estuaries	1800	1	0.56
Lakes and streams	100	0.02	0.04

Source: Adapted from Whittaker, R. H., and G. E. Likens. *Human Ecology* 1 (4):357–369, 1973.

at different light spectra (Stomp et al. 2004; Figure 2.3). As such, each taxon can use a unique arrangement of these pigments to access their niche in an ecosystem (Rockwell et al. 2014). Rockwell et al. (2014) showed that diverse algal taxa have high plasticity in the ability for *phytochromes* (i.e., light receptors) to respond to variable wavelengths of light. Thus, autotrophs have evolved plasticity to enhance photosynthesis under a variety of light conditions.

Production by autotrophs is not uniformly distributed across the planet. In terrestrial ecosystems, the highest autotrophic production occurs in wetlands, with tropical and temperate forests also having high plant turnover (Table 2.1). Even though temperate zones experience winter, the summer pulse in production is sufficient to overcome the pause in autotrophy. The primary factor limiting terrestrial production is water, with deserts on Earth having the lowest rates of carbon fixed. In aquatic systems, where water is obviously plentiful, the coastal zones of oceans are most productive; however, the open ocean, which dominates the surface of Earth, is similar to the terrestrial deserts in autotrophic production (Table 2.1). Here, nutrients like elemental iron are major factors limiting photosynthesis by phytoplankton (Johnson et al. 2011; Chapter 12). Areas around the sea ice in the Arctic Ocean also support some primary production of up to 15 g C m^{-2} year^{-1} (Gosselin et al. 1997). Remote sensing with satellites and spectrometers now makes it simpler to measure global

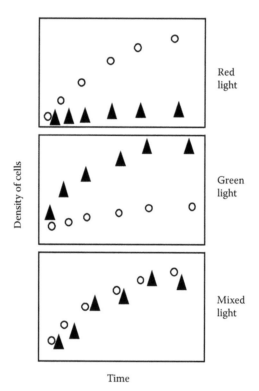

FIGURE 2.3   Hypothetical results of an experiment showing how two species of algae coexist under different wavelengths of light. When grown in the same medium, species a (white circle) grows better than species b (filled triangle) under long light wavelengths (red). In green light (medium wavelengths), species b outcompetes species a. When light is mixed wavelengths, both species grow. (Based on actual results from Stomp, M. et al. 2004. *Nature* 432 (7013):104–107.)

changes in productivity through time as Earth's surface colors (e.g., the spectral imagery of chlorophyll) change (Box 2.1).

   With the advent of deep-sea exploration by ecologists a few decades ago, an alternative pathway to photosynthesis as a base of the trophic pyramid that had long been predicted called chemoautotrophy was discovered. Geothermal vents on the seafloor were found to support complex ecosystems far out of the reach of sunlight (Corliss et al. 1979). Symbiotic bacteria in the tissues of organisms such as the vent shrimps (*Rimicaris* spp.) oxidize molecules (e.g., hydrogen sulfide) emanating from the vents, providing an alternative energy source to sunlight (Van Dover et al. 2002). *Chemosynthesis* has been well documented in terrestrial and aquatic ecosystems. But the idea that it forms the foundation for ecosystem processes was certainly novel when it was discovered. Not all of the production at the vents is chemosynthetic. Some photosynthetic bacteria are also present and capable of fixing inorganic carbon into sugars by using the glow of the vents as a light source (Beatty et al. 2005). The biofilms created by chemosynthetic bacteria living directly on the seafloor also are

BOX 2.1

Satellite remote imagery of net terrestrial primary production (i.e., the pro-
duction that results in accumulation of plant biomass) on the continents of
Earth during June (top) and December (bottom) 2015. The satellite is the
NASA Terra and the data are taken from the Moderate Resolution Imaging
Spectroradiometer (MODIS) on board. Monthly data are available for down-
load on the NASA web site. Darker areas are near the maximum of 6.5 g C
m^{-2} day^{-1}. Because these data are averaged across the month, negative values
can occur if NPP declines through time. MODIS quantifies the intensity of 36
spectral bands, of which a subset corresponds to the absorption and reflectance
of light from chlorophyll. This spectral radiance is correlated with chlorophyll
concentration on the planet surface.

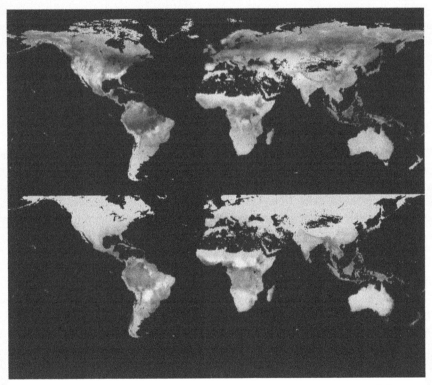

gC/m^2/day

−1.0 0   6.5

consumed by heterotrophic organisms. Of course, none of these interactions would be possible without the creation of Earth via the sun. So, similar to life on the sunlit surface of Earth, the energy source is ultimately solar in nature. Interestingly, many scientists believe that similar chemosynthetic pathways may support life at subsurface thermal vents on other planetary bodies such as the water-covered moon Europa circling Jupiter (Greenberg 2011).

2.4 PRIMARY CONSUMERS

Any organism consuming a photosynthetic or chemosynthetic autotroph and deriving energy from it is a primary consumer (Figure 2.1). Some organisms may inadvertently consume autotrophs but are incapable of digesting them and thus are not primary consumers. In fact, some putative herbivores are inefficient at digesting plant matter and rely more on gut endosymbionts or bacteria on plant substrates (Findlay et al. 2002; France 2011) as a source of energy. In this section, we focus on herbivores, organisms that derive their energy directly from eating plants. After exploring some of the many ways herbivores eat, we will determine how herbivory has evolved through time and evaluate the impact of these organisms on plants. Ultimately, the plants themselves are largely responsible for influencing much of the diversity of life on Earth through the ways they manipulate their consumers.

2.4.1 Modes of Herbivory

The impact of herbivory on plants varies in many ways. One of the most fundamental relationships between plants and consumers is apparently symbiotic, with arbuscular mychorrhizal fungi providing an important role by improving nutrient and water uptake at roots. In legumes, captive bacteria (*Rhizobia* spp.) fix unavailable atmospheric nitrogen gas in root nodules. Nitrogen is often a limited nutrient in soils. By providing organically available nitrogen to the plant, the bacteria gain a source of carbohydrates and other materials from the plant host (Table 2.2). Other microbial relationships with plants are not mutualistic, with many fungal, bacterial, and viral pathogens having devastating impacts on hosts (Table 2.2).

Grazing and browsing are two very common feeding tactics. Grazing is a method by which herbivores consume plants indiscriminately (Table 2.2). This feeding mode is likely to occur in terrestrial systems when herbivores are hungry and attempting to maximize energy input. Black-tailed deer (*Odocoileus hemionus*) at high densities can create distinct grazing lines within forests as they forage widely on all available plant matter with little regard to quality (Martin et al. 2011). In aquatic systems, many crabs, snails, and crayfish can be considered grazers. Rusty crayfish (*Orconectes rusticus*) can wipe out extensive beds of vegetation in lakes in which they invade (Lodge et al. 1994). In contrast to grazing, browsing is discriminate foraging, which is more likely among terrestrial herbivores and can in fact allow many specialists to coexist (Gordon and Prins 2008). In aquatic systems, some grazers will be less likely to switch to browsing due to morphological constraints. For example, mouth structures (i.e., gill rakers) that filter the water of planktonic particles make

TABLE 2.2

Different Modes of Herbivory and Their Potential Impact of Autotrophs

Mode of Herbivory	Impact to Autotroph	Example
Mutualism		
Microbial	Increase nitrogen	Legumes and *Rhizobia*
Pollen	Enhance sexual reproduction	Honeybees
Nectar	Attract pollinator	Hummingbird
Seed	Dispersal of offspring	*Pacu* fish in Amazon
Grazing	Indiscriminate loss of tissue	Deer
Browsing		
Folivory	Loss of leaves	Giraffe
Granivory	Loss of seeds	Golden mouse
Frugivory	Loss of fruit	Owl monkey
Rooting	Possible plant death	Pocket gopher
Pathogen/parasite	Possible whole plant death	Dutch elm disease
Piercing and sucking	Reduced growth (often minor)	Aphid
Galling	Reduced growth of structures; deformation	Gall wasp
Mining	Reduced growth of leaves	Leaf miner insect
Boring	Inhibit sapflow, increase infection, possible plant death	Pine beetle larvae
Filter feeding	Death to particulate autotrophs	Freshwater mussel

it impossible for silver carp (*Hypopthalmichthys molitrix*) to browse efficiently on rooted aquatic plants. No terrestrial counterpart to filter feeding exists (Table 2.2).

Many other modes of herbivory exist. Similar to aquatic filter feeding by fish, some aquatic invertebrates like caddisfly larvae build silk nets that allow them to gather drifting particles including suspended algae (Table 2.2). In terrestrial ecosystems, leaf feeding, also called folivory, is a specialized form of browsing. A possible adaption is the long neck of the African giraffe *Giraffa camelopardali*, which facilitates its ability to feed on leaves of tall trees (Table 2.2). There is some debate about the function of the tall neck of this species and its relationship to foraging. Some ecologists argue that neck length may be related to sexual selection, where males with longer necks have greater mating success (Pratt and Anderson 1985). Some modes of herbivory like seed feeding (granivory) and frugivory (fruit eating) as well as pollen and nectar collecting may have some advantage to the plant (Table 2.2), although any biomass consumed by the herbivore constitutes an energetic cost to the plant and thus is going to be minimized through natural selection (Pellmyr 2002). Several species of insects burrow into various plant structures and induce galls, which are enlarged portions of plant tissue in which the insect larvae reside (Table 2.2). Mining and boring are other herbivorous tactics used by insects to reach specific plant tissues (Table 2.2). In some herbivores, complex behaviors especially related to spatial memory arise. A good, flexible memory ensures that an herbivore can respond rapidly to changes in resources. Social herbivores such as honeybees (*Apis* spp.) can communicate spatial information to conspecifics about the location of rewarding foraging locations. Honeybees communicate the location of flowers producing pollen or nectar by using complex waggle dances that relate the direction and distance of the food source relative to the position of the sun.

2.4.2 HISTORY OF HERBIVORY

Herbivory is very old, likely arising not long after the first multicellular algae evolved. In aquatic systems, fossil evidence suggests that an arms race occurred between macroalgae and their herbivores about 540 million years ago at the Proterozoic–Phanerozoic boundary (LoDuca and Behringer 2009). The concept of an evolutionary arms race is simple (Van Valen 1977). As the prey evolves a defense, the predator evolves a countermeasure. Thus, interesting characteristics of both arose rapidly. Planktonic algae were also readily consumed in the ancient oceans. Filter feeding occurred in brachiopods as early as 535 million years ago (Figure 2.4). These organisms, some of which persist to the present, look like bivalve mussels but have a very different phylogeny and feeding morphology, filtering water with tentacle organs called lophophores. The fossil record suggests that marked herbivory by fish occurred in the early Cenozoic (63 million years ago), likely having strong impacts on the structure of plant communities (Bellwood 2003). Paleolimnologists studying the deep cores of the bottom of lakes have documented robust grazing effects of zooplankton on phytoplankton dating centuries in the past (Carpenter and Leavitt 1991).

Herbivory quickly followed plants as they moved onto land, with arthropods and their damage to plant tissue appearing as early as 430 million years ago (Labandeira 2007; Figure 2.5). Patterns of herbivory arose at different times in the past with boring

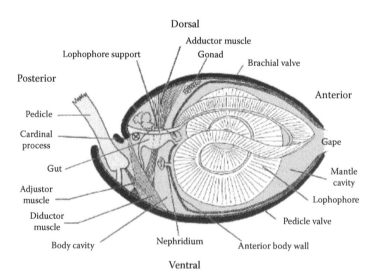

FIGURE 2.4 Anatomy of a brachiopod, once the most abundant herbivore on Earth. Phytoplankton are concentrated by an extensive lophophore in the chamber between the valves.

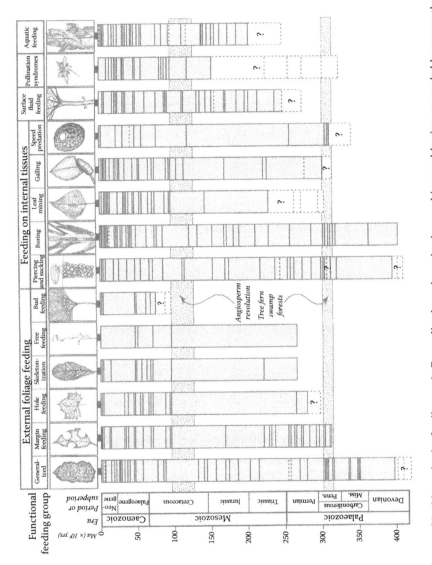

FIGURE 2.5 History of herbivory in the fossil record. Generalized grazing, piercing, sucking, and boring are probably among the oldest forms of herbivory. (Adapted from Labandeira, C. 2007. *Insect Science* 14(4):259–275.)

into tissue and piercing/sucking of internal plant fluids arising first (Labandeira 2007). Evidence of generalized feeding also appeared early in fossils, although it is difficult to differentiate between tissue loss due to herbivory versus damage from other sources such as weathering and tearing. Other methods of feeding like leaf mining and pollen feeding developed more recently, about 175–150 million years ago (Labandeira 2007; Figure 2.5). Fossils of galls, the insect larvae that induced them, and potential parasitoids that attacked the insect larvae provide important information about trophic interactions that occurred in deep history (Stone et al. 2008). Herbivory by vertebrates on land began much later and likely evolved first in carnivorous amphibians (Sahney et al. 2011). Similarly, phylogenetic analysis of microbes in mammalian guts suggests that carnivory in mammals was the earliest feeding mode and that herbivory evolved later, with greater complexity of microbial taxa occurring in herbivores (Ley et al. 2008). Many mammals such as *ruminants* are obligatory herbivores and cannot function on other diets. Ruminants such as domesticated cows have complicated, multichambered digestive tracts that require extended fermentation times.

2.4.3 IMPACTS OF HERBIVORY

Plants must contend with a constant assault by herbivores. In terrestrial environments, about 50% of all mammals are herbivorous (Lindroth 1989). As they did in the ancient past, arthropods such as beetles and lepidopterans comprise some of the major herbivores on land. In aquatic ecosystems, the diversity of herbivorous vertebrates is high including damselfishes in coral reefs, cyprinid fishes in freshwater rivers, and amphibian larvae in temporary pools. Marine iguanas *Amblyrhynchus cristatus* are an interesting example of herbivorous reptiles only found in the Galapagos Islands (Figure 2.6). Invertebrates that rely on plants to provide energy in oceans and in

FIGURE 2.6   Marine iguana shows the unique characteristic of a terrestrial reptile making forays into the ocean for herbivory. (Courtesy of Shutterstock.)

freshwater are exceedingly numerous and include just about every major phylum, including mollusks, arthropods, echinoderms, cnidarians, rotifers, and sponges.

Although herbivory is a critical pathway allowing energy captured by autotrophs to move up to higher trophic levels, the transfer of energy is very inefficient. As Lindeman (1942) aptly pointed out, only about 10–15% of the energy assimilated by the autotroph trophic level is available to herbivory. The remainder of the energy is either used for maintenance metabolism or lost to other ecosystem compartments such as detritivores (Figure 2.1). Generally, herbivores are inefficient at accessing plant standing biomass, probably only defoliating a total of 15% of plant matter annually (Crawley 1989). In terrestrial ecosystems, excess plant matter is lost in soil and can be washed into streams, lakes, and oceans. Some of this production washes downstream from headwater streams into rivers to drive biological interactions in these flowing aquatic systems. This forms the foundation of the *river continuum concept*, which predicts that terrestrial vegetation drives production upstream, whereas internal autotrophic production drives interactions downstream (Vannote et al. 1980). Marine algal and plant production can be lost in the dark, cold depths of the bathypelagic. However, nutrients from decomposed algae can be resuspended in areas of ocean upwelling such as at the coasts of continents, stimulating primary productivity and driving trophic interactions in coastal ecosystems (Figure 2.7). The

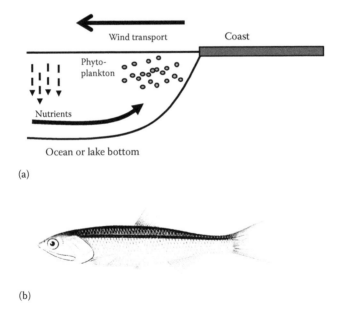

(a)

(b)

FIGURE 2.7   (a) Process of upwelling and its relationship to nutrient resuspension. Wind transports water offshore, causing deep, cold, nutrient-rich water to flow upward. Nutrients plus sunlight stimulate primary production, which supports high secondary production. Wind transports some phytoplankton offshore, which die, decompose, and elevate nutrients in the deep, cold, dark bottom waters. (b) Peruvian anchoveta is a species of fish that is supported by high plankton productivity at upwelling areas off the western coast of South America.

anchoveta fishery off the coast of Peru is supported by these processes. At a smaller scale, biomass produced by primary production in freshwater lakes can be lost in the dark, cold strata at the bottom, and may be resuspended during seasonal turnover events (Figure 2.7).

Hairston et al. (1960) hypothesized that the world is green. What they meant was that herbivores and their consumptive effect on autotrophs should be held in check by predation by the secondary consumer trophic level. Without carnivory, herbivores would completely consume all the plant matter on the planet and the world would cease being green and ecosystems would collapse. A debate ensued about this issue that continues to the present. Crawley (1989) wrote a comprehensive review of invertebrate herbivory and concluded that invertebrate herbivores are limited more by the defenses of plants and inefficiencies of the herbivores than top-down control by secondary consumers. However, herbivores have strong impacts under certain contexts. Leaf-cutting ants are well-known agricultural pests, decimating crops to grow their fungal gardens (Currie et al. 1999a,b). Zooplankters in lakes are capable of grazing phytoplankton to the point of clearing the water (Strong 1992; Borer et al. 2005). Many vertebrate examples of overgrazing plant resources exist as well, typically when predators are absent and grazers are confined within an area (Martin et al. 2011). The effects of herbivores are typically spatially isolated. A global onslaught of herbivores is kept in check by a combination of predation, environmental heterogeneity, and ultimately the broad diversity of plants and their associated defenses.

As the fossil record suggests, plants continually evolve ways to offset herbivory through an arms race between the two. Physical ways to reduce herbivory are numerous. Thorns and spines deter herbivory by vertebrates. Plants can invest energy into large indigestible carbon compounds such as lignins to reduce their digestibility. Plants also can allocate nutrients differentially to structures, reducing the nutritional value of exposed structures and deterring choosy herbivores. Another option available to plants and algae is poison. Chemical defenses were brought to the fore by Ehrlich and Raven (1964), who suggested that plants accumulate these compounds to avoid being consumed. Many of the substances isolated from plants are alkaloids, which are carbon molecules with nitrogen in their structure that range from being mildly toxic to deadly. All of these defenses are energetically costly to the plant and should negatively affect reproductive output.

2.4.4 TRADE-OFFS

Herbivory always incurs some cost to plants. The concept of *compensation* suggests that plants reallocate energy biomass to structures that are lost to herbivory (Belsky et al. 1993; Figure 2.8). When this happens, the plant may appear to compensate for tissue loss by increasing aboveground biomass. The real cost of this apparent compensation may be reduced belowground biomass or perhaps lost seed production. Although some investigators have suggested that plant growth could be stimulated by herbivores allowing them to overcompensate for herbivory, this idea has largely been discredited (Crawley 1989; Belsky et al. 1993; Ward 2010). Similar to compensation is the idea that plants must contend with herbivory by either tolerating or resisting tissue loss (Stevens et al. 2007; Figure 2.8). Tolerating loss occurs as plants allow some burden of herbivores to occur

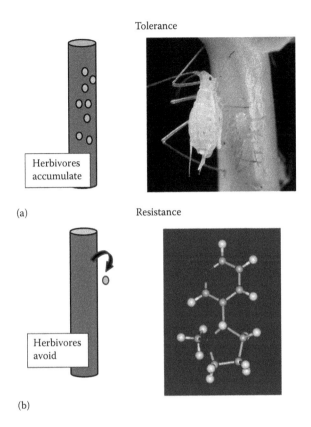

FIGURE 2.8 (a) Tolerance of herbivory on a plant, with a stem having a large number of herbivores (represented as dots) present (left). A plant uses energy to compensate for the loss of tissue to herbivores such as aphids (right). (b) Resistance of a plant to herbivores by generating a toxin or producing indigestible parts (left). Alkaloid chemicals such as nicotine (molecule on the right) are produced to deter herbivory, at an energetic cost.

on their structures (Figure 2.8). Plants tolerate loss by regrowing tissue at an energetic cost. Resisting loss occurs by increasing physical and chemical defenses, although again at some energetic cost (Figure 2.8). These two tactics can be considered complementary ways to deal with herbivory (Trudgill 1991), with both occurring to offset the effects of herbivory while still ensuring successful, albeit reduced, reproduction.

Herbivores can become quite specialized on plants, relying on them not only for food but also as habitats for rearing their young. Thus, it is in the interest of these herbivores' fitness to not kill their host. Aphids are a good example of insects that rarely kill their hosts, and in fact are conspicuous prey for predators. To persist, aphids must rely on factors such as mutualisms with ants, where the ants provide protection while the aphids produce honeydew for the ants to eat. *Galling* insects have similar dependencies on their plant hosts and are often a target for parasitoids. Herbivore interactions may eventually evolve into mutualisms, such as those between fungi and roots (Kumari 2011) and bees and their flowers (Armbruster 1992). The degree of coevolution between plants and herbivores can vary dramatically among taxa (Sachs et al. 2011; Figure 2.9). If strong linkages between coevolved species are lost, then trophic transfer should decline. Honeybees and other pollinators are declining worldwide (Kearns et al. 1998). Loss of the associations may reduce plant productivity, potentially devastating agriculture (Gallai et al. 2009; Winfree et al. 2011).

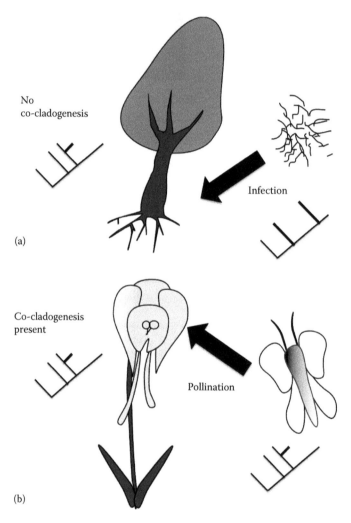

FIGURE 2.9 Plants and their symbiotic organisms vary in the degree of coevolution. (a) Most fungi that form mycorrhizal associations by infecting the roots of trees live independently in soil. Thus, natural selection between the two groups occurs independently, with co-cladogenesis unlikely. Co-cladogenesis is the process by which clades (i.e., diagrams of evolutionary relationships) are similar between taxa with coevolved histories. (b) Some species of plants and their symbionts are quite interdependent, with both requiring the other to successfully reproduce. For example, many species of moths and orchids have coevolved closely, with co-cladogenesis likely.

2.4.5 DIVERSITY AND ECOSYSTEM EFFECTS

Although the idea that the world is green because herbivores are controlled by their predators is likely not universally true, herbivores can have marked, geographic-scale effects on patterns of plant diversity and ecosystem structure (Gossner et al. 2014). Herbivores thus influence the *shades of green* on the planet, when their consumers are rare or absent. Perhaps the greatest effects of herbivory arise when unchecked plant pathogens exert strong, unilateral effects on their hosts (Loo 2009). A notable example is Dutch elm disease caused by fungi (*Ophiostoma* spp.) and their nearly global effect on some elm species. Chestnut blight caused by *Cryphonectria parasitica*, an introduced fungus from Asia, has nearly caused the extinction of American elm *Castanea dentata* in North America. Many uncontrolled invading insects with no apparent predators can have similar devastating effects on specific plant taxa, thereby influencing patterns of diversity across broad geographic extents. A recent example in North America is the emerald ash borer *Agrilus planipennis*, which arrived from Asia in the 1990s likely in wooden shipping crates and threatens ash trees (*Fraxinus* spp.), a major component of forests throughout the continent. It is currently found in states throughout the northeast United States, with occurrences as far west as Colorado, and provinces of eastern Canada.

At the scale of plant populations and communities, herbivores may exert positive effects. American bison (*Bison bison*) promote grassland plant diversity through multiple processes that advance spatial heterogeneity of resources and recruitment of native plants. These processes include selective grazing, seed dispersal, and nutrient excretion (Rosas et al. 2008; Burns et al. 2009). In coral reefs, grazers can mediate interactions between competing coralline algae, enhancing their diversity (Steneck et al. 1991). Herbivory also can increase the turnover of plant matter (i.e., enhanced productivity) in some ecosystems by shifting plants to more rapidly growing juvenile forms (Craig 2010). Foraging by some filter-feeding fishes like silver carp and gizzard shad (*Dorosoma cepdianum*) can favor nutrient resources that support blue-green algae over green algae (Lamarra 1975; Vanni and Layne 1997). Thus, from the perspective of certain plant species, herbivory can be a boon.

2.4.6 COMPLEX RESPONSES

Herbivores can dramatically affect temporal and spatial patterns within plant communities. Some cycles of herbivory can be long and difficult to assess, such as the 17-year emergence cycle of some periodical cicadas, which can lead to large pulses of carbon and nutrients from the soil into terrestrial and aquatic ecosystems (Whiles et al. 2001). Thus, long-term monitoring is necessary to track the potential impacts of cyclical events such as these through time. Some temporal patterns of herbivory are seasonal, such as the spring through summer succession of zooplankton grazers on phytoplankton in freshwater lakes (Sarnelle 1993). The presence of zooplankton can deter the emergence of encysted algae from the sediments of lakes, thereby slowing the algal successional process (Rengefors et al. 1998). Similar successional effects of herbivores on algae occur in streams (Dudley and Dantonio 1991). Seasonal migrations of large vertebrate herbivores like African elephants (*Loxodonta africana*) in the Serengeti influence patterns of plant succession (McNaughton 1985). The loss of herbivores within ecosystems can dramatically alter the diversity and function of plant communities.

Herbivores and plants interact at many spatial scales. The physical structure of plants influences spatial patterns of herbivory. In the Great Barrier Reef, the spatial patterns of the size of macroalgae influence the distribution of herbivore taxa present (Hoey 2011). Termite mounds in the African savanna create hotspots of plant production, thereby structuring herbivore distributions (Levick et al. 2010). Fragmentation of plant communities also influences herbivores, typically in negative ways (Searle et al. 2011). For example, experiments showed that growth of migratory grasshoppers is negatively affected by fragmentation of resources, even when the abundance and quality of the resources are unchanged (Searle et al. 2011).

Understanding how herbivores depend on plants in a spatial context is important for designing reserves and corridors between them. Also, when attempting to hinder the movement of nonnative invading herbivores, understanding how to manage the plant landscape is necessary. For example, plants that are undesirable to invading herbivores may be planted to create dispersal barriers, protecting desirable plant hosts behind them (Jonsson et al. 2010).

2.5   HIGHER-LEVEL CONSUMERS

Herbivores may have been the first consumers, but natural selection ensured that they would be consumed by organisms at higher trophic levels. In this section, we focus primarily on carnivores, organisms that consume the tissues of other consumers. The benefit of carnivory is that digestive assimilation efficiency is often very high, reaching 90%. However, several costs of carnivory arise. Energy available to these trophic levels is limited by the inefficiencies of trophic transfer between the autotrophs and the primary consumers, as well as energy losses to heat within the autotroph trophic level (Elton 1927; Figure 2.1). Also, many prey are not defenseless and can incur substantial costs. Broken teeth of carnivores are common both in the fossil record and the present (van Valkenburgh 2009). For many species, a broken tooth can mean infection and possibly death. Other taxa such as some sharks can regrow teeth, although maintaining additional teeth is a cost in itself. Prey can cause life-threatening injuries to carnivores with spines, horns, bites, kicks, and scratches. Prey also can avoid being consumed by hiding, running, and crypsis, causing carnivores to expend energy during searching.

2.5.1   HISTORY

Carnivory is not a recent phenomenon. Shells likely evolved as a way to deter the consumption of soft-bodied herbivores by early carnivores. However, shells were quickly overcome. Borings by predators occur in fossil shells dated as early as the Precambrian–Cambrian boundary, about 542 million years ago (Bengtson and Zhao 1992). Interestingly, many examples of trophic interactions between carnivores and their herbivore prey exist in the fossil record (Martin et al. 2016). Fossilized diet contents of a chiton showed that they consumed crinoids about 440 million years

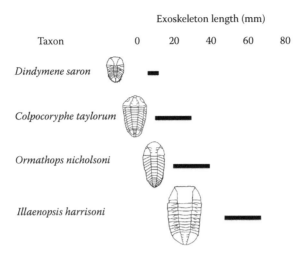

FIGURE 2.10 Common trilobite predators of the Cambrian. The size distribution likely reflects different foraging niches within the trophic level. (Adapted from Fortey, R. A. 2004. *American Scientist* 92 (5):446–453.)

ago (Donovan et al. 2011). In trilobites, a common taxon of the Cambrian seas, carnivory was common (Figure 2.10), with Fortey (2004) describing fossilized records of trilobite and worm tracks that suggest that trilobites departed from encounters but worms did not. Herbivorous trilobites were likely important prey for carnivores as evidenced by injuries in fossilized individuals (Babcock 1993). Although fossil records are scant for most cephalopods, they likely preyed on herbivores. Vertebrates in water and on land were typically carnivores and then evolved herbivory later (Sahney et al. 2011). Terrestrial plants such as pitcher plants and flytraps also have adopted carnivory, as apparent adaptations to low nutrient availability in their sunny, waterlogged bog habitats (Ellison and Gotelli 2001). Animals are rich in nitrogen, which these carnivorous plants cannot obtain from their low-nitrogen environment.

2.5.2 MODES OF CARNIVORY

As with plant defenses, there are many different modes of carnivory (Table 2.3). We characterize them by the ways that carnivores can detect, capture, and consume their prey (Table 2.3). A seemingly simple, yet very complex way that carnivores find and capture their prey is through the pressure-induced sensation of touch. In aquatic systems, the diverse taxa in the phylum Cnidaria use pressure-sensitive structures called *nematocysts* on their bodies and tentacles to attach to prey and inject poison in them (Figure 2.11a). Many predators rely on the sensation of touch, where pressure on their bodies causes neurons to fire. These neurons integrate with a neural net or central nervous system, allowing the predator to respond by pursuing, engulfing, grasping, or biting. Many fishes have a lateral line system, which responds to changes in the pressure of the water around them and can be used to detect prey (Figure 2.11b;

TABLE 2.3
Morphological and Behavioral Adaptations in Carnivores

Prey Detection and Capture	Method	Examples
Touch	Nematocysts	Jellyfish, corals
	Lateral line	Fish
	Somatosensory	Wolves
Heat	Pit organs	Pit vipers
Chemical	Taste	Catfish
	Smell	Lion
Electrical	Electroreceptors	Shark
Hearing	Ears	Bats
	Weberian apparatus	Fish
Vision	Compound eyes	Dragonfly
	Complex eyes	Squid
Mechanical	Grasping	Venus fly trap plant
	Constriction	Rat snake
	Piercing	Aquatic beetle
	Venom	Wasp
	Infection	Komodo dragon
	Web	Spider
Learning	Cognition	Humans
	Social interaction	Hyenas

Bassett et al. 2007). No other group of organisms has a similar system for sensing pressure changes in the environment. Somewhat similar to the sense of touch is the ability to detect temperature differences in the environment to increase encounters with warm-blooded prey. A good example is the ability for pit viper snakes to detect very small gradients in temperature with specialized sensory pit organs in the head (Table 2.3). Many predators have well-developed hearing that allows them to detect prey and consume them. Cyprinid fish have a special system called a Weberian apparatus that amplifies sound and identifies prey (Holt and Johnston 2011).

Predators also can use chemicals and electricity to detect prey in the environment (Table 2.3). Slime molds can use organized chemosensation among cells to organize movement to microbial prey without the aid of a nervous system (Boisseau et al. 2016). The chemosensory systems known by humans as smell and taste are common in carnivores and allow these organisms to make decisions about foraging choices. For example, carnivorous fishes have a highly developed ability to taste specific amino acids found in the bodies of herbivores (Carr et al. 1996). The presence of these molecules stimulates feeding. Many predators are capable of cuing in on the odor of their prey as the prey use chemicals to signal reproductive readiness to mates (Zuk and Kollru 1998; Hamer et al. 2011). Prey also can detect their predators through smell, leading to a complex interplay of odor production and suppression depending on predation risk (Hamer et al. 2011). In aquatic systems, some species of fish such as sharks and eels have electroreceptors that allow them to detect the weak electrical fields generated by

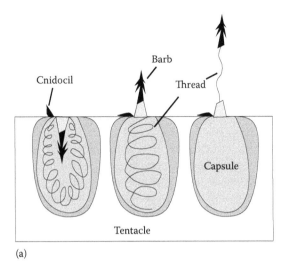

(a)

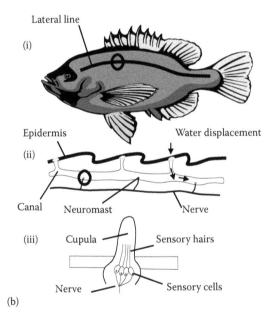

(b)

FIGURE 2.11 Pressure sensitivity in predators. (a) Nematocyst is a common structure in cnidarians. The fluid-filled capsule bursts when the cnidocil is triggered. A barb affixed to a thread is thrust into the prey and injects a toxin. (b) Lateral line system in many teleost fishes extends along the side of the body (i). The lateral line is a canal (ii) that integrates the displacement of water through a series of holes. Water displacement is sensed at neuromasts (iii), which are structures with hair cells that sense the direction that water is flowing in the canal.

their prey in the water (Table 2.3). There is a vast landscape of chemical and electrical signaling occurring between herbivores and their consumers at multiple trophic levels.

Vision is another way that predators can acquire their prey (Table 2.3). Although such a complicated sense as sight may seem to be a relatively recent biological phenomenon, a Cambrian predator, *Anomalocaris* spp., probably related to the arthropods, found its prey with large compound eyes (Whittington and Briggs 1985; Figure 2.12). Compound eyes continue to be typical in many arthropods with photoreceptors specialized for differentiating prey (Land 1997) and, because of their superior ability to detect motion, have been recently incorporated into robotics (Floreano et al. 2013). Similar to compound eyes, complex eyes with lenses have evolved many times since the Cambrian. In the complex eyes of fish, the photoreceptors on the retina are sensitive to the light spectra produced by their prey's skin (Chiao et al. 2011). Thus, carnivore sight and prey visual characters are often morphologically linked and important in moderating trophic interactions.

Successful carnivores must possess the ability to capture and consume prey, which requires morphological characteristics to restrain prey and the intelligence to respond appropriately to them. Characteristics such as sharp teeth, claws, constrictor muscles, barbs, strong jaws, nails, venom, and even infectious bacteria are ways that predators overcome their herbivore prey (Table 2.3). Prey can be quite clever (Burns and Rodd 2008). Learning and cognition are ways that the nervous system of carnivores allows them to overcome adaptations of intelligent prey in complex environments. Arguably, humans have evolved large brains and the ability to use them to overcome their prey (Brain 2000). The ability to coordinate in a complex fashion to hunt and consume herbivores occurs in other taxa as well (e.g., hyenas; Drea and Carter 2009). The characters and behaviors seen in higher-level consumers allow these predators near the peak of the trophic pyramid to persist even in the face of declining resources.

FIGURE 2.12 Note the large compound eyes of the Cambrian predator *Anomolocaris*. Vision is not a new phenomenon. (Courtesy of Shutterstock.)

2.6 OMNIVORY

Omnivory, the consumption of food at multiple trophic levels, happens. However, the degree of omnivory and its impact on trophic interactions are still open to debate. Omnivory is problematic because it challenges the simplicity of the trophic pyramid. Omnivory can arise in many ways. Organisms can occupy different trophic levels during different life stages or perhaps as a function of internal state and food availability during a single life stage. From our perspective as humans, omnivory seems reasonable. However, many investigators argue that natural selection should favor specializations in nature, thereby eliminating the generalist strategy of omnivory (Yodzis 1984; Loxdale et al. 2011; Figure 2.13).

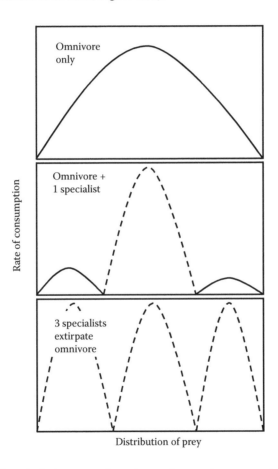

FIGURE 2.13 Hypothetical consumptive patterns of an omnivorous species (solid line) as a function of the distribution of prey. In the absence of other consumer species (top panel), the omnivore has high consumption rates across the distribution of prey. If one specialist species arrives (middle panel, dotted line), it should be a better competitor for a subset of the prey, forcing the omnivore to only consume prey at the ends of the tails of the prey distribution, potentially reducing consumptive rate. If three specialist species arrive (three dotted humps), the omnivore will be outcompeted across the entire distribution of prey (bottom panel).

Although theory may dispel omnivory, it does occur in nature and can dramatically affect trophic structure (Lodge et al. 1994). Polis (1991) pointed out that much of the research done to identify the commonality of omnivores was conducted by using food web inventories, where investigators attempt to tally all the species in a community and draw lines of trophic interactions between them. The problem with this approach is that inventories are artificial and may miss important species and their interactions (Polis 1991). Experimentation is necessary to identify important food web linkages and determine whether omnivory is playing a role. The frequency of omnivores in nature is still not well known and will require techniques outlined later in this book to quantify. To illustrate, techniques such as stable isotope analysis mixing models may be used to resolve some level of an omnivory in consumers, but these models are limited when many prey items are consumed across trophic levels (Chapter 13).

By definition, omnivores should have physical and behavioral characteristics that are intermediate between those of herbivores and carnivores. These characteristics should render these species able to switch among food items and perhaps successful invaders in novel ecosystems. However, they should fare poorly when competing directly with foraging specialists for limited food because they are unable to consume prey as efficiently and quickly. Some intermediate characteristics of omnivores should include a shorter gut than herbivores but a longer gut than carnivores, mobility to access plant and animal prey, large mouths for accessing a variety of prey sizes or the ability to chew/disassemble large prey, a diverse gut microbial community for digesting plant and animal tissue, and strong chemosensory ability for learning whether prey are palatable. Humans are an obvious example of an omnivore, with our gut flora being consistent with that expected for this trophic mode (Ley et al. 2008). Crayfish that consume both vegetation and aquatic invertebrates indiscriminately also may be considered omnivorous (Lodge et al. 1994). A fish, gizzard shad (*Dorosoma cepedianum*), consumes both phytoplankton and zooplankton and can strongly affect trophic interactions (Vanni et al. 2005). Most species probably could be considered omnivores depending on the resolution of trophic groups being consumed. The degree to which different food items are consumed and how labile the consumer is at switching diets needs to be better defined to precisely define this trophic strategy.

2.7 CONCLUSIONS

Trophic pyramids are important, albeit simplistic constructs in ecology, providing a useful way for us to think about the structure of ecosystems and participate in debate. One view is that natural selection favors specialization and should lead to distinct trophic roles that can be categorized into discrete trophic levels (Figure 2.13). In fact, the variation in feeding modes that may occur in one apparent species might reveal the presence of a host of *cryptic species*, each with a discrete foraging specialization (Loxdale et al. 2011). A contrasting view is that omnivory and other generalist foraging tactics are common in ecosystems and that it is misleading to cram species into false trophic categories. Other important ecosystem processes such as detritivory and ecological subsidies also are missed in trophic pyramids, although a large proportion of autotrophic production can move down this pathway.

An iconic ecosystem that does not fit well into the trophic pyramid conceptual model is a coral reef. The algal base of these highly diverse and productive biological systems lies in the bodies of cnidarian carnivores (Wooldridge 2010). In fact, these algae called zooxanthellae may provide as much as 90% of the energy consumed by corals, making the corals indirect herbivores as well. This enormous energy consumption by coral consumers led Wooldridge (2010) to suggest that corals are parasites of these *symbiotic* algae, likely reducing algal reproductive fitness. Regardless of how the interaction is viewed, the relationship between zooxanthellae and corals blurs the distinction among trophic levels and makes the trophic pyramid concept tenuous in this ecosystem.

The terms used by ecologists to categorize trophic interactions are often as constraining as the trophic pyramid concept itself. Terms such as *autotroph, consumer, herbivore, carnivore*, and *omnivore* are useful for communicating key ways that organisms capture energy. However, ecologists must recognize that this lexicon only may hold true under specific circumstances or for particular life stages. We should be particularly vigilant about avoiding these strict categories when empirical evidence suggests otherwise.

QUESTIONS AND ASSIGNMENTS

1. What are alternative reasons for a trophic pyramid to be inverted?
2. Do autotrophs really exist?
3. How is primary production distributed across ecosystems on Earth? What is the major limiting factor?
4. What is a primary consumer? Are they always herbivores?
5. What is the difference between grazing and browsing?
6. When did herbivory first arise? How did algae and plants respond to herbivory?
7. At what time did herbivory start appearing on land?
8. What are the different modes of herbivory and what is their chronology through evolutionary time?
9. Why is herbivory typically so inefficient at removing autotrophs? Do exceptions exist?
10. What are the two general ways that plants and algae resist herbivory?
11. How does the spatial distribution of plants and algae influence herbivore distributions in space and time?
12. Describe the evolutionary history of carnivory.
13. Why are certain adaptations for carnivory useful in some environments but not others? Use examples.
14. Should omnivory be rare or common in ecosystems? Why?
15. In what ecosystems might trophic pyramids be particularly problematic as a conceptual model?

3 Scavenging and Decomposition

3.1 APPROACH

We introduce the concept of the detrital web, showing that it is driven by complex interactions that are not well understood in most ecosystems. Rates of scavenging by large organisms and consumption by bacteria and fungi vary, and these groups may either work as *consortia* to decompose dead organisms or compete for resources. We will show how important decomposition is to ecosystem processes and biogeochemical cycling at the local and global scales, with broad implications for issues such as global climate change.

3.2 INTRODUCTION

Death, waste, and decay surround us. Yet, rarely do we see much evidence of those processes unless we go looking for them. A dead animal on the highway, decaying logs on a walk in the woods, a pile of manure in a field, and fallen leaves blowing in the wind are evidence that death and decline are occurring and that plants and animals are inefficient at using materials they produce and consume. But after a few short moments in our life, those things disappear into a seemingly invisible world that exists beyond the traditional Eltonian trophic pyramid (Chapter 2). Unlike other trophic levels that have distinctive names, this trophic category is so complex and varied that it has never received a proper name. The end products are detritus, scavenger biomass, microbial biomass, carbon dioxide, heat, and *mineralized* nutrients, so the *detrital web* is probably the best term to use (Figure 3.1).

The accumulation of the products of death and waste in the environment can be quantified in many ways. Probably the most common is the contribution of dead organisms and their waste to the carbon cycle. The net carbon produced on Earth each year by autotrophs has been estimated at about 104 gigatons (Field et al. 2005), which is a fraction of the 2000 gigatons of standing organic matter on the surface. Most of the organic matter is found in soil (1500 gigatons) and is nonliving. This highlights that organically derived living carbon is relatively rare on the planet's surface. Most of the 104 gigatons of fixed carbon does not remain in living organisms for long. Some of it persists for a short while as individuals grow to adulthood, reproduce, and finally die. A small fraction of fixed carbon is captured through herbivory or predation. Most of it senesces or dies and thereafter follows many complex consumptive pathways, which include consumption by large animals, fungi, microscopic organisms, and eventually, once again, autotrophs as carbon dioxide or organically derived carbon (Figure 3.1).

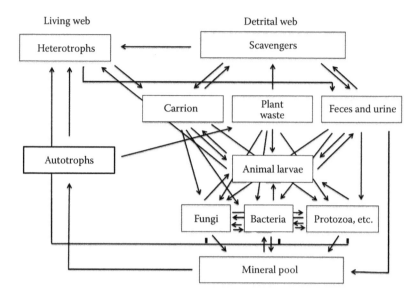

FIGURE 3.1 Comparison of hypothetical energy flow between compartments within organisms consuming living matter (left) and those consuming dead materials (right). Arrows show the direction of flux of energy and nutrients between compartments.

Organisms that consume dead organisms, plant waste, and animal waste are surprisingly understudied given their importance to carbon cycling and other ecosystem processes such as the nitrogen cycle, the hydrological cycle, and other biogeochemical pathways (McGuire and Treseder 2010). Because humans are releasing large amounts of carbon dioxide into the atmosphere through consumption of fossil fuels, the fate of the dead organisms and the carbon and other pollutants (e.g., nitrogen and sulfur) they contain is very important to know. The carbon will eventually end up in the atmosphere as carbon dioxide via respiration or in some sink that may be unavailable to primary producers and consumers for millions of years. About half of the carbon dioxide produced by humans is incorporated into the Earth's biota (Field et al. 2005); the remainder contributes to climbing atmospheric concentrations. Nitrogenous and sulfuric compounds produced by burning fossil fuels contribute to acid rain and enrichment of ecosystems through atmospheric deposition.

Accumulation of dead organisms is rare on many parts of Earth because the scavengers and decomposers are efficient at consuming the waste and breaking it down further to organic matter. This chapter will distinguish between decomposition pathways for animals and plants. The composition of dead organisms influences how they are consumed and the relative amount of time they remain intact in the environment, before becoming mineralized (i.e., converted back to their component elements). The fate of these materials including carbon, nitrogen, phosphorus, and sulfur drives many ecosystem processes. We will assess how processes such as ecosystem production and biogeochemical cycling of materials at the local and global scales are driven by the dynamics of death and decay.

3.3 ANIMALS, SCAVENGERS, AND DECOMPOSERS

Ecologists have spent much time and effort quantifying and modeling foraging and, more specifically, predation. The general assumption is that the animal prey are alive when consumed. The reality is that many times the food is either dying or dead. In Chapter 5, we will show that the act of searching, pursuing, and handling or overcoming prey can be dangerous and energetically costly (Figure 3.2). The benefit of predation is a fresh, nutritious meal with potentially little energetic cost (at least initially) from competitors (Figure 3.2). Coming very close to predation in terms of profitable foraging is scavenging for carrion. Because in this case prey are already dead, the cost of pursuit and overcoming them is minimal. However, the killer may still be lurking and waiting on another meal via the scavenger. Other scavenging competitors for the carrion may increase the chance of injury and loss of foraging opportunities. Many organisms have specific physical and behavioral characteristics that aid in foraging on dead bodies (Tamburri and Barry 1999). For example, different species of Old World vultures have unique physical characteristics that allow them to specialize in consuming different body parts (Houston 1983). Different species of birds and mammals partition carrion resources in space and time, likely to avoid competition and *intraguild* predation (Selva et al. 2005; Table 3.1).

Scavenging is very common in nature, even among animals thought to be apex predators (Table 3.1). Our human ancestors were likely opportunistic scavengers, searching the sky for vultures and running to gather fresh carrion (Ruxton and Wilkinson 2013). In western culture, the act of eating a live animal is typically not considered palatable. In other words, many humans continue to be scavengers, preferring to eat fresh carrion rather than a struggling prey item. The dinosaurs known as velociraptors, made famous in the *Jurassic Park* movies, also scavenged when possible (Hone et al. 2012), as do great white sharks today (Fallows et al. 2013).

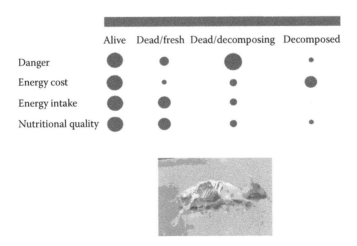

FIGURE 3.2 Costs and benefits, as depicted by the size of each dot, of predation and scavenging. Scavenging is broken down into consuming fresh carrion, decaying carrion, or decayed carrion (detritus).

TABLE 3.1

Characteristics of Some Carrion Consumers

Scavenger	Character	Utility
Vulture	Naked heads	Reduce infection
	Hooked beaks	Break bones
	Olfaction	Detect carrion
Hyena	Strong jaws	Break bones
	Olfaction	Detect carrion
	Aggression	Reduce competition
	Pack behavior	Reduce competition
Crayfish	Olfaction	Detect carrion
	Sharp mandibles	Tear apart carrion
	High density	Reduce competition
	Aggression	Reduce competition
Hagfish	Slime	Reduce competition
	High density	Reduce competition
Humans	Sight	Detect carrion
	Running	Access carrion
	Tools	Access carrion/reduce competition
Bacteria	Toxic chemical	Reduce competition
	Rapid reproduction	Reduce competition

Hagfishes are an exceptional example of scavengers (Figure 3.3). These eel-like marine fish have no jaws and produce copious amounts of mucus, swarming in a mass of slime to stave off competitors (Tamburri and Barry 1999) and reducing fish and marine mammal carcasses to bones within a day.

Vertebrate scavengers have limited time to consume carrion and waste. Insects, aided by their capacity for flight and keen senses of smell, rapidly colonize dead bodies and begin consuming them. Many insect species also use carrion as rearing substrata for immature stages (Putman 1977; Rivers et al. 2011), a process that can rapidly render a carcass useless as a nutritious food source to vertebrate scavengers. However, some

FIGURE 3.3   A hagfish.

predation occurs on the insect larvae, pupae, and emerging adults, but the density is usually very high; the predators accumulating around the corpse cannot overcome the high rate of consumption of the carrion by insect larvae (Putman 1977). Simultaneously, bacterial and fungal assemblages accumulate. Interestingly, insect larvae are often able to consume the flesh rapidly before it becomes infested with bacteria, keeping the rate of decomposition in check. Doctors have known this for centuries, using fly larvae (i.e., maggots) to keep human wounds clean as they heal using maggot debridement therapy. Once microbial activity dominates, carrion is shunned by many scavengers. Even most insects will not use the body as rearing habitat after advanced microbial colonization (Rozen et al. 2008). Lack of palatability is not the only reason for avoidance. Many microbes have infectious and toxic characteristics, which Janzen (1977) hypothesized were ways that these tiny organisms are able to outcompete scavengers for carrion. Bacteria genera that are harmful in carrion include *Clostridium*, *Escherichia*, *Staphylococcus*, and *Shigella*. Many foul-smelling amine compounds like amylamine, putrescine, cadaverine, and tyramine are produced during composition (Lovenberg 1973), and these can be associated with illnesses caused by consuming decomposing carrion. As such, scavengers often avoid consuming carrion with these odors.

Janzen's hypothesis appeared to receive little attention until recently. Burkepile et al. (2006) tested whether scavengers in the ocean preferred fresh versus decomposing carrion. They set traps for stone crabs baited with either fresh or decaying fish in coastal Georgia (Figure 3.4). Stone crabs consumed more than two times the fresh fish bait than the decomposing carrion. Bacterial analysis of the fresh carrion showed that it was dominated by aerobic, nontoxic taxa. Aged carrion was dominated by anaerobic, toxic genera such as *Bacillus* and *Clostridium*. Thus, ocean scavengers, like their terrestrial counterparts, are deterred in similar ways by bacteria they compete with for carrion.

FIGURE 3.4 Deron Burkepile (left) and Brock Woodson (driving) transporting crab traps that will be baited with either fresh or rotten menhaden and deployed in tidal creeks near Skidaway Island, Georgia, USA. (Courtesy of J. D. Parker. Reprinted from Ecological Society of America. With permission.)

Once the usable carbon and other nutrients are eliminated from a dead animal, the body may require a long time to further decompose. Insufficient energy is available to make consuming the remainder of the organism energetically feasible (Figure 3.2). Non-water-soluble refractory minerals and substances like phosphate and calcium carbonate may remain for a very long time, eventually becoming fossils under the proper conditions. Coral reefs depend on the durability of calcium carbonate deposits from dead corals for their structure and function. Carbon and nitrogen trapped in large, stable molecules found in tissues such as chitinous exoskeletons, cartilage and other connective tissue, scales, and fur may remain in the environment for a long time, but will eventually be consumed as long as the material is limiting in the environment. A sad example of this was shown by Robert Ballard. Upon sending a remotely operated vehicle down to the hundred-year-old sunken Titanic in 2004, he discovered only the chemically tanned leather shoes of the victims on the ocean floor. The remainder of the bodies were completely processed by scavengers and microbes in that nutrient-poor environment.

3.4   PLANTS AND DECOMPOSERS

Like predators, some herbivores do not eat plants while they are living. Many plants, like animal prey, have defenses that make it difficult or perilous for herbivores to consume them while they are living (Chapter 2). When plants die or are damaged, or parts are removed, the edible tissues that were once protected by physical and chemical defenses are exposed, allowing scavenging herbivores to consume them successfully. Humans speed up this early decomposition process in plants, most notably through cooking, to enhance the plants' nutritional content and reduce their potential toxicity. Many organisms have unique ways to capture and process discarded or dead plant matter into edible forms (Table 3.2). Associated with this is the perishability hypothesis that suggests that herbivores store food based on the relative perishability of items (Hadj-Chikh et al. 1996). Perishable items are consumed, while those that

TABLE 3.2
Scavengers of Plants and Their Characters

Scavenging Herbivore	Character	Utility
Goat	Jaws, flat teeth	Break down plant matter
	Ruminant stomach	Chemical digestion of cellulose
Leaf-cutting ants	Cut leaves	Remove leaves for collection
	Ferment leaves with fungus	Convert leaf into food
Mussels	Filter feeding	Gather particles
Woodlice	Hard bodies	Difficult to eat
	Chewing	Process wood
	Unpalatable	Reduce predation
Earthworms	Digging	Reduce predation
	High reproductive rate	Quickly access plant biomass
Caddisfly larvae	Shredding/nets	Gather particles

tolerate storage by resisting decay are stored. Tannin content affects this in seeds. Those that have low tannin contents are often consumed first. Those with high tannin contents are cached. Through time, the tannin breaks down making these seeds more palatable.

Because plants are lower in nutrients such as nitrogen than animals, the rate of decomposition and potential toxicity of dead matter to consumers are low initially. However, as with animal carrion, pathogenic organisms will colonize and use chemical deterrents to reduce scavenging (Janzen 1977). The process of decomposition usually reduces the nutritional content of the plant by simplifying complex, nutritionally valuable molecules (see Chapter 9). Limited nutrients like nitrogen are often preferentially removed by decomposers. By-products of decomposition include oxalic acid, methanol, patulin, aflatoxin, and botulin, all of which are toxic to vertebrates at low concentrations. Production of these by-products by decomposers is likely not accidental, but intended to reduce competition with scavenging herbivores (Janzen 1977).

Many bacterial antibiotics have been isolated from fungi that coexist with bacteria in decaying plants. Penicillin is the most common example. Bacteria and fungi often compete for the carbon and nutrients in the plant litter. Thus, the most parsimonious explanation is that the fungi produce toxins to kill bacteria to gain access to more resources. Janzen (1977) posited that the fungi may have a more dubious intent for herbivores. Many herbivores rely on specific gut microflora to properly digest plant matter (Chapter 2). Fungal antibiotics may upset these stomach assemblages, harming the herbivore competitor and reducing its competitive ability.

Humans, being the resourceful creatures we are, have learned to take advantage of the potential toxic defenses of bacteria and fungi in decaying plants through controlled fermentation (Janzen 1977). The strong smells of acids produced in pickling may have been meant to be a deterrent to herbivory because they suggest anaerobic processes and the presence of potentially deadly bacteria such as those that produce botulism. Fermentation of grains and fruit produces ethanol, again a by-product of anaerobic respiration, indicating potentially lethal conditions. Most animals avoid food or juices with high ethanol concentrations. The intoxicating effect of ethanol may be enjoyable for some humans, but it is probably not useful for most animals trying to avoid predation and find resources in a dangerous world.

As with animals, the final stage of decomposition produces materials that are not easily digested and provide little energy to most consumers. These refractory materials, like lignin and cellulose, sometimes collectively referred to as *lignocellulose*, are long chains of phenolic (ring-like) carbon molecules that are not easily broken down. Some fungi and bacteria produce enzymes capable of cleaving these molecules into digestible compounds. In the absence of these organisms, lignocellulose (e.g., wood) can persist for centuries, hence its use as a building material for humans and other organisms such as beavers (Naiman et al. 1999).

It is important to note that not all interactions among animals, fungi, and bacteria consuming decaying plant matter are antagonistic. In both soils and aquatic ecosystems, scavengers, fungi, and bacteria may work in concert through ecological *consortia* to increase the digestibility of the organic matter. Leaves falling in a stream may be processed by larval insect shredders such as caddisflies (Figure 3.5), which

FIGURE 3.5   Larval caddisfly, an important shredder of decaying leaf matter in streams.

tear the leaves into small pieces that have higher surface areas for microbial and fungal activity. Fungi produce hyphae that penetrate the leaf matter and cause it to break down further. Bacteria can grow on exposed surfaces. Hieber and Gessner (2002) found experimentally that *shredders* account for up to 64% of leaf matter breakdown. Fungi and bacteria contributed about 18% and 9% to breakdown, respectively. This will be explored further in the following section.

3.5   ENVIRONMENTAL EFFECTS

Decomposition rates and the relative contributions of scavengers, microbes, and the physical environment to decomposition depend both on the type of dead organism and the environmental conditions. The percentage of annual net primary production and the amount of it that enters the detrital pool varies both temporally and spatially. Most of the energy used by humans at present derives from nonrenewable legacy detritus, some having been fixed and stored in the Earth's crust for more than half a billion years. Thus, historically, the carbon produced was not perfectly recycled but lost under the sediments of the ocean and shallow inland seas (Woodwell et al. 1978). Anaerobic conditions and, much later, tectonic forces helped convert this organic carbon into oil, coal, and natural gas, all of which are likely to be exhausted as carbon dioxide into the atmosphere within the next century. Fossil carbon is greater than the standing biomass of organic carbon, likely approaching 4000 gigatons.

Many physical factors affect the fate of the carbon and other materials produced in organisms. One of the most important factors affecting decomposition in terrestrial ecosystems is water availability. As with all organisms, the bodies of scavengers and decomposers rely on water to function. If the carrion is dry, it will not be readily consumed—the reason why drying is an effective preservation technique for human food. In most aquatic systems, decomposition is not limited by water availability. However, if pH is low such as in a bog ecosystem or salt concentration is high such as in a saline lake, decomposition will again be tempered because of limited microbial activity in these conditions. Temperature affects the metabolic rates of scavengers and decomposers. Endothermic scavengers have a competitive advantage over microbial decomposers at cold temperatures, because the metabolic rates of bacteria and fungi are depressed. At warm temperatures, microbial activity will be high, giving decomposers a competitive edge. Depressed oxygen concentrations also limit decomposition rates. Hypoxic or anoxic conditions may occur in soil and some aquatic ecosystems such as the bottom of lakes and wetlands. In some cases, such as many wetlands, anoxic conditions and suppressed decomposition rates are natural. However, anoxia coupled with reduced decomposition rates in aquatic habitats are often symptomatic of cultural eutrophication, which results from excess inputs of limiting nutrients such as nitrogen and phosphorus associated with agricultural activities and urbanization (V. H. Smith et al. 1999).

Organisms present in the environment may interact with environmental conditions to affect the rates and fates of decomposing organisms. Makkonen et al. (2012) were interested in whether microbial activity differed across latitudes to influence decomposition potential of ecosystems. They conducted a reciprocal transplant experiment using decomposers from multiple latitudes and found no difference of origin on the potential for breaking down carrion and plant matter. Thus, environmental conditions interacting with bacteria and fungi, not the species composition of the microbial assemblage, likely affect decomposition at that stage.

Animals ranging from tiny invertebrates to large vertebrates can influence decomposition directly by feeding on materials, or indirectly through physical breakdown from trampling, excretion, and similar activities. The presence of arthropods such as leaf-litter dwelling ants accelerates the decomposition of leaf litter (McGlynn and Poirson 2012). The worldwide decline of scavenging species such as wolves and vultures negatively influences the rates of decomposition of animal carcasses (Sekercioglu et al. 2004). Some animals can both enhance and depress decomposition. For example, forest floor amphibians can speed decomposition through production of nitrogen-rich excretory products that enhance decomposer activity, but they can also indirectly slow decomposition by feeding on detritivorous arthropods (Beard et al. 2002; Davic and Welsh 2004).

In soils and many aquatic habitats, entire assemblages of invertebrates are responsible for converting leaf matter and other senescent materials from whole parts, often referred to as coarse particulate organic matter (CPOM), to smaller particles (fine particulate organic matter [FPOM]). This breakdown process generates food for groups that feed on smaller particles and increases surface area for microbial colonization. In streams, assemblages that feed on coarse particulate materials, such as senescent leaves from the surrounding forest, are referred to as shredders.

Shredder feeding activities produce smaller particles, which are in turn fed on by filter feeders such as net-spinning caddisfly larvae, active filter feeders such as mussels, and collector gatherers such as midge larvae (Wallace and Webster 1996). This breakdown process from CPOM to FPOM also promotes the export of energy from headwater streams, where detritus inputs can be quite high, to downstream food webs because smaller FPOM particles are more easily entrained in the water column (Vannote et al. 1980; Box 3.1). As materials are processed by these groups, they are continuously colonized and further broken down by bacteria and fungi.

As discussed earlier in this chapter, microbes may sometimes compete with scavengers and render dead materials less palatable or even dangerous to consumers. However, microbial colonization can also enhance the palatability and quality of nonliving plant materials, which, without microbial colonization, may have little nutritional value. Microbial biomass, which is generally consumed along with the dead plant materials, is much higher in nutrients such as nitrogen and phosphorus compared to the mostly carbon substrates of decomposing plant materials. The process of microbial colonization enhancing nutritional quality of detritus is sometimes termed the *peanut butter (microbes) on the cracker (detritus) concept* and it ultimately lowers the carbon to nitrogen (C/N) or phosphorus (C/P) ratio of the material (France 2011). Recent studies using stable isotopes and fatty acids as tracers demonstrate that many detritivores derive much of their nutritional needs from microbes (see Chapter 13). Thus, an organism's functional role (e.g., detritus shredder) may be different from its trophic status (e.g., microbivore).

BOX 3.1

The river continuum concept was developed by Vannote et al. (1980) to describe longitudinal patterns of community structure and energy flow along streams from headwaters to large rivers. Although most ecosystems are considered driven by autotrophic production from the base of the trophic pyramid, small- and medium-sized streams neither receive enough sunlight nor produce enough algae to support the communities of primary and secondary consumers in them. Stream ecologists have long known that some of the energy in streams was from outside (allochthonous) terrestrial inputs and that these systems were therefore heterotrophic in that most energy is derived from outside sources. Vannote et al. (1980) formalized this idea, demonstrating how energy derived from surrounding terrestrial habitats falls into headwater streams and is subsequently processed by shredding invertebrates and microbes that break it down into simpler forms that are used by other groups such as filter feeders.

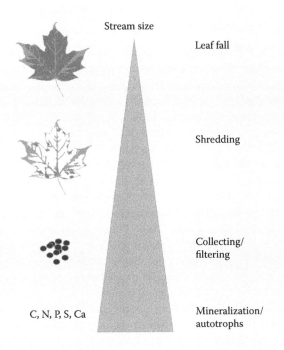

As the organic materials are processed, they also become smaller in size and are more easily transported in the moving water, fueling downstream food webs. As the stream becomes wider and receives more sunlight because of reduced forest canopy cover along the middle reaches, primary production increases and the stream is more autotrophic, whereby autochthonous (from within) sources of energy are increasingly important. At some point along the continuum, the river channel becomes deep and the water is turbid, with minimal light penetration to the substrata. Here, the system shifts back to a more heterotrophic state, with important detrital energy inputs from upstream and floodplain sources. Along with invertebrates, larger detritivores such as many fishes are present in the large river habitats. This conceptual model was one of the first to show how the relative importance of detrital and autotrophic processes influences ecosystem structure and function.

The above figure shows the fate of a leaf falling into a small stream that flows into a large river, as per the river continuum concept (Vannote et al. 1980). Heterotrophic processes of consumption dominate initially where the leaf is processed by invertebrate shredders, bacteria, and fungi. The *waste* produced flows downstream where it is collected and filtered by other organisms such as filter-feeding mussels. Nutrients released support both autotrophic production and detritivory. The width of the triangle along the continuum corresponds to stream size.

3.6 COMPARATIVE ECOSYSTEM CONSEQUENCES OF DECOMPOSITION

As demonstrated in Figure 3.1, the detrital food web is terribly complex and generality seems implausible. However, some patterns do exist for trophic ecology that are unique to aquatic and terrestrial ecosystems. Aquatic systems like lakes, streams, and the ocean should have *loose* associations between scavenging and decomposition because water moves decaying matter (Beasley et al. 2012). The process of organic matter transformation in streams described earlier is an example of how a dead organism or body part falls upstream and is converted to simpler forms as it moves downstream through the stream continuum (Vannote et al. 1980; Box 3.1). In marine ecosystems, a marine mammal may serve as prey for sharks and other fish while it floats near the surface. Once it sinks into the cold, abyssal depths, it can provide prey for thousands of scavengers and eventually bacteria, lasting for months to decades (Smith and Baco 2003; Figure 3.6). Similar processes occur in deep freshwater lakes with minimal mixing. Warm, light water accumulates near the surface called the *epilimnion*. Cold, dense water becomes trapped at the bottom. Microbial activity rapidly exhausts oxygen in this zone called the *hypolimnion*. Organisms sinking into this zone may remain intact for a long time because of reduced microbial activity. In both oceans and freshwater lakes, upwelling is the process by which nutrients trapped in the depths are transported back to the surface via wind- and temperature-dependent currents. This water, rich in nutrients and organic matter, stimulates primary production (Figure 3.6). In temperate latitude lakes, this often occurs in the fall when cooling temperatures break down the epilimnetic and hypolimnetic layers, resulting in mixing. In the ocean, changes in wind-driven currents interacting with coastal zones and islands create upwelling conditions (Figure 3.6).

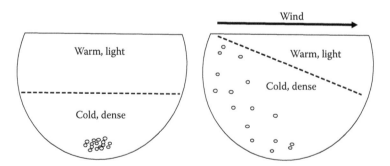

FIGURE 3.6 Stratification of water that occurs in lakes and the ocean. Warm water of low density floats above cold, dense water (left). With no water movement, dead animals and algae drop into the cold zone and settle in the bottom, where decomposition is slow. Wind at the surface pushes water and allows cold, dense water to well up toward the surface. This process suspends nutrients and organic matter trapped in the bottom, allowing it to be consumed by autotrophs.

Although the nutrients in a dead organism in water may be transported many kilometers from the place of its demise, carrion on land typically rots where it drops (Beasley et al. 2012), although scavengers may move sections across the landscape. Terrestrial organisms are scavenged and decomposed more efficiently than in water, because a stationary carcass can be rapidly consumed by specialized scavengers, insect larvae, and then microbes (Coe 1978; Carter et al. 2007). The distribution of carrion on the landscape affects heterogeneity in arthropod assemblages and nutrient availability to plants that persists for long times. Similarly, areas where animals drop feces and urine create a mosaic of nutrients and organic matter on the landscape. A simple look at the heterogeneous patches of green in the backyard shows where the family dog does its business and promotes primary production in the lawn.

These differences between processes of decomposition in water and land explain in part why so much of the mineralized fossilized body parts we see in museums and the fossil fuels we depend on for energy derive from aquatic systems, particularly the ocean. The ocean and swamps of Earth have been and continue to be sinks for organic matter produced by autotrophs and incorporated into dead consumers. The decomposers in water become limited by depth, oxygen, and temperature allowing carbon and minerals to accumulate in the Earth's crust.

3.7 ECOSYSTEM PROCESSES DEPEND ON DECOMPOSITION

Most conceptual and computational models of food webs fail to fully account for the energy that is consumed and transformed by the detrital food web. Wilson and Wolkovich (2011) conducted a review that showed that scavenging is underestimated by 16 times in food web studies. Most predators also are facultative scavengers, taking advantage of an energetically inexpensive meal when possible. Thus, a significant proportion of energy entering the consumer trophic levels may be derived from the detrital web (Figure 3.1), with predation and its associated energetic costs being overestimated. This is a problem for trophic ecology because the energy budget for ecosystems depends not only on the energy flux between trophic levels but also on the associated costs of foraging. Predation may have a much different cost per foraging patch than scavenging, thereby affecting models of dispersal (see Chapter 4). Indirect effects of foraging activities are discussed in depth in Chapter 10. These effects are related to direct foraging but not the actual foraging event. A predator may selectively consume one prey organism that is a competitive dominant in a community. By reducing the population size of this competitor, the predator may indirectly help another species that was being outcompeted by the competitive dominant. Predators are considered important community components, promoting diversity (Chapter 5). However, Wilson and Wolkovich (2011) aptly noted that this assumption may be wrong if the predator is actually scavenging the competitive dominant. In this case, some other unknown factor, maybe disease, is controlling the competitive dominant and the scavenger is getting undeserved credit. This is a particular problem if predation is inferred from diet studies rather than experiments or direct observation of consumption (Chapter 7).

As the children's book by Gomi Taro *Everyone Poops* (1977) proclaims, all organisms produce waste. While solid (egested materials or feces) and liquid (excretory products such as urine) waste materials are produced by terrestrial organisms, most aquatic organisms do not produce liquid waste (with the exception of excreted water). Rather, the nitrogen produced by aquatic organisms as ammonia is excreted directly into the water via gills or other highly permeable organs and tissues. Like carrion and plant matter, solid waste produced via inefficient digestion (feces) is typically processed rapidly in the environment and may play an immediate role in ecosystem processes. Groups like dung beetles and many true flies are specialized at processing feces. All the organisms in an ecosystem are producing waste continuously, in contrast to the occasional dying individual, which is a discrete event producing a relatively small amount of organic material. Nutrients in feces and excreta are more labile than those in carrion because waste matter has already gone through digestion, where molecules have been converted to simpler forms.

Living bodies of organisms—consumers and scavengers alike—must contend with ammonia as an end product of protein metabolism. Ammonia is toxic to living tissue and must be excreted. It does not build up in the environment because it is the preferred nitrogen source for most plants, algae, and many bacteria. Nitrogen rarely is available from abiotic sources because it is tied up in its gaseous form in the atmosphere, so nitrogenous waste from consumers is the primary source in all ecosystems (Vanni 2002). Until relatively recently, the primary source of nitrogen in terrestrial and aquatic ecosystems was believed to be from the waste produced by the microbial *ooze* (Lindeman 1942; Figure 3.7). The nitrogenous waste produced by multicellular organisms including large vertebrates is now recognized as being important in nutrient cycling (Vanni 2002). The movement of these organisms organizes both terrestrial and aquatic communities by enhancing primary production around them (see Chapter 12). Daily movement of fish between the shoreline and the open water of lakes transports nutrients. Sunfish consuming microscopic zooplankton in the open water of lakes during the day excrete and egest nutrients in shoreline habitats at night (Vanni 2002). Thus, feces and excreted nitrogen promote shoreline plants through an indirect pathway.

Nitrogen is only one of the many components of waste produced by all organisms. Phosphorus, sulfur, carbon, calcium, and many other elements are released present in waste products. Nitrogen and the other materials are often rapidly consumed by autotrophs through a process called *nutrient recycling*. As we will show in Chapter 12, the ratio of nutrients released back into the environment will affect the types of autotrophs and consumers that arise in ecosystems. Nutrients can be temporarily *stored* in soil and sediments as detritus. These elements can be cycled back into the environment via consumption by detritivores (Vanni et al. 2005). Detritivorous fishes, including mullets in marine ecosystems and gizzard shad in freshwater lakes and rivers, specialize on consuming highly decomposed organic matter on the bottom, and in doing so they release nutrients in it back into the open water for algae to utilize (Winemiller 1990; Vanni 2002; Figure 3.8). Earthworms in soil serve a similar purpose by releasing nutrients in detritus back to highly available forms for autotrophs. This process has been called a nutrient *pump* (Figure 3.8).

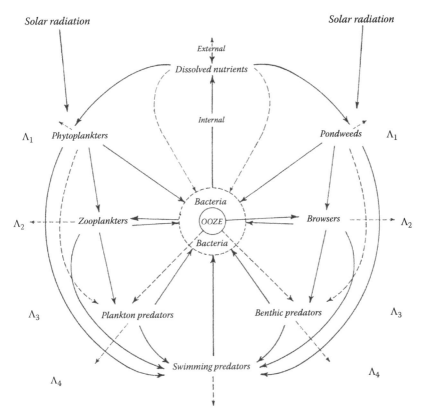

FIGURE 3.7 Trophic structure of a lake food web as envisioned by Lindeman (1942). At the center of the wheel of energy transformations is the microbial loop (ooze) that recycles nutrients and supports the ecosystem. (Reprinted from the Ecological Society of America. With permission.)

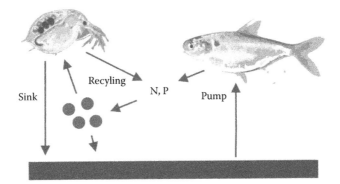

FIGURE 3.8 Simplified nutrient recycling and pumping in a freshwater ecosystem. A zooplankter consumes phytoplankton (left) and excretes nitrogen (N) and phosphorus (P). Some of these nutrients are consumed by the phytoplankton. Both the zooplankton and phytoplankton die and sink into the lake sediment. A detritivorous fish like gizzard shad (right) consumes the organic matter in the ooze and resuspends it for use by the phytoplankton. Arrows indicate the direction of flux of nutrients.

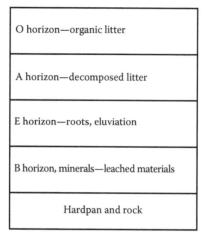

FIGURE 3.9 Simplified structure of soil.

Both the soil in terrestrial ecosystems and the sediments of aquatic ones are structured by complex interactions between decomposition and the physical environment. Soil ecologists divide its layers into horizons (Jenny 1980; Figure 3.9). Organic material from decomposition is limited to the top layers. At the surface is litter (i.e., the O horizon), which is composed of plant and animal matter that is beginning the process of decomposition. Temperature and water determine how long this layer persists. Under winter snow, the litter layer will remain but will be rapidly consumed once the spring thaw occurs. This top layer is rapidly incorporated into the next layer, the A horizon, by microbial decomposition and invertebrate consumption and egestion. This detrital material is called *humus* by soil ecologists. The process of eluviation or leaching occurs in the E layer, where most plant roots occur. Water percolating down through the soil carries minerals down into the next layer (B horizon), where clay accumulates. Thus, most biological decomposition and recycling processes are limited to these areas where carbon and nutrients occur. Once the carbon is removed and only water-insoluble minerals remain, biological activity stops. Without disturbance, these layers will become quite pronounced. In tropical forests, the O horizon may only be a few centimeters deep, because organic matter and nutrients are rapidly recycled back into the aboveground biomass. Conversely, in temperate zones where organic matter accumulates seasonally and in some regions benefits from a legacy of glacial deposition, the O horizon can be deep and productive, making these soils attractive for agriculture.

In lakes, wetlands, ocean sediments, and the floodplains and backwaters of rivers, the processes of decomposition and nutrient recycling are less structured than those that occur in soils. In the open water, nutrients are mostly consumed by phytoplankton and autotrophic bacteria. Ammonia can be converted to nitrite and nitrate through aerobic biological processes called *nitrification* (Figure 3.10). Most of the aerobic activity in the sediments in aquatic ecosystems including nitrification is restricted to shallow depths. Lack of mixing and relatively impermeable materials like clay inhibit the physical penetration of oxygen, which is relatively insoluble in water. In response to this, most vascular aquatic vegetation usually has no roots,

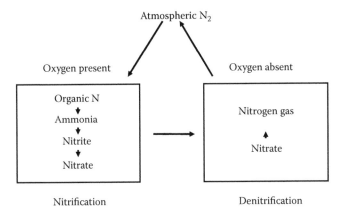

FIGURE 3.10 Simplified nitrogen cycle that is applicable to terrestrial and aquatic ecosystems. Nitrogen gas is fixed into organic (living) forms via fixation by certain autotrophic microorganisms. Ammonia excreted as waste is converted to nitrite and nitrate by bacteria through a process known as nitrification. Nitrate can be converted back to gas under anoxic conditions by anaerobic respiration of certain bacteria.

very shallow roots, or accessory roots that extend upward into the water column to acquire oxygen (e.g., mangroves) (see Lodge et al. 1988). Decaying organic matter that is buried in aquatic sediments is consumed by a specialized class of decomposers called *denitrifiers* (Figure 3.10). These bacteria use the bonds in nitrate, the byproduct of aerobic bacterial nitrification at the surface of the sediment, as an energy source. The waste product of denitrification is elemental nitrogen, which returns to the atmosphere as gas. This is a very important process of nitrogen cycling in aquatic ecosystems (Lijklema 1994). Many anaerobic denitrifying and aerobic nitrifying bacteria may form *consortia*, whereby the anaerobic bacteria benefit from the reduction of oxygen by the aerobes and their production of nitrate as an energy source (Paerl and Pinckney 1996). Another anaerobic process of microbial organic matter processing in aquatic sediments produces methane as a by-product.

As we noted earlier, water moves materials around, decoupling the initial process of scavenging and eventual nutrient recycling. In temperate lakes, the process of seasonal turnover dominates nutrient cycling from sediments to the open water. In the coastal zones of marine ecosystems, upwelling can be important. However, waves and tidal movement are probably more common at physically disturbing the sediments and recycling nutrients in most areas. In rivers, flooding of extrachannel areas such as side channels, oxbow lakes, wetlands, and floodplains suspends organic matter and nutrients trapped in anoxic sediments and transports them downstream.

3.8 DECOMPOSITION AFFECTS GLOBAL CYCLING ACROSS ECOSYSTEMS

Detritus and nutrients produced by decomposition do not always remain in the ecosystem in which they were produced. They are transported among ecosystems via many climatic, geologic, and biological pathways. Rivers transport materials to estuarine and coastal ecosystems. Islands and coastal ecosystems are supported by the influx of marine detritus (Chapter 10). Organic matter suspended in the ocean is rapidly and efficiently filtered by the mussels, barnacles, and other filter feeders in coastal ecosystems. Aquatic insects emerging as adults from streams and lakes move nutrients into terrestrial ecosystems after they reproduce and die or are preyed upon by riparian predators. As we will show in Chapter 10, salmon produced in the ocean transport marine-derived nutrients when they migrate to spawn and are consumed or die in their natal streams.

Cycling of detrital materials at a global scale may be affected by many factors (see Polis et al. 1996 and Chapter 10 for review). Damming large rivers reduces the transport of organic matter on a continental scale, potentially reducing productivity downstream, including in coastal delta areas. The loss of floodplains through damming has a similar effect. Climatic events that affect the direction and intensity of currents in the ocean influence patterns of upwelling and nutrient availability in coastal regions. Marine birds move tons of phosphorus from the ocean to coastal areas annually via feces called *guano*. Losses of these birds through pollution and hunting reduce this lateral movement of nutrients. Detritus that accumulates in the O/A horizon of the soil can be transported long distances. During the Dust Bowl in the central United States in the 1930s, poor land use coupled with intense drought literally took the O/A horizon airborne, transporting it via wind for continental distances. The soil in the wind was experienced far into the Atlantic coastal region of the continent. Detrital material from soil in Africa is regularly transported via wind across the Atlantic Ocean to South America (Chapter 10).

The oceans of Earth are vast and deep. Productivity in most of these systems is very low (Chapter 2). Lack of uptake of nutrients and consumption of detritus means that there is a net flux of particulate organic materials, often collectively called *marine snow* (Turner 2002), into the abyssal zone. Most of the detrital materials produced on land eventually move toward this zone via runoff and airborne particles. The remainder enters the atmosphere as carbon dioxide where the autotrophs in the open ocean cannot consume all of it because of other limiting nutrients, such as iron, which is typically insoluble in seawater (Kerr 1994). The question facing biogeochemists today is, what is the eventual fate of these by-products? It is a race between the ability for the microbial assemblages on the continents and in coastal regions to consume the matter and convert it back to carbon dioxide and other minerals versus the detritus entering the ocean as unprocessed organic matter and sinking to the depths.

This question is far from answered, but is critical for determining how much of the carbon produced by the burning of fossil fuels will persist in the atmosphere and likely warm the planet. Beasley et al. (2012) showed that temperature is one of the primary factors influencing the balance between particulate organic matter making it into the ocean and carbon getting pumped into the atmosphere. Elevated temperature gives the bacteria and fungi a competitive edge over the scavengers. If atmospheric temperature was to increase by 10°C, carrion availability to macrofauna such as vultures, insect larvae, and many of the world's apex predators might be reduced by 10–40% (Beasley et al. 2012). This change in food availability may jeopardize the success of many macro-organisms on Earth. Perhaps more ominously, organic matter flux into the marine environment where it is eventually lost in the abyss may decline by as much as 50% (Smith et al. 2008). If true, then the amount of carbon dioxide waste in the atmosphere will greatly increase because it never makes it into the Earth's ultimate organic sink.

3.9 CONCLUSIONS

The business of death, excretion, and defecation is important to trophic ecology and to the broader disciplines of ecosystem ecology and biogeochemistry. How carrion and waste are consumed by multiple trophic levels ranging from large vertebrates to single-celled bacteria determines the distribution of resources on Earth, including the chemical composition of our atmosphere and perhaps the behavior of our climate.

Ecology is typically defined by textbooks as the study of the distribution of living animals and plants. The general assumption by many ecologists is that food webs driven by living organisms are most important to study. However, even apex predators scavenge when possible, meaning that fresh, dead organisms contribute substantially to the dynamics of these species. Changes in the physical environment that affect the availability of carrion to obligate and facultative scavengers may determine the stability of entire food webs and ecosystems.

Scavengers and decomposers compete for access to carrion and waste. How this competition interacts with environmental factors such as water availability, temperature, weather patterns, ocean currents, and river flooding plays an important role in ecological interactions. Scarce water limits the rate of decomposition. However, lack of water greatly increases the chance of detrital organic matter trapped in soil being moved by wind at global scales. Temperature, which is expected to rise in many regions of the globe, should give microbes a competitive advantage, leading to more rapid production of carbon dioxide, methane, and other chemical waste. Changes in the movement of water both on land and in the ocean as affected by human and climatic factors will further determine the fate of decomposed materials.

The future of trophic ecology will have to account for the multitude of detrital pathways, linking with other disciplines like microbiology, climatology, and geology to answer questions about how consumption drives ecosystem processes at a global scale. Many of the fundamental approaches described in this book are already being applied to this problem and need to be refined.

QUESTIONS AND ASSIGNMENTS

1. How much of the production of live organic matter makes up the total organic carbon on Earth?
2. Why are decomposers important to determining the fate of the organic matter that dies after it is briefly alive?
3. Is predation or scavenging by purported predators more common? Why?
4. How do microbes deter the consumption of carrion by scavengers?
5. What is the perishability hypothesis?
6. Why is lignocellulose so persistent in the environment?

7. Give an example of a consortium of decomposers consuming plant matter.
8. How does temperature mediate the interaction between scavengers and microbial decomposers? Tie this in with predictions about global climate.
9. What are some of the differences between the fate of decaying matter in water and terrestrial ecosystems?
10. Why is ammonia such an important commodity deriving from decomposition?
11. What are denitrifiers? How do they differ from nitrifying decomposers?
12. Where is the ultimate sink for decomposed organic matter?

Section II

Mechanisms at the
Organismal Scale

4 Foraging in Patches

4.1 APPROACH

The intimate trophic interaction between a consumer and its prey is the foundation of trophic ecology. Foragers typically live in environments that have distributions of food that are patchy both spatially and temporally. To persist under these conditions, the internal state of organisms must interact with the abundance, distribution, and variance of food in the environment to ensure survival and reproduction. This chapter will explore several theoretical and simulation approaches used to assess how organisms make foraging decisions to ultimately influence the flux of energy and materials through ecosystems. As the environment becomes increasingly unpredictable through climate change, fragmentation, altered habitats, and other factors, the ability for foragers to successfully respond to changing information will determine whether they persist.

4.2 INTRODUCTION

Food resources on Earth are not distributed homogenously. Rather, food typically is found in patches of differing size, shape, density, and distance (Figure 4.1). The same landscape may look very different depending on the foraging species (Figure 4.1). Foraging organisms must search the aquatic or terrestrial landscape looking for places to eat and then decide how long they should stay before searching again. Unlike humans riding down a highway, foragers do not have the luxury of referring to billboards or GPS devices that announce the distance of the next restaurant on the trip. Ultimately, the decisions foragers make about what patches to occupy will determine whether they thrive or eventually starve to death.

From an ecosystem perspective, the ability for organisms to find their food resources and consume their prey affects how efficiently energy is transferred to higher trophic levels (Figure 4.2; Schmitz et al. 2008). Inefficient foragers will do a poor job of matching their resources in the environment, leading to inefficient ecosystems. For example, imagine a population of herbivorous insects distributed in a field with different levels of primary autotrophic production (Figure 4.2). If the insects distribute themselves randomly or erroneously choose the patches of low primary production, then capture of energy will be low (Figure 4.2a) and the ecosystem will likely be limited in trophic levels and perhaps diversity. In contrast, efficient insects that closely track the availability of food in the environment will transfer more primary production to higher trophic levels (Figure 4.2b), presumably supporting more trophic interactions and more species. This assumes that the insects are available to insectivores. Thus, as foragers within ecosystems become more specialized through natural selection, the efficiency of trophic transfer and

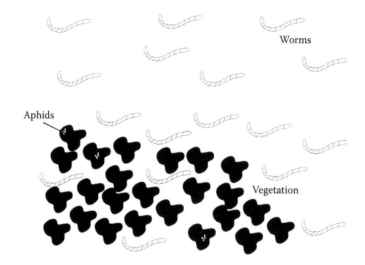

FIGURE 4.1   The foraging landscape will look different depending on the organism. For a robin searching for worms, food patches of worms are homogeneously distributed. For a grasshopper looking for patches of vegetation to consume, the food source is patchy, distributed in the lower left quadrant of the landscape. For an ant searching for aphids in the vegetation, food is very rare and patchily distributed.

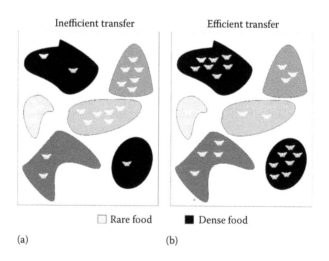

FIGURE 4.2   Concentration of food (level of shading) differs among patches. If foragers do a poor job of tracking the food concentration in the environment, then their distribution will be random (a). Conversely, if foragers track food availability, their density should correlate positively with food availability (b).

complexity of ecosystems should increase. Some plants and algae respond by resisting herbivory, which will reduce the efficiency of transfer and lead to the evolution of plant-defense-herbivory trade-offs.

An improved ability of foragers to track prey in patches should have many additional ecological consequences. The ability for foragers to control their prey via direct and indirect pathways (see Chapter 6) is largely driven by foraging efficiency. These *top-down* effects influence many aspects of prey populations such as their size structure, morphology, and standing biomass (Terborgh and Estes 2010; Chapter 6). Natural resource managers of ecosystems interested in predicting whether novel species will become successful invaders need to understand the relationship among patch choice, patch suitability, and the invader's foraging needs (Box 4.1). Efforts to conserve or restore native species in ecosystems require similar information.

Foraging is a complex business. Over the past few decades, theoretical and empirical research has grown substantially, occasionally generating considerable debate, most notably over the term *optimality* (Pyke 1984; Pierce and Ollason 1987;

BOX 4.1   INVASIVE SPECIES AND FORAGING

Predicting the ability for a new species to successfully invade and then thrive is difficult (Lodge 1993). Foraging opportunities contribute to the potential success of a species and allow it to spread from its point of introduction. If contiguous patches of food are limited, then the ability for the invader to spread is limited. Thus, a management goal is to reduce the food source around the invading species to discourage its spread.

Emerald ash borer *Agrilus planipennis*, which threatens to spread and decimate forests throughout the continent, is a recent invader in central North America. When ash borer adults or their larvae are found, a common management action is to destroy any trees that are infected. Because the larvae of this species consider the space under tree bark a foraging patch, patches and their larval occupants can move in firewood and lumber. Thus, management needs to curb the spread of this species by discouraging the movement of these materials. A campaign in the United States and Canada is underway to educate the public about the danger of moving firewood and spreading this invader via its food source.

Humans can facilitate invasions of species by creating large plots of homogeneous foraging patches. Agriculture is undeniably important for human survival. However, much of agriculture has moved toward creating huge plots of single species of plants. This practice makes farming very efficient, but it also increases the probability that a pest species will find a vast unimpeded habitat patch in which to forage and spread. Agriculture combats this with pesticides and pest-resistant crops. However, new resistant pest species can invade at any time and known pests can develop resistance through natural selection. A way to naturally reduce the ability for invasive pests to spread is by farming using a diverse array of plants. However, the trade-off is loss of efficiency.

Bull and Wang 2010). Optimality is likely not attainable in most ecological systems because it requires nearly perfect information about the environment, which is not possible, perhaps with the exception of highly specialized foragers such as parasites within their hosts (Lozano 1991). We start with the simplest theoretical models and then build up from these to more complex approaches. These models deal with patch choice of foragers on prey without considering specific interactions between the two. We will explore specific predator–prey interactions later (see Chapter 5). At the end of the chapter, we will determine the efficacy of the optimality concept and determine whether it holds any validity for trophic ecology.

4.3 GENERALITIES

Some general factors influence the decisions of foragers. Note that at this juncture, we are only considering within-population factors influencing the motivation and ability to feed. We do not consider how various external factors such as interspecific competition and predation affect foraging decisions (Chapter 6). Stephens and Krebs (1986) noted that there are three general factors that motivate foragers as they search for patches. The first factor revolves around the decision process. Foragers must decide where to eat. This is a complicated issue involving direction and duration. When a forager settles into a patch, it must choose what items to consume and when it is time to leave as the prey items decline. Prey items decline for a variety of reasons. They may be reduced by being eaten, fleeing the area, hiding, reducing activity, or ending up as a meal for competitors (Chapter 6). Without understanding what causes a forager to give up, it is difficult to make predictions about its residence time in patches. Experimental approaches and direct observation are needed to evaluate factors influencing patch choice.

In addition to the prey base of the patch, the currency used to assess the value of a patch may be important in driving foraging decisions. Organisms can be motivated to use a patch based on maximizing energy intake (such as total kJ, biomass, carbon intake, or some other factor). A forager may stay longer if the return for each feeding bout is near its maximum capacity, which is often determined by mouth or gut size. The quality of prey such as energy density or nutrient content may be as important as quantity. High nitrogen content in prey often stimulates foraging because this is typically a limiting nutrient in ecosystems (Pastor et al. 1993; Pretorius et al. 2011; Chapter 2). Foragers might use a patch in order to minimize some biologically meaningful factor such as energy expended or time spent in that area. The more time spent in a particular patch may mean the forager is missing a better foraging opportunity in another patch. Also, more time spent eating means that a competitor may be exploiting limited prey more quickly or that a predator may be able to strike. Another measure of patch currency is its stability. If a patch provides a stable, predictable supply of a resource, it may be considered valuable to a forager. In contrast, if resources are fluctuating, the uncertainty may deter the organism, causing it to seek more reliable patches. As we will see, the variance in return on foraging investment depends on the perspective of the forager. To make the situation more complicated, foragers may be evaluating some ratio of the currencies, such as the ratio of food intake relative to time spent foraging. Obtaining a large amount of food per unit foraging effort should encourage an organism to linger in a patch. In contrast, if each bite yields less reward, it may be time to leave. A challenge that still remains unresolved is translating these motivational factors to a biologically relevant feedback mechanism, which could be physiological or behavioral. The decision to leave a patch can be influenced by complex internal factors such as the sensation of hunger or the visual perception of declining food availability (Chapter 9). The presence of conspecific competitors and changes in prey behavior or morphology also may reduce foraging time in a patch. For example, prey may become defensive (e.g., scratching, biting, clawing) causing the forager to reassess the potential gain versus cost, such as a nasty, perhaps life-threatening wound.

The third factor that motivates organisms to occupy patches is associated with foraging abilities as well as contrasting constraints. The primary factor affecting foragers is the information that they have available to make decisions. Organisms have a variety of cues in the environment they might use to improve their choices. However, not all this information might be perceptible. Factors constraining the information gathered or the ability to process (and evaluate) the information play a large role in determining whether a forager successfully exploits patches. While many physical characteristics aid an organism in finding foraging areas, they also may constrain foraging. For example, small body size or low mobility may limit the ability for an organism to sample multiple patches in its environment. Small mouths may limit the size of prey available within a patch. The inability to detoxify defensive chemicals in plants may also constrain the availability of patches to herbivore foragers. Of course, characteristics that limit foraging under one set of conditions may facilitate foraging in another situation. The large mouths of largemouth bass limit their ability to feed on small food items such as most invertebrates. However, their large gapes allow them to consume very large prey items that are unavailable to invertivores (Hambright et al. 1991).

Ultimately, most foragers are equipped with a set of tools that allow them to perform better than expected based on random expectations. We can envision the foraging abilities of organisms across patches as a continuum. If foragers are unable to access resources, they will cease to exist either through direct starvation or competition with other better-equipped species (Hutchinson 1959; Figure 4.3). On the opposite end of the continuum, it is impossible for any organism to have perfect knowledge of the environment because the environment is continually changing (Figure 4.3). Natural selection should favor individuals that have increased sensory capacity to evaluate their environment and rapidly find food. Selection should also favor individuals with the ability to respond to a changing environment through learning (Figure 4.3). The result is foragers with differing levels of information about the environment rather than perfect knowledge (Stephens and Krebs 1986), unless a GPS unit, the Internet, and the ability to use them are handy. An unspecialized forager will only overlap well with its food resource when the resource is widely distributed (Figure 4.4a). If prey are limited to a small proportion of patches, the forager able to respond to the changed environmental variation will be more successful (Figure 4.4b). Under most circumstances, even the best foragers will be unable to completely track their prey (Figure 4.4b).

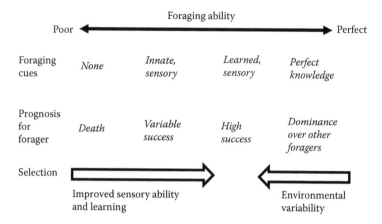

FIGURE 4.3 Continuum of foraging abilities and their relationship to ecological and evolutionary processes. A poor forager is unable to identify important cues and therefore starves to death. It will be selected out of a population. Natural selection should select individuals with a better ability to forage in their environment. However, unpredictable environmental variability will make selection for a perfect forager impossible.

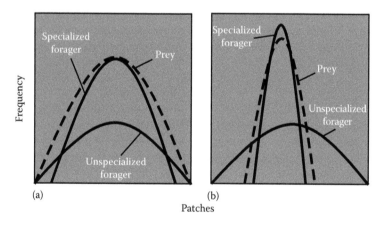

FIGURE 4.4 Overlap of foragers with their prey in patches. (a) If prey are widely distributed across foraging patches, then both general and specialized foragers will do well. (b) If prey are narrowly distributed in only a few patches, then a specialized forager will overlap well with prey and a generalist will not.

4.4   IDEAL FREE DISTRIBUTION

The *ideal free distribution* (IFD) is a deceptively simple concept in trophic ecology that continues to cause debate (Abrams et al. 2007). According to the IFD, consumers should distribute across patches in the landscape in proportion to the availability of resources (Fretwell and Lucas 1970). Thus, the IFD predicts that any ecosystem should have strong linkages between consumers and their prey, leading to efficient trophic transfer (Figure 4.1).

The IFD provides a useful springboard for developing theoretical models for trophic interactions. In its simplest form, the value of a patch W to an individual i is

$$W_i = Q_i/N_i$$

where Q is the food input into the patch and N is the number of conspecifics in the patch. Conspecifics are considered to be competitors and will reduce the value of the food resources in proportion to their abundance (Figure 4.5). If the patch was a pie, the pieces would become disproportionately smaller as more mouths are present. The distribution of individuals across patches should be such that W_i is equal among all of the feeding locations. Of course, this means that the foragers have knowledge of what their average W_i should be. Natural selection may lead to some average internal value that organisms target depending on the environment in which they evolved. They may have a sense of how much pie they should get. If they do not receive the expected return, then they will move on to other areas. Tools for determining how organisms determine their expected energy return are discussed in Chapter 8.

The IFD model can be modified in many ways to account for variables that affect the perceived value of the patch. For example, Sutherland (1983) added an interference competition term m to the model

$$W_i = Q_i/N_i^m$$

that incorporates physical interactions between foragers in patches. The presence of conspecific competitors is disproportionately greater in this model

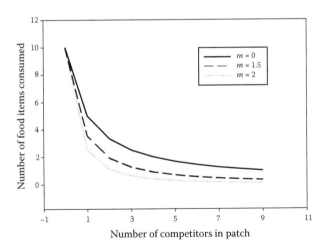

FIGURE 4.5 IFD results where the number of food items per patch is 10. If no competitors are present (number of competitors = 0), then under all scenarios, a forager will receive the maximum reward. As competitors increase within patches, the average expected reward per patch declines. The variable m increases the intensity of interference competition within patches, thereby reducing the value of the reward to additional individuals.

(Figure 4.5) because they are actively preventing other foragers from accessing the pie. Interference can occur in many ways. Species with a dominance hierarchy such as wolves or some trout may access resources differently depending on their status. Increased dominance of some individuals in patches will increase the value of m in the model.

Obviously, the IFD model is simplistic and has many assumptions. The value Q must be constant, meaning that prey do not decline as they are consumed. This is not impossible. Prey production may be sufficiently high to match their removal by foragers. The IFD model as described does not account for external factors affecting foraging decisions. Resources can come from outside the patch, influencing the value of Q (Polis et al. 1996). Predation has long been recognized as an important force affecting the choice of patches (Sih et al. 1985). Foragers avoiding predators will likely avoid or limit time spent in patches that pose a greater risk of mortality, even if the energy gain in these areas is high (Chapter 6).

Definitive tests of the IFD are rare in the literature. Tregenza (1994) cited several early studies of the IFD and noted that they typically failed to meet critical assumptions. For example, most studies could not confirm that the reward per patch was constant. Also, the influence of other factors such as interference and the fear of being consumed was missed (Tregenza et al. 1996a). Tregenza et al. (1996b) conducted research on parasitoid wasps laying eggs on moth larvae. Wasps initially spent time searching for host sites. Once they began ovipositing, the interference IFD model was generally supported.

In our view, the IFD concept may hold better under some circumstances than others. Mobile foragers should be more likely to quickly sample across patches and assess their relative value, thus distributing as per the IFD. A simple IFD without interference probably is quite rare because foragers residing in a patch will be aware of competitors and likely will defend their space through contest competition. Organisms that have conspicuous defenses against predators may be more likely to distribute per the IFD because they are not at risk of being consumed while searching across patches. Animals that are able to communicate information about foraging locations should be more successful at distributing across patches. Eusocial insects that transmit information about foraging through chemical and behavioral cues such as many ants, termites, and bees are likely very adept at distributing themselves across resources via the IFD (Shellmanreeve 1994; Figure 4.4).

4.5 MARGINAL VALUE AND PATCH USE

Not long after the IFD concept arose in ecology, the *marginal value theorem* (MVT) was developed (Charnov 1976). Again, the concept is simple, determining the time an organism should spend foraging in a patch before giving up and moving to a new, more profitable location. Rather than the value of the patch being continuous as in the IFD, the cumulative value decelerates as the forager stays there (Figure 4.6a). The forager has some knowledge about the time necessary to move to the next patch of food (Figure 4.6b). The time foraging in the patch should depend on the time necessary to move between patches and the gain from the patch, which can be determined by drawing a tangent line from the expected transit time to the intake curve

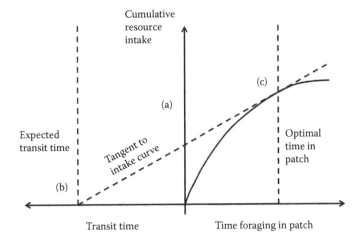

FIGURE 4.6 Graphical depiction of the predictions of the MVT. Time on the *x*-axis can either be spent foraging in a patch or moving to another patch. Time costs energy and needs to be balanced between the two activities. Energy intake (*y*-axis) only occurs in the patch and declines the longer the forager remains (a). The graphical solution to how long to remain in the patch is determined by the transit time between patches (b) and the shape of the intake curve (c).

(Figure 4.6c). The point where the line intersects the intake curve indicates the time when the forager should give up and move to the next patch (Figure 4.6c). It is easy to visualize how changing the parameters of the model affects the giving up time in patches. A slower deceleration of resources will lead to a longer time in patches. Similarly, a longer transit time between patches will lead to longer foraging times. Many organisms such as hummingbirds (Pyke 1978) and ducks (Tome 1988) appear to forage in a manner consistent with MVT predictions.

Although the MVT accounts for new parameters relative to the IFD, it requires that foragers have some knowledge of the mean value of patches as well as the distances between them. Transit time is particularly difficult to assess and needs to incorporate predation risk. Kotler (1997) assessed the giving up times of gerbils in patches with different amounts of seed and exposed to owl predators. With more food and no owls, the gerbils spent less time in patches foraging, which is consistent with the MVT. The owls also reduced time in patches, confirming that predation risk must be considered a foraging cost (Chapter 6).

Foragers that follow patterns as per the MVT must be able to gather and integrate a large amount of information about their environment. Many fish, birds, and mammals integrate long-term patterns in food availability and, through learning, better exploit their environment. Many invertebrates also perform well in their foraging environment by integrating information (van Alpen et al. 2003). Parasitoid insects initially respond to olfactory cues via using their antennae, positively responding to the concentration of odor molecules. When they arrive at the location of their host, they integrate additional cues from the frass, chemical compounds, excreta, and other materials produced by their prey, which are often related to the prey density in the patch (van Alpen et al. 2003). If a host is selected, the parasitoid will often mark it with an olfactory compound as a place-marker. Thus, the odor compounds of conspecifics can be used to assess cumulative foraging in a patch.

For all foragers within patches, the marginal decline in value must be assessed. How this is sensed is probably as complicated as defining, finding, and then initially evaluating the overall value of a patch. If food intake is positive, then the process keeping the forager in the patch is a form of incremental decision making, which will continue as long as intake is valuable as per the MVT. As several investigators note (Mangel and Clark 1989; Kotler et al. 1997; Noonburg et al. 2007), the internal physiological state of the organism and the impact it has on the forager's perception of the foraging environment may well alter the use of the patch.

4.6   STATE DEPENDENCY

Understanding how a forager perceives its environment is important for predicting patch choice and residence behavior. A host of physiological factors influence internal state. Hunger and satiation may seem to be simple processes. However, they are affected by a host of interactions between various organs and, in many organisms, the brain (Brodin and Clark 2007; Figure 4.7). The only known peripheral signaling

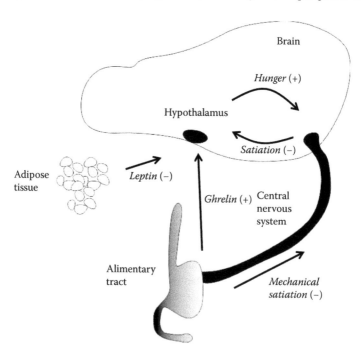

FIGURE 4.7 Simplified mechanisms of hunger and satiation in a typical vertebrate. Information triggering hunger can be produced both in the alimentary tract as well as in the brain. Dropping blood sugar (and a corresponding drop in insulin) also may trigger hunger. Satiation is also influenced by external factors such as an increase in adipose tissue and the expansion of the gut as well as within the brain.

compound produced in the gut that stimulates feeding is a hormone called *ghrelin* (Cummings et al. 2001). In the nervous system, several substances stimulate anabolic (body building) activity such as cortisol, norepinephrine, and neuropeptides. The energy content of the forager integrates with internal chemical cues to stimulate hunger and to initiate the behavioral search for patches in which to forage. Of course, external factors such as the presence of prey, environmental conditions, meal size, competitors, and predators determine whether foraging by a hungry organism actually occurs.

Processes leading to satiation appear to be directly linked to physiological status. The stomach or gut pouch expands during a meal and sends nerve signals back to the brain to gauge fullness (Brodin and Clark 2007; Figure 4.7). Another group of signals is related to the energy stores or adiposity of the forager. Insulin generated by the pancreas and leptin produced by fat cells are two well-recognized hormones that reduce hunger in direct proportion to energy stores. Within the brain, many compounds exist including serotonin and oxytocin, which promote satiation. Thus, as a forager remains in a patch, it is assessing both the cumulative value of the patch as well as its own internal sense of energy stores and gut fullness.

Foraging decisions in patches also are affected by long-term environmental factors. Seasonal changes in resources, temperature, or photoperiod affect foraging behavior. Preparation for scarce winter food at temperate latitudes or high altitudes leads to caching in many organisms, where the foragers create their own food-rich patches (Figure 4.8). Organisms can scatter-hoard, spreading their food stores over a large area. This can cost energy in terms of increased brain matter needed for memory and then searching for stores (Volman et al. 1997). Conversely, larger hoarders place their stores in one large patch, such as a burrow. This may seem a more cost-effective solution, but the forager risks losing its entire stores to a raid or spoilage. Energy stores may be accumulated internally as fat in preparation for winter scarcity (Garvey et al. 2003). Other resources can affect foraging behavior in a spatial context. Scarcity of water in the environment such as on the African savanna greatly affects the movement of organisms and their perception of foraging patches

FIGURE 4.8   Western scrub jay (*Aphelocoma californica*) is a classic scatter hoarder.

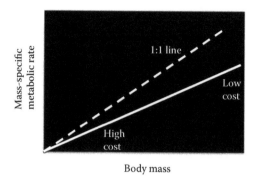

FIGURE 4.9 Body size and related metabolic needs affect the time that foragers remain in patches. Large organisms have lower metabolic requirements relative to their mass than small ones. Small organisms must have better access to foraging patches than large ones to meet their energy needs.

(Ogutu et al. 2011). To some organisms, the lack of water may make foraging perilous, while other foragers such as carnivores may use the tight spatial packing of prey around water holes as an opportunity to capture vulnerable, exposed prey.

The influence of internal state on foraging choice and ultimately ecosystem processes will depend on the physiology of the organism and the spatial scale of the foraging patches relative to body size. Small organisms, particularly those that are endotherms, have high energy demands relative to their body size (Figure 4.9). For example, shrews must eat nearly constantly and thus cannot be far from foraging patches at any time. In contrast, large organisms with low metabolic needs relative to their mass (e.g., *poikilotherms*) may have the luxury of living in an environment with fluctuating patch value and availability (Figure 4.9). A crocodile in a small lake may consider its foraging patch to be a small portion of shoreline as it waits for its prey to arrive for a drink. Because of its low energy needs relative to its large body size, a single encounter with a large, energy-rich prey is sufficient to meet its long-term energy needs. Predictions about energy flow in ecosystems must account for evolutionarily significant matches and mismatches between foraging constraints and scale-dependent patch availability (Figure 4.1).

4.7   VARIANCE SENSITIVITY

In addition to the relative value of patches, foragers may be able to perceive the variability in reward. This concept is often referred to as *risk sensitivity* in the literature, although it is unrelated to the risk of predation (Stephens and Krebs 1986). Rather it refers to risk in an economic sense, whereby a forager invests energy into seeking food with the potential value of the food reward being linked to its variable energetic return. As with economic investing in financial markets or gambling in casinos, a *safe* bet yields a small, yet relatively dependable return. A *risky* return may yield a large reward or nothing at all.

Foraging variability is perceived by foragers as a function of their internal state. Thus, the utility of a foraging patch may not equal its reward. An organism that is not hungry may perceive a food-rich patch as having relatively low utility. Conversely, a very hungry forager may perceive a low food patch as having a high utility. The shape of the relationship between reward and perceived utility lends insight into the risk sensitivity of the forager (Stephens and Krebs 1986). A linear relationship suggests that the forager is risk-neutral when selecting patches because it has perfect perception of value (Figure 4.10a). A risk-averse forager is one that perceives the utility of the reward as declining as it seeks more (Figure 4.10b). Thus, the reward is increasing but the perception of its utility declines. This may happen as a forager becomes sated. A risk-seeking (or risk-prone) forager is one that perceives the utility of the reward as accelerating with more foraging effort (Figure 4.10c). It is important to note that the value of reward increases the same in all these cases. Rather, it is the perception of the forager that drives behavior in the patch. Perception is difficult to measure directly and requires experimentation.

A simple experiment to assess variance sensitivity to foragers involves providing them with two foraging options with the same mean reward but different variance. A low variance treatment leads to a small but consistent return (Figure 4.11a). A high

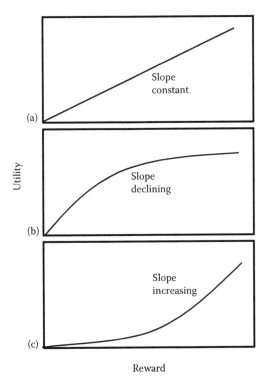

FIGURE 4.10 Relationships between reward and perceived utility of the reward within foraging patches. (a) Reward and utility are the same. (b) Risk-averse foragers perceive less utility of reward as reward increases. They will be less inclined to keep foraging in a specific patch. (c) Risk-seeking foragers perceive exponentially increasing utility in a patch, although it is only increasing linearly. They may continue foraging in a patch even when its average return is diminishing relative to its perceived intake.

variance treatment may provide either a large reward or nothing at all (Figure 4.11b). In experiments, juncos with a positive energy balance chose the low variance option (Caraco et al. 1980). However, when energy balance was negative, the birds chose the high variance treatment. The juncos perceived the occasionally large reward as accelerating utility causing them to keep seeking the riskier option. Humans behave similarly in risk taking (Symmonds et al. 2011).

A similar experiment can be conducted on the temporal delay in prey return. If the reward is delayed, foragers will tend to choose the more variable award now than wait for a more consistent, yet lower reward later (Buchkremer and Reinhold 2011). This is related to the concept of discounting, where foragers will choose to forage now because of perceived or real uncertainty in reward in the future. Organisms preparing for migrations, winter scarcity, and reproduction may well choose to forage even if rewards vary in the present in preparation for a major change in internal state in the future.

Further complicating a forager's perception of risk is the need to provision for offspring. Foraging parents must balance their energy state as well as attend to the

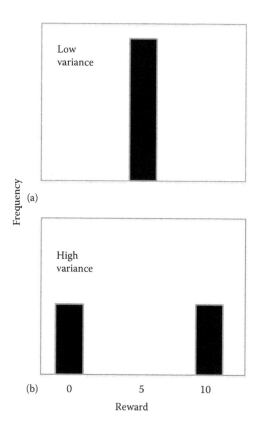

FIGURE 4.11 Classic design of a foraging experiment. In both treatments, the average return is the same ($n = 5$). However, the variance in reward differs. (a) In the low variance treatment, the reward is always the same as the average. (b) In the high variance treatment, a forager may receive either a large reward (double the average) or no reward at all.

needs of growing young. A study of red-winged blackbirds showed that the variance sensitivity of parents depended on the energy state of the nestlings (Whittingham and Robertson 1993). When offspring were in a negative energy balance, adults chose relatively small, easily obtainable food items in patches close to the nest. When energy balance of young was positive, adults flew longer distances to return with larger prey from distant patches.

Finding generalities in the risky business of foraging is difficult, because foragers are often able to make sequential dynamic decisions rather than following simple rules of thumb. Dynamic decisions may be driven by learning. Spatial memory, typically processed and stored in the *hippocampus* of vertebrate brains, is very important in influencing future foraging decisions (Sherry and Hoshooley 2009). A similar structure called the *mushroom body* is found in insects (Sherry 2005; Figure 4.11). The hippocampus can increase in volume during seasons of high caching behavior to allow foragers to find their food stores (Volman et al. 1997). The memory of past energy intake influences future foraging decisions in hummingbirds (Bacon et al. 2011). Thus, patterns of foraging that occur in ecosystems are driven not only by internal physiological cues and the landscape but also by the dynamic *memoryscapes* that exist in many foragers' brains. A recent study with ants showed that they indeed can rely on memory to find food when their pheromone trails are broken, with private information from individuals being communicated to other ants to reach high-quality food caches (Czaczkes et al. 2016).

Risk-taking has a heritable component (Dingemanse et al. 2012; Naguib et al. 2011), meaning that some individuals within populations have a genetic predisposition to approaching risk-sensitive foraging in different ways. Marchetti and Drent (2000) found that tits differed in their foraging exploration behaviors. When aggressive *producer* tits were present, *scrounger* conspecifics would wait to forage until the producers found productive forage sites. In the absence of producers, scroungers would actively search for foraging areas. Risky individuals should make risk-prone foraging decisions more frequently, which may benefit them when resources are low, but serve to their detriment if the risky behavior incurs an energetic cost when resources are available.

4.8   SEARCHING FOR OPTIMALITY

Although foraging by most or all organisms is by no means optimal, natural selection should favor characteristics that improve foraging efficiency in a very complex and variable world (Stephens and Krebs 1986). Thus, foragers should perform better than random expectation; rules such as the IFD, MVT, and variance sensitivity appear to provide some insight into the motivations behind many foragers. Sequential decision making can be explored further using dynamic programming models (Mangel and Clark 1989; Hilborn and Mangel 1997). Unlike the IFD and MVT models, which use analytical methods to find the *best* solution for foraging in patches, dynamic models begin with a function that maximizes some variable related to fitness, such as food intake rate. The models then account for the sequential environmental and state-dependent variables that maximize the output of this function (Box 4.2). The options for these models are practically unlimited. The downside is that these models are often context-dependent with limited generality for understanding foraging

BOX 4.2   DYNAMIC PROGRAMMING MODELS

Dynamic programming models are a useful way for assessing the most parsimonious solution for a foraging problem in ecology. These models work backward to maximize the output of some function related to the fitness of the forager. For the uninitiated in math or computer programming, a function is a mathematical equation that produces some biologically relevant output. For example, a dynamic programming equation may be generated that produces the best energy return from foraging across a landscape of stable patches. At a time step before the fitness function is maximized, the program determines which course of action based on internal state variables and environmental variables maximizes the function in the future time step. This is a clever way to assess the past selective factors affecting current foraging patterns. In the example of foraging across a landscape of food items, patch distance and patch quality might be considered external variables, while the energy state of the forager also will affect the decision about which patches to select and how long to stay there. The model then continues to work backward at each time step, with each determining the decision that maximizes the energy return in the future terminal time step.

The dynamic foraging decision model is asking the same fundamental ecological question that the MVT asks. However, the conditions of the model can be very specific to the system of study. Further, the dynamic programming model provides a sequence of decisions rather than a single solution. Thus, the spatial and temporal patterns of foraging predicted by the model can be compared to empirical patterns. If the patterns match, then the model may be capturing fundamental processes affecting foraging. If the patterns do not mimic those in the field, then new questions about internal and external variables affecting foraging need to be quantified. Foraging may be affected by the presence of predators, which needs to be incorporated into the model.

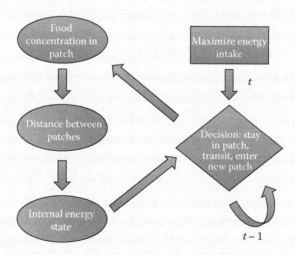

The figure shows the flow of a dynamic programming model evaluating the best decision during each time step that maximizes future total energy intake (rectangle). The first model time step (t) starts at a function that maximizes energy reward. At each time step ($t - 1$), the model determines which patch should be occupied based on the patch food concentrations, transit time between patches, and the internal energy state of the forager. The patch decision that maximizes future energy intake is the one chosen (triangle). The model then continues to iterate backward by $t - 1$ time steps. A typical time frame would be the growing season of a forager, where t is days.

decisions beyond the system of study. Other models that can be used to assess foraging decisions include those that use differential equations to solve for *optima* in foraging models (e.g., find the point where energy intake is maximized using derivation) and genetic algorithms. A genetic algorithm starts with random distributions of food resources, patches, or prey and then finds the optimal solution for foraging on these items through series of decisions made by computer simulations, similar to the way natural selection preserves genes that are adaptive and removes those that are not (Figure 4.12). Wajnberg et al. (2012) used this approach to find out how parasitoid wasps allocated energy to their foraging forays in fragmented habitats.

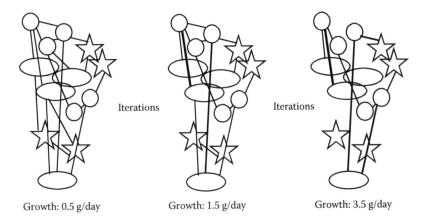

Growth: 0.5 g/day   Growth: 1.5 g/day   Growth: 3.5 g/day

FIGURE 4.12 Simplistic model of using a genetic algorithm to determine the best foraging pathways to patches in the environment. Patches are depicted as different shapes. Foraging pathways (or corridors used) are the connecting lines. The pathways and the frequency that they are used are considered units of selection and change (i.e., mutation) in the model. Before the first model iteration, all possible pathways are present with the same frequency of use, depicted by the thickness of the lines. These pathways generate growth of the forager of 0.5 g/day (left). Several iterations where the pathways are either selected or eliminated. They also are *mutated* to become stronger. Halfway through the simulations, the pathways selected increase modeled growth to 1.5 g/day (middle). Further iterations lead to a set of pathways with different strengths that maximize expected growth at 3.5 g/day (right).

4.9 CONCLUSIONS

The foraging theory provides a useful foundation for assessing patterns at broader spatial scales. Decisions made at the level of the forager can scale up to population, community, and ecosystem levels. For example, the spatial distribution of an herbivorous pest and its choice of plant patches can determine whether an economically important crop is lost or saved (Chapter 2). The impact of top-down predators on ecosystem production will be affected by their ability to seek out prey in lower trophic levels and successfully control their populations (Chapter 10). Understanding the response of ecosystems to environmental change will require a grasp of the foraging capabilities of its component species. Hence, foraging models based on sound theory are critical for connecting mechanistic interactions with phenomena such as materials and energy flux in ecosystems.

Although many models, both theoretical and simulation, exist for predicting the distribution of organisms ranging from viruses to whales relative to food (or host) availability, surprisingly few experimental or field tests have been conducted to directly test theoretical predictions such as the classic foraging experiments and field observations conducted by Gilliam and Fraser (1987) with fathead minnows and Hixon and Beets (1993) in coral reefs; many pollinator examples exist (e.g., Real 1981 and its predecessors). Because optimality became a dirty word in ecology by the early 1990s, the search for decisions that organisms make to distribute themselves across the landscape may have fallen out of favor. However, many advances in understanding the impact of internal state and how it affects decisions continue to be made using simulations, where the decisions are realized to lie somewhere between imperfect and optimal. Uncertainty can be built into the models and used to determine how organisms make the best decisions in an imperfect environment.

Fragmentation of the Earth's habitats coupled with changes in climate at small and large scales (Halpin 1997) necessitates that organisms have the ability to reliably respond to changes in the food within their environment. In the North Sea, up to two thirds of marine fish have moved north, likely as a function of climate warming and changes in prey distributions (Perry et al. 2005). Conservation and management must recognize how organisms perceive and respond to changes in food in the environment to ensure survival. Application of theories and models such as those in this chapter is critical for making sound decisions.

QUESTIONS AND ASSIGNMENTS

1. How are forager distributions in food patches related to trophic transfer in ecosystems?
2. What are the problems with the concept of optimal foraging?
3. What are the three factors influencing foraging decisions made by organisms?
4. Why are most foragers not random searchers for food?
5. How does the ideal free distribution work? How might foragers be distributed across the landscape of prey based on its predictions?
6. When should a forager give up eating in a patch according to the marginal value theorem?

7. What is meant by an internal state and how does it affect foraging decisions? How does size and metabolism affect state and foraging decisions?

8. Do all organisms perceive the risk of foraging in patches the same way? Explain factors that affect the perception of risk.

9. If a reward is delayed, will foragers choose riskier patches or wait for a more certain yet lower quality patch?

10. How might a genetic algorithm or state-dependent dynamic programming model be designed to assess the best foraging decisions in uncertain environments?

5 Predation

5.1 APPROACH

Predators are a special category of consumer. We define what predators are, showing that these organisms have evolved to be specialized and efficient. With specialization comes fragility, especially regarding susceptibility to human consumption. Events of predation can be generalized, with predators likely minimizing energy costs and maximizing energy intake across these stages. We will show that predators probably do not make optimal foraging decisions in the environment but that their foraging choices are not random. We conclude by showing why predators rarely hunt their prey to extinction. Humans, having shown a strong proclivity in the past several millennia for causing mass extinctions of top predators in ecosystems, are the exception.

5.2 INTRODUCTION

Predation is an ecological process that thrills and frightens. Humans may not like to consider it, but we occasionally fall prey to predators; this certainly was a real threat in our evolutionary past (Berger 2006) and occasionally occurs today. Humans have excelled at being predators (Brain 2000; Stiner 2002), with hunting and fishing continuing to be common human pursuits. Commercial fishing is an example of predation at a massive, industrial scale, with global harvest hovering around 92 million metric tonnes per year (FAO 2016). Domestication of animals in agriculture and aquaculture simplifies the predation process for humans. People have successfully preyed on species to extinction, such as the North American passenger pigeon *Ectopistes migratorius* that once had a population in excess of 3 billion (Conrad 2005). The impact of this mass extinction on lower trophic levels and perhaps biodiversity will never be known.

Predation can have multiple definitions. We broadly define it as the consumption of living organisms. The processes of mass consumption such as herbivory and filter feeding are predation but typically are considered separately because of their unique characteristics (see Chapter 2). For this chapter, predation is defined as the intimate pairwise interaction between a predator and a single prey. A pack of lions attacking a heard of wildebeest, a python constricting an antelope, and a shark consuming a mackerel are all instances of predation. A term related to predation is *depredation*. Depredation is typically used to describe the outcome of a terminal predatory act, such as the complete removal of eggs from a nest. In contrast, predation is a process where some prey may survive the encounter.

Predation is a complex, dynamic process involving the interplay between a predator and its prey. In this chapter, we focus primarily on the predator response to the prey and the importance of predation to community and ecosystem integrity.

TABLE 5.1

Common Large, Apex Predators in Ecosystems

Predator	Potential Impact
Large predatory birds (eagles, hawks)	Loss of control of prey populations such as small rodents
Polar bears	Increase in seal populations
Wolves	Increase in elk or moose, with decrease in plant diversity
Shark	Loss in coral reef fish diversity
Sea otters	Reduced kelp bed diversity
Jaguar	Herbivore overgrazing of plants

In Chapter 6, we consider the rich ways that prey respond to their consumers. Ultimately, the success of a predator depends on its ability to capture and consume prey to maintain a positive energy balance and successfully reproduce. Natural selection is operating at the level of the individual predator and will lead to traits that improve the predator's performance. Predators do not exist alone. They must contend with conspecifics. Intraspecific interactions may be beneficial through shared efforts such as cooperative hunting. In contrast, exploitative and interference competition reduce the value of meals, particularly as the predator population grows and resources become limited. At the level of the community and ecosystem, predation becomes more complicated as populations compete for prey, influence the distribution of resources, and alter the flow of energy.

Understanding predation is critical for general ecology as well as for the management and conservation of economically important species. Many of the world's most imperiled species are large predators (Table 5.1; Terborgh and Estes 2010). The conservation and restoration of these species require an understanding of their foraging needs and factors that facilitate their success as predators. Large predators like the African lion, salmon, and wolves have long been recognized as important in structuring the diversity and function of communities and ecosystems (Table 5.1; Chapter 10). Their loss can reduce diversity and hobble ecosystem function (Terborgh and Estes 2010).

This chapter follows a hierarchical structure. We start from the perspective of individual predators, identifying what characteristics make them successful and exploring any generalities that might exist. We scale up to predator populations, reviewing how populations behave in unique and sometimes surprising ways and are intimately linked to prey population dynamics, and then consider how these predatory processes scale up to ecosystem function, such as nutrient cycling.

5.3 INDIVIDUALS

Predation occurs among organisms at all spatial scales. Gause (1934) conducted his famous predator–prey experiments at the microscopic scale using *Didinium* predators consuming *Paramecium* (Figure 5.1). Predation also occurs at colossal scales. The largest predator to walk the Earth was probably *Spinosaurus*, an 18-m-long theropod that lived about 100 million years ago (Dal Sasso et al. 2005; Figure 5.2). Predator–prey interactions between this species and its prey may have been a bit

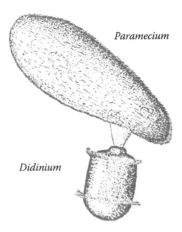

FIGURE 5.1 Predator–prey interaction between the predator *Didinium* and its prey *Paramecium*. (From Gause, G. F. 1934. *The Struggle for Existence*. Dover Phoenix.)

FIGURE 5.2 *Spinosaurus*, possibly the largest terrestrial predator in the history of Earth. (Courtesy of Shutterstock.)

more dramatic than those in Gause's test tubes. As we pointed out in Chapter 2, predation is likely common across so many taxa and ecosystems because conversion efficiency of prey flesh is high. Trade-offs do exist. The act of finding and consuming unwilling prey has significant energy costs.

Predation involves a sequence of events (Figure 5.3). Natural selection should minimize the amount of energy spent at each point along the sequence. This can be achieved by either reducing the time spent in the activity or reducing its energy cost. As with foraging across patches, a predator must spend time searching for its desired

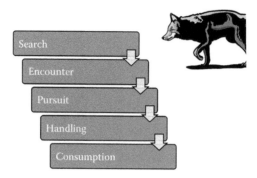

FIGURE 5.3 Generalized predation sequence. Organisms must balance the energy and time spent in each of these behaviors.

prey (Figure 5.3). The predator may have a *search image*—an internal representation of the prey that it seeks and the landscape on which it forages (Chapter 4). The term *image* is misleading because the cues may not be visual. They can involve many senses such as olfaction, electricity, touch, gustation, and sound. When a predator encounters a potential prey item, it must be able to recognize it as prey. Many prey have ways to hide in plain sight (crypsis or inactivity; Chapter 6). If the prey is recognized and the predator decides that it is energetically beneficial and nonharmful, then it will pursue the prey (Figure 5.3). Pursuit can be energetically taxing if prey are fast-moving or able to effectively use cover. If the predator is able to make contact with the prey, it needs to be able to effectively capture and handle it, getting it into the digestive system (Figure 5.3). Handling can be a dangerous and costly event for the predator, particularly when prey have characteristics such as large size, armor, venom, teeth, claws, and strength. Assuming that the predator consumes the prey, it must be able to digest the prey organism (Figure 5.3). Prey can be toxic, difficult to digest, spiny, or poorly nutritious. These characteristics can pose additional hazards to the predator. The final steps of predation are excretion and egestion, which also incur energy costs depending on prey characteristics (see Chapter 8).

Predators can adopt various strategies for capturing prey based on selection across the range of predatory events. One group along a continuum (Figure 5.4) remains relatively stationary. The energy cost of searching is reduced by this tactic. However, these predators must make the most out of encounters with prey. Generalities are hard to make, but under most circumstances, the predator needs to be large-bodied with a big mouth to consume a wide size range of prey. A good example would be a grouper (*Epinephelus*) in a marine reef, which is stout-bodied with a huge mouth and rarely moves until prey are near. It lunges, using its mouth to suck in its prey in an instant. Species such as these are often called *lie-in-wait* predators. Other tools that stationary species may possess to make their encounters successful are big teeth or grasping organs, suckers, barbs, large guts, low metabolic rates, the ability to suck prey into the mouth, sticky tongues or tentacles, camouflage, stealth, venom, and lures. For the stationary tactic to be successful, encounters with prey must be above some frequency threshold to ensure that predators have a positive energy budget. If prey are rare, stationary predators cannot make a living.

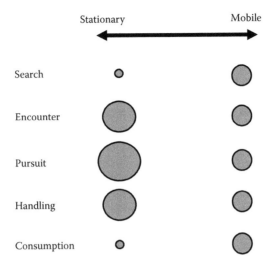

Stationary   Mobile

Search

Encounter

Pursuit

Handling

Consumption

FIGURE 5.4 Relative energy consumption (size of circles) spent by a stationary, Lie-in-wait, predator versus a mobile predator. The total energy spent in predation is about the same, but distributed differentially among the sequence of events.

On the opposite side of the continuum of predation, predators can be mobile, expending energy in search and pursuit (Figure 5.4). Many predatory birds overcome the energetic cost of searching by gliding on warm, rising air currents and are able to rapidly dive toward their prey when they are sighted. If the prey are fast, the predator needs speed, agility, and grasping ability to overcome the prey if capture occurs. Sharp teeth or beaks are useful for quickly killing prey and then tearing them into bite-sized pieces. Some species such as scale-eating cichlids tear pieces from their prey rather than consuming the entire organism (Van Dooren et al. 2010). Many species that must capture speedy prey have acute eyesight, elevated metabolisms, and intelligence. Although fishes are poikilotherms, tuna are a highly mobile fish that warm their muscle tissue to make them faster swimmers (Shiels et al. 2011). Cooperative hunting evolved in many taxa such as wolves, killer whales, and lions to overcome the challenge of pursuing fast prey. If prey are slow or stationary, then mobile predators need not expend much energy in pursuit but may have to contend with consuming smaller, less mobile prey that may be less energetically profitable (Figure 5.4). Many insectivorous birds have adopted this approach. When consuming prey organisms with indigestible armor (e.g., chitin) or low-quality flesh, some energy may have to be invested by the predator in digestion such as chewing, grinding, and lysing. The point is that all of these tactics balance energy costs across the sequence of predation (Figure 5.4). If prey density or quality falls below a positive energy balance, then the predator population will decline or move.

A tremendous amount of research has been expended on determining the *optimal* decisions that predators make about the prey they should choose. As we noted in Chapter 4, the term *optimal* is controversial (Pierce and Ollason 1987). However, when an individual predator is confronted with multiple prey items, the choice is typically not random (Schoener 1974). Most predators show preference, and it makes sense that the choice depends on maximizing energy intake and minimizing the cost of energy acquisition. Stein (1977) was among the first to experimentally assess the energetic decisions that predators make when choosing their prey. He used a cost/benefit equation to determine whether smallmouth bass (*Micropterus dolomieu*) were selecting *Orconectes* crayfish prey in ways that minimized energy costs and maximized gains (Box 5.1). The goal was to minimize the function $(H + P)/O$, where H was the handling time in seconds, P was the pursuit, and O was the digestible organic content of the crayfish (Figure 5.5). He found that smallmouth bass consistently chose the sizes and life stages of crayfish that followed this simple function. Similar research followed on food choice in other organisms such as bluegill (Mittelbach 1981) and small mammals (Brown 1988).

Many predators are limited by the size of prey they can consume. Thus, the size distribution of prey in the environment will interact with the size limitations of the predators to influence patterns of predator growth (Figure 5.6). Matches or mismatches between prey and predator through time may influence the flow of energy through ecosystems. These interactions are dynamic. Many predaceous larval fish

BOX 5.1   DESIGNING A SELECTIVE PREDATION EXPERIMENT

Experiments are required to determine predator choice (Sih et al. 1985). Inferences can be made from field observations, but it is impossible to control for all the factors influence foraging choices. When designing an experiment, do not expect the predator to be acting in an optimal manner. Rather, choices made by a predator about which prey items to consume are a combination of innate behavior, learning, prey response (see Chapter 6), and chance. Hypotheses to be tested need to be developed from field observations and theory. The experimental venue should be as realistic as possible, accounting for the proper spatial scale at which the predator–prey interactions occur. Experiments conducted at multiple spatial scales can provide insight into the validity of the results.

Mittelbach (1981) was interested in the predation choices that different sizes of bluegill sunfish (*Lepomis macrochirus*) made in inshore and offshore lake habitats. In laboratory feeding trials, he determined the relative costs of handling and energy gains of consuming common prey, including small insect larvae and zooplankton. Similar to Stein (1977), he found that bluegills chose the most energetically profitable prey items. However, in the field, the laboratory patterns for the smallest bluegills did not hold. The experiments and models that Mittelbach used predicted that these fish should feed on zooplankton in the open water, but they remained in the lake inshore areas. He predicted that this was due to the high risk of predation in the offshore areas, which was later confirmed. Mittelbach (1981) showed that identifying the inconsistencies between small-scale experimentation and field observation generated new hypotheses and future knowledge.

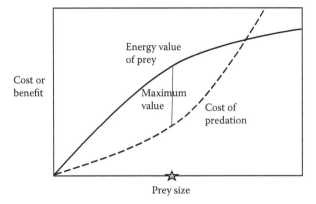

FIGURE 5.5 Hypothetical model of *optimal* foraging of a predator on a specific size of prey. Prey items have greater energy content as they become larger (solid line), although this declines as a greater proportion of the prey includes indigestible components such as chitin in insects or bones in vertebrates. The cost of consuming prey (dashed line) increases exponentially with prey size. Given a distribution of prey sizes in the environment, a predator should select the size that maximizes the difference between the energy intake and the cost of predation. This also can be addressed as a function minimizing the ratio of cost of predation over the energetic benefit of the prey consumed.

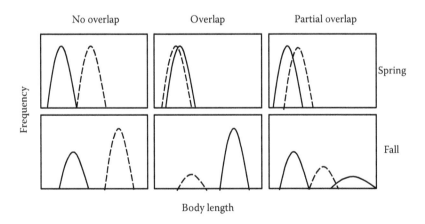

FIGURE 5.6 Size overlap between growing predators (solid line) and prey (dashed line) when the predator is limited by the size of prey it can consume. Overlap is shown during spring and fall, assuming that growth and density of the predator and prey depend on each other during the summer. If prey are larger than the size that the predator can consume (no overlap), then prey grow uninhibited through the summer and predator growth is poor and density declines. If size overlap is nearly perfect, then the predator should grow and reduce the density and perhaps growth of the prey. If only a subset of the predators can consume prey, they will grow rapidly, while the remaining predators will not (partial overlap).

and amphibians are gape limited (Hambright et al. 1991; Wilbur 1997), because they consume their prey whole. The growth *race* between these predators and their prey greatly affects the outcome of their interactions (Adams and DeAngelis 1987; Persson et al. 1996; Wilbur 1997). Because gape increases with body length, slight increases in length provide distinct advantages in predation ability. Increasing length also increases swimming ability and presumably improves capture efficiency. From the prey's perspective, rapid growth is advantageous under these circumstances. If predators overlap poorly with prey, then predator growth will be low, prey will grow beyond vulnerable sizes, predation declines, and predators starve (Figure 5.6). In the unlikely case that overlap with prey is perfect, then predators will grow rapidly and keep prey in check (Figure 5.6). If some predators overlap well and others do not, then the growth of the predators will be skewed as the larger individuals are able to outgrow the prey (DeAngelis et al. 1993). In these cases, the large predators may grow to sizes where they can cannibalize conspecifics (Forney 1977; Wahlstrom et al. 2000; Figure 5.6). Thus, factors affecting starting sizes of larvae such as birth date and hatch size may greatly affect the success of cohorts of predators within populations, population size structure, and ultimately the impact of predator populations at the community and ecosystem scales.

5.4 POPULATIONS

To understand the ways that predator populations respond to prey, it is important to review the processes and patterns of population growth. Unlike individuals for which growth is typically measured as a change in body size, populations grow in density of individuals. Predator populations follow a logistic pattern of growth, where they grow slowly when they are sparse or near their carrying capacity. They grow most rapidly at an intermediate density (Figure 5.7). Prey populations also follow a similar pattern of growth. Thus, how predators and prey affect their growth rates along the continuum of densities affects the dynamics of the predator–prey system through time.

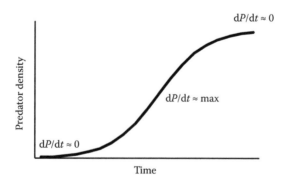

FIGURE 5.7 Logistic growth pattern for a predator population based on the assumption that growth rate (dP/dt) of the population declines as the population reaches its theoretical carrying capacity.

In isolation of each other, unperturbed populations of predators and prey grow to their carrying capacity and then theoretically should remain at these equilibrium densities. When the two populations interact and their equilibrium densities depend on each other, the expected population densities of predator and prey at equilibrium are depicted as *isoclines* on graphs (Figure 5.8). Mathematicians and theoretical ecologists are interested in the behavior of these isoclines when predator and prey interact (Berryman 1992). These isoclines occur when the two interrelated equations describing the instantaneous change in the populations are set to zero (Figure 5.8). Two of the simplest versions of these change equations are

$$\frac{\partial N}{\partial t} = aN\left(1 - \frac{N}{K}\right) - bNP$$

$$\frac{\partial P}{\partial t} = cP\left(1 - \frac{eP}{N}\right)$$

where N is the prey population density and P is the predator population density. Growth of the prey population in this model follows a logistic pattern, with the population growth limited by two factors. First, the prey population is limited by its carrying capacity, K. As N approaches K, the population's growth rate declines. The constant a affects the rate of increase. Prey population growth is further curtailed by the predator density P and a coefficient b. Predator growth also follows a logistic pattern, with a growth rate of c. The coefficient e is the marginal subsistence value of the prey to the predator. The ratio of N/e is the carrying capacity of the predator

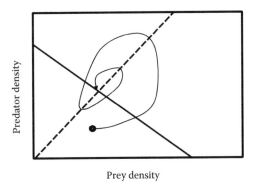

Prey density

FIGURE 5.8   Predator (dashed line) and prey (solid line) isoclines, where the density of each affects the carrying capacity of the other. An isocline is the density where a population will remain if unperturbed. In this case, the intersection between the prey and predator isoclines is the joint, stable equilibrium density for the two populations. As the prey density increases, the equilibrium density of the predator increases. Similarly, as the predator density declines, the prey equilibrium density will increase. Starting at any joint prey–predator density, the populations will fluctuate through time (follow the curved arrow) until they stabilize at the intersection point.

population. If the prey population N is large, then the population grows rapidly. As N declines, the ratio of P/N increases, reducing predator population growth. Berryman (1992) called the processes incorporated in these equations ratio-dependent predator–prey interactions. More simply put, these are predator–prey interactions that incorporate both intra- and interspecific factors that limit populations.

Like all models, these differential equations may be modified in many ways to explore the behavior of predator–prey systems. In fact, unlike the model presented above, the earliest equations developed by Lotka (1925) failed to incorporate logistic constraints on population growth. In other words, only interspecific interactions were included, and the models led to unstable systems, predicting that predator or prey extinction should be common in nature. In reality, predator–prey systems can be quite stable (Box 5.1). The model by Berryman (1992) in its simplest form incorporates what is known as a numerical response in predator populations (Solomon 1949). In an ecosystem sense, as the carrying capacity of the prey is elevated, for example by adding a limited nutrient to the system, the predator population will respond by increasing its carrying capacity. This does not happen immediately. Numerical predator responses lag because it takes time for predators to ramp up reproduction or migrate to the patch from other areas.

Predators can respond behaviorally to changes in prey that affect the predator–prey system. This is called the predator's *functional response*, which leads to many shapes of consumption depending on the conditions of the system (Solomon 1949). These patterns were predicted by Holling (1959), who conducted experiments to confirm how predators react to prey. As prey density increases, the number of prey consumed by the predator increases. Prey consumed can be displayed as either the total number or as a proportion. In a Type I functional response, the number of prey consumed increases linearly as prey density increases (Figure 5.9). At some point, the increasing prey density overwhelms the predator, and the proportion prey consumed steadily declines (Figure 5.9). These patterns might occur in a filter feeder, such as a whale shark or a web hunter, like a spider. At some threshold prey density, the prey items clog the feeding apparatus, preventing the predator from collecting additional items.

In a Type II functional response, predator efficiency decelerates as prey density increases. Predators are most efficient at removing prey when their prey are rare (Figure 5.9). Predator efficiency declines as prey density increases due to handling constraints. This can be described using a typical rate-limiting equation:

$$P(N) = \frac{aN}{1 + ahN}$$

where P is the predation rate, N is the prey density, a is the predation rate, and h is the handling time. Sea otters consume sea urchins by bringing them to the surface, lying on their backs, tearing them open, and lapping out the soft innards. The rate a in this model is determined by the average time (and related energy) necessary for the otter to travel to the bottom, capture an urchin, and return to the surface. Handling time (h) depends on the otter's skill at eviscerating the urchin. From the

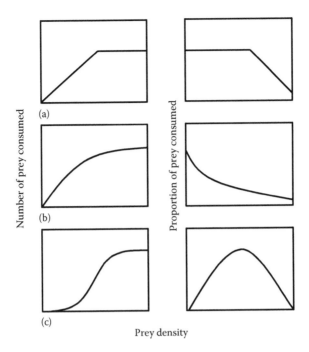

FIGURE 5.9 Type I (a), Type II (b), and Type III (c) functional responses of predators to their prey. A Type I consumer will respond immediately to prey availability until its ability to consume prey is overwhelmed. Type II predator is most efficient at consuming prey at low densities, which may lead to the extirpation of the prey and the collapse of the predator population when prey are rare. A Type III functional response is likely most typical in ecological systems. Predators become less efficient at consuming prey when prey are scarce, causing predators to switch to other prey species until the prey recover.

urchin's perspective, h provides a density-dependent refuge, because proportionally in less individuals are vulnerable at high densities (Figure 5.9). A fun experiment to use ecology labs follows the lead of Holling (1959) (Box 5.2).

A Type III functional response occurs when the predation rate accelerates to a maximum at an intermediate prey density and decelerates as prey either become rare or dense (Figure 5.9). These patterns have important implications for predator–prey systems. As with the Type II model, there is safety in numbers as predator efficiency declines with increasing prey density. Unlike the Type II model, prey find a refuge at low densities. This may occur when alternative prey are available. This is a process called *prey switching* (Figure 5.10). As one prey item becomes rare and less frequent in the diets, the predator switches to the more abundant species. Switching is a fairly simple concept and should be detected with the following relationship:

$$C = (P_A/P_B)/(N_A/N_B)$$

where P_A and P_B are the relative proportions of prey items A and B in the forager's diets, and N_A and N_B are the relative densities of the prey in the environment,

BOX 5.2   HUMAN FUNCTIONAL RESPONSE

Holling (1959) was interested in the functional response of predators to their prey. He quantified the response of deer mice to European pine sawfly pupae in cocoons and found that it followed a Type III pattern. Later, he experimented with blindfolded human subjects searching out sandpaper discs, finding their response to be similar to a Type II response.

A fun and tasty experiment similar to that conducted by Holling (1959) can be conducted in an ecology laboratory session. Blindfolded students are asked to search for pieces of candy, unwrap the pieces, and eat them. As the density of the candy prey increases, the act of handling and consuming each item should follow a Type II pattern. Hiding some of the candy pieces, making them more difficult to find, should generate a Type III pattern, because the candy prey have a refuge from the students. After a set time, the students are unmasked and then tally the consumed and remaining, unconsumed prey.

Many statistical packages allow students to enter an expected model such as a Type II (see equation on page 90) or Type III. A Type III model is sigmoidal and could be structured similarly to a logistic model

$$P(N) = \frac{Ke^{rN}}{K + e^{rN} - 1}$$

where P is the predation rate, N is the density of prey, r controls the rate of the functional response, and K is the asymptotic prey density. The parameters of the model are then estimated iteratively using a statistical computer package. The package SAS has a nonlinear function (Proc NLIN) where the model forms (e.g., Type II versus Type III) are included in the program. The program minimizes the sum of squares to determine how well the data fit each model. Another option would be to use a maximum likelihood in program R (package friar) to estimate the models and then use Akaike's information criterion to find the model with the best fit (Burnham and Anderson 2002).

---

respectively. C is the index of switching. If C is near 1, then the two prey are consumed in proportion to their density in the environment, meaning that foragers feed on the most abundant species, providing a density *refuge* for the less abundant prey. If C deviates significantly from 1, then selection for one of the prey is occurring and that prey is consumed when rare, violating the assumption of a Type II model at low densities. Without an alternative prey species present, a prey item may become inconspicuous, inactive, or cryptic at low densities, falling below the predator's search image (Relyea and Werner 1999; Relyea 2001). Switching and rarity in predator–prey systems may serve to stabilize predator–prey interactions. Prey can

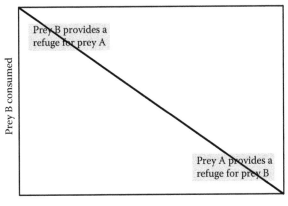

Prey B provides a
refuge for prey A

Prey B consumed

Prey A provides a
refuge for prey B

Prey A consumed

FIGURE 5.10 Predator switching. When prey B is rare in the environment, a high proportion of prey A are consumed. As prey A declines in the environment (and thus in the diets),
then prey B is consumed at higher rates.

recover from intense predation, preventing their extinction. This also ensures that
the predator population does not completely exhaust its food supply and starving
itself to extinction.

The ultimate outcome of all these interactions is complex systems with potentially simple underlying processes. These processes can be teased apart in laboratory experiments. However, in the field, other factors can mask underlying
dynamics. Probably one of the best-documented, long-term examples of predator–
prey cycles is between wolves and moose in Isle Royale National Park, a series
of islands in Lake Superior of North America (Nelson et al. 2011). Attempts to
accurately predict their dynamics have proven difficult. Not only do the wolves and
moose fluctuate depending on their relative abundance, but also the food resources
available to the moose and other factors such as disease affect population dynamics (Nelson et al. 2011).

Understanding predator–prey cycles is critical for sound conservation and
management. In many ecosystems, the apex predators have been removed.
Wolves, natural predators of species such as elk and deer, are examples of predators that have been extirpated from much of their range, allowing populations
of their prey to increase dramatically (Nelson et al. 2011). Other factors such as
human-induced changes to the landscape such as increasing *forest edge* habitat that deer use also likely increased population density of these prey species
(Ruzicka et al. 2010). To manage the prey, humans must act as the predator
through controlled hunting.

Humans are undeniably the dominant predator on Earth. Humans typically do
not abide by predator–prey cycles and are quite adept at hunting prey to extinction. This is not a new phenomenon. Evidence is mounting that along with climate
change following the end of the last glaciation in North America, overhunting by

humans led to the decline of many large species (Ripple and Van Valkenburgh 2010), although other researchers suggest that many of these animals went extinct before humans arrived (Boulanger and Lyman 2014). The recent loss of species such as the passenger pigeon and American buffalo in North America is testament to the ability for humans to hunt species very efficiently, without regard to fundamental ecological mechanisms such as search, handling, or prey switching. The goal of modern natural resource management is to impose limits to hunting and fishing to prevent prey extinction because natural ecological mechanisms do not exist in this context. These efforts have had mixed success. The western black rhinoceros recently was declared extinct in the wild (Emslie 2012), largely due to poaching for the species' horn, which is believed to have medicinal properties in some cultures.

5.5 MULTIPLE PREDATORS

Within species, some predators cooperate (Drea and Carter 2009; Yosef and Yosef 2010). This is problematic for ecologists. Natural selection should favor the most *selfish* individuals in a population. Thus, evolutionary biologists cite the *prisoner's dilemma* when describing the emergence of cooperative hunting or other seemingly selfless behaviors in populations (Nemeth and Takacs 2010). If two prisoners are given independent choices, they will pick the most selfish one, even if it works to their joint disadvantage. To illustrate, the prisoners are told, without being able to communicate with each other, that their sentence will be reduced by two years if they agree to a deal where the other prisoner does not get a reduced sentence. They are given another option that involves cooperation. If they both agree to a deal, they can leave prison immediately, but if one prisoner does not agree then they will experience an extended sentence. Because neither prisoner can guarantee that the other will choose the jointly beneficial decision, they will choose the two-year reduction, which ends up in both losing a very beneficial opportunity. Similarly, within a species, natural selection should lead to selfish, nonbeneficial predation decisions, because of the risk that the other individual will not hold up their end of the deal. In other words, there is a risk of the other predator not cooperating and then taking advantage of the free meal (Figure 5.11).

The fact that cooperative behavior is found in many predators suggests that the prisoners' dilemma does not always apply to ecological systems and has arisen as an evolutionary strategy (Stevens et al. 2005). Many social insects cooperate as predators. Army ants (*Eciton* spp.) hunt in cooperative groups that result in them overcoming prey many times their size (Kaspari et al. 2011). Social cooperation that involves learning and some level of cognition also can arise. Killer whales hunt sea lions and other species cooperatively (Ward et al. 2011). Birds also can cooperate (Peron et al. 2010). The prevailing explanation is that cooperative hunters are genetically related and kin selection is responsible, although this view is not holding up to genetic evidence (Mitani et al. 2000). Evolutionary game theory, which tests scenarios like the prisoners' dilemma under different ecological conditions, can be used to assess how

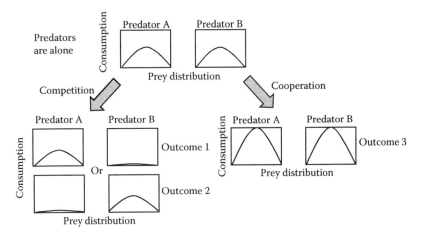

FIGURE 5.11 Illustration of why cooperative behavior should not evolve as an ESS (see Box 5.3). Two predators (species A and species B) in isolation should have similar consumptive intake across a distribution of prey when the predators are alone (top panel). As per the competitive exclusion theory, when the predators overlap (i.e., are in sympatry) and consume the same distribution of prey, one species should outcompete the other (left, bottom). The outcome of the competitive interaction is not known until the encounter occurs. Outcome 1 is where predator A is successful, and outcome 2 is the other possibility. Evolutionary theory predicts that both predator species should opt to avoid contact with each other or divide prey distributions to avoid competition. A third outcome, cooperation, may lead to a surprising, different result, whereby both species A and B *increase* their consumptive intake. Cooperation does occur among some individuals within and among species, even though theory predicts that this involves some *a priori* information that one of the cooperators will not attempt to outcompete the other during the interaction.

cooperation arises among individuals within populations (Nemeth and Takacs 2010; Box 5.3).

For most predators, cooperation does not occur. Rather, competition for prey is commonplace. According to the *competitive exclusion principle*, it is impossible for predators to completely overlap in prey consumption (Gause 1934). However, some *symmetric* or *asymmetric competition* is possible and its influence on population growth of a predator species can be described by the following equation:

$$\frac{\partial P_1}{\partial t} = r_1 P_1 \left(1 - \frac{P_1}{K_1} - \frac{a_{1,2} P_2}{K_1} \right)$$

where P_1 and P_2 are the numbers of two species of predator, r is the rate of growth of species 1, K is the maximum sustainable number of species 1, and $a_{1,2}$ is a coefficient of competition between the two species. In this case, competition is asymmetric, with predator species 2 having a competitive effect on species 1. This is a logistic growth equation with a term added that slows the growth of predator 1's population when either the competitor species is abundant or the competition coefficient is very

BOX 5.3   GAME THEORY

Evolutionary games can be developed to determine the best strategy a predator may use to capture its prey. These are mathematical models where a series of predatory strategies are pitted against each other. Each has an outcome with the strategy leading to the most successful outcome *winning* the game. For example, two strategies, lie-in-wait versus pursue, may be adopted by two genotypes of predator in a population. Based on the conditions of the model, one strategy will be more successful and will dominate. This will lead to the emergence of an *evolutionary stable strategy* (ESS). The ESS will be maintained in the population and will resist the invasion of other strategies into the population (Maynard Smith 1982).

In the example of hiding and attacking prey versus actively pursuing prey in a population, the energetic costs of pursuing a prey versus the encounter rate with prey may drive which ESS arises. In a large-bodied, ectothermic predator with relatively low energetic costs like a crocodile, the game between waiters and pursuers will likely lead to an ESS of waiters because attacking and consuming a large prey item and then digesting it over a long period is more energetically profitable. In a small mammal with a high metabolic rate like a shrew, the ESS of pursuit will arise, because the individuals adopting the alternative ESS of wait will starve to death while waiting for prey to arrive.

Game theory is a very powerful tool for determining solutions to difficult questions in predator–prey systems and ultimately in trophic ecology. Evolution has led to systems with a high proportion of ESS that, although not perfect, have complex and efficient trophic interactions. As humans selectively harvest particular genotypes from populations, specific ESS may be disrupted or lost, leading to reduced reproductive viability and loss of stability of predator–prey interactions. The challenge for conservation biologists and ecosystem ecologists is determining when ESS are present in populations and their impact on ecosystem function.

high (Figure 5.12). If the competition coefficient a is 1, then each individual in predator species 2 has the exact same effect on resources as each individual in species 1. When predator species 2 reaches densities the same as the carrying capacity of species 1, species 1 will stop growing. The model can be modified in many ways to explore joint, interactive effects of competition among multiple predators.

Competition between predators can be complex depending on factors such as size, experience, and timing of appearance. *Priority effects* often give one predator an advantage over another, in that the first predator species to arrive at a foraging location will have greater access and perhaps superior interference competition over later arrivals. This has been shown in fish recruiting in coral reefs (Poulos and McCormick 2015), larval amphibians in small ponds (Lawler and Morin 1993), and many other ecological systems where communities assemble frequently (Drake 1991; Garvey and Stein 1998).

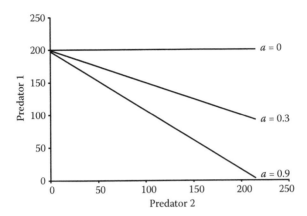

FIGURE 5.12 Theoretical reduction of a superior predator (species 2) on the equilibrium density of another predator (species 1) as a function of a competition coefficient in a simple competition equation.

5.6 COMMUNITY AND ECOSYSTEM

Predation is undeniably a critical structuring force in communities and ecosystems. In the discipline of limnology—the study of freshwater ecosystems—predation was typically omitted from discussion about driving factors in lakes in favor of physical processes such as nutrients and temperature. A seminal paper by Brooks and Dodson (1965) demonstrated that size-selective predation by land-locked zooplanktivorous alewife influenced the size structure of zooplankton, reducing the average body size of these prey and potentially the ability for the zooplankters to graze algae. This contributed to the development of the trophic cascade hypothesis, which will be covered in Chapter 10 when discussing food webs.

Predation has been shown to promote the diversity of organisms in some ecosystems. The classic study by Paine (1966) in the rocky intertidal zone of the Pacific Northwest was the first to illustrate experimentally how predation maintains diversity. Paine (1966) manipulated the density of a predatory seastar (*Pisaster* spp.) and found that it selectively preyed on the competitive dominant (a mussel, *Mytilus*) in the community. When *Pisaster* were removed, the competitive dominant took over, excluding many other species. Paine (1969) coined the term *keystone species* to describe the influence of *Pisaster* and similar species, which have effects that are disproportionate to their density in the environment.

Following Paine's research, a debate about how competition and predation structure communities began. Field observation rather than experimentation was typically used to conclude that predation was important. The problem is that causation could not be established, leading ecologists to call for more controlled experiments (Sih et al. 1985). Menge and Sutherland (1987) developed a formal model of predation depending on the trophic level of the prey species. They predicted that basal prey should be most susceptible to predation, particularly when environmental conditions are stable (Figure 5.13). Basal prey are abundant, are exposed to many predators,

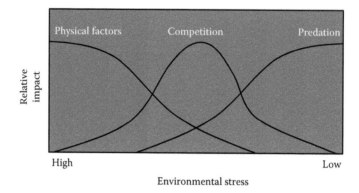

FIGURE 5.13 Response of prey populations to physical stress, competition, and predation as a function of an environmental stressor. Predation will likely have the greatest impact of prey when environmental conditions are mild, allowing predator populations to expand and control prey. (Adapted from Menge, B. A. and J. P. Sutherland. 1987. *American Naturalist* 130:730–757.)

are typically fast growing, and dominate communities. Stable conditions allow these prey and their predators to build populations and interact intensively, with predatory control of prey occurring. In contrast, predation on top consumers (i.e., carnivores) will be insignificant at the highest trophic level given that there are no higher consumer levels (Figure 5.13). Some predation is still possible within trophic levels such as cannibalism or between predator species (i.e., intraguild predation) (Polis et al. 1989). Predictions by Menge and Sutherland (1987) hold well in some aquatic systems (Wellborn et al. 1996) but less so in terrestrial ecosystems (Strong 1992). We will explore reasons for this in Chapter 10.

Predation is an emergent property of ecosystems. As we note in Chapter 2, predators at higher trophic levels are limited by energetic limitations of the trophic pyramid. Many predators thus fall in the realm of *k*-selected species, with typically low reproductive rates, long lives, high investment in individual offspring, and large bodies (MacArthur and Wilson 1967). These characteristics make predators and their control of ecosystems especially vulnerable to environmental stressors (Menge and Sutherland 1987). The world is in the middle of a mass extinction (Dirzo et al. 2014) in what is classified now as the Anthropocene (Zalasiewicz et al. 2015), which is unique in that the extinction is not uniform among species. Rather, large, predatory species occupying upper trophic levels (i.e., apex predators) are being lost at a high rate (Terborgh and Estes 2010). Because so many of these predator species play a critical role in promoting diversity at lower trophic levels, their loss propagates throughout entire ecosystems (see Chapter 10). The end result will be lost biodiversity as well as the elimination of unique trophic interactions that maintained this biodiversity (Terborgh and Estes 2010). Conservation and restoration of ecosystems should pay special attention to the importance of predation by reintroducing or protecting large predators when possible (Gil et al. 2015). Unfortunately, many of these species are economically valuable or feared as pests making effective policy difficult. To illustrate, large felids such as jaguars in Central America and lions in Africa are considered important species that likely exert important top-down effects on herbivores, promoting biodiversity. They are considered a threat to domestic livestock and human safety, although they often have less impact than perceived. Options such as reserves (Roberts et al. 2001) can be beneficial to maintaining predator populations. However, many reserves are incapable of providing adequate protection, especially as predators move due to factors such as changes in prey distribution.

5.7 SUMMARY

Predation is an active, top-down force in communities and ecosystems, usually involving species with unique behavioral and morphological features that allow them to specialize on prey. Although we avoid using terms like *adaptive* and *optimal* to describe predators, successful predators have characteristics that typically match those needed to efficiently capture their prey. Most top predators (e.g., secondary or tertiary consumers) have well-developed senses such as hearing, vision, and olfaction. Predators typically trade off strategies of either staying put and waiting for prey to arrive or actively tracking prey and overcoming them, depending on the energetic balance of the predatory sequence. Learning is an important component in many predator species.

The efficiency of predators to capture their prey has both top-down and bottom-up effects. Predators are typically selective in prey choice and can affect the diversity, size distribution, and behavior of their prey, either promoting or reducing biodiversity in communities. The efficiency by which predators consume their prey affects the flow of energy and materials to their predators, including humans. Predator–prey systems may be stable, particularly if predators are able to switch among different prey. If predators are too efficient, then prey extinction may occur, leading to the extinction of the predator and potentially the collapse of the food web.

Most of the species going extinct on Earth are *tertiary* or top predators in food webs. These species are particularly vulnerable to human activities including harvest and habitat loss, because they have low densities, low production rates, and are typically large-bodied and conspicuous. The loss of these species typically causes changes in lower trophic levels that further compromise biodiversity and lead to less efficient trophic interactions (Hooper et al. 2005). If conservation is emplaced, many predators will likely have low recovery rates or fail to recover at all, given constraints both within populations and changing ecosystem characteristics, requiring active recovery efforts by humans (Gil et al. 2015).

QUESTIONS AND ASSIGNMENTS

1. What is the difference between predation and depredation?
2. Why are humans considered the ultimate predator on Earth?
3. How does the predation sequence of events differ between a lie-in-wait and pursuit predator? How are energetic investments distributed differently among these events?

4. What is optimal foraging by predators? Is it possible? In your answer, explain how predators become specialized.

5. How does the size of prey relative to predators affect patterns of predator growth and their impact on prey populations?

6. Explain how the theory of predator and prey dynamics allows them to coexist without one causing the other to go extinct. Explain how theoretical models by Berryman and others allow predators and prey to coexist. Hint: It involves density-dependent processes in the logistic model of population growth.

7. What are the differences between functional and numerical responses of predators?

8. Explain prey switching and how this is important for stabilizing predator–prey interactions.

9. Why is cooperation among predators improbable?

10. What is a priority effect?

11. How does selective predation promote biodiversity in ecosystems?

6 Prey

6.1 APPROACH

This chapter reviews the characteristics of prey that affect their vulnerability and ultimately their contribution to energy and matter flux within food webs of ecosystems. We will learn that one of the most important factors affecting the susceptibility of prey is age and development, with the relative production of biomass at vulnerable life stages influencing stability of food webs. Prey are not helpless and have many physical, behavioral, and chemical characteristics that may reduce predation. Some of these traits are permanent. Others are induced by the presence of predation. These trait-mediated interactions are an important frontier in the study of trophic ecology. Humans are able to work around prey defenses and often target life stages that are typically well protected from predators. This leads to instability in food webs.

6.2 INTRODUCTION

Almost all organisms are prey during some or all of their existence. Being prey does not mean that an organism is helpless. Rather, many species have evolved unique defenses and offenses to reduce the individual probability of being eaten during vulnerable periods of life (Marrow and Cannings 1992). The consumption of prey items is a major conduit by which energy moves through ecosystems. Thus, the distribution, physical characteristics, and behavior of prey and their interaction with consumers will greatly affect ecosystem structure and function (Werner and Peacor 2003).

At the simplest level, prey may be viewed as nothing more than discrete packets of energy and nutrients to be transported up the trophic pyramid. This is typically not the case. Herein, we explore ways that prey need to be considered as complex components of ecosystems in order to better understand their contribution to trophic fluxes. Prey may encounter predators without being consumed, but the encounter itself may start a series of events that have unanticipated consequences. Surviving prey that suffer injury may die later (e.g., Bowerman et al. 2010) and enter detrital or scavenging pathways rather than consumer trophic levels. A small deer bitten by a monitor lizard may develop an infection or succumb to the lizard's venom, and die a week later. Not only may the monitor lizard track down and consume the deer, but the downed prey also serves as a host of food for other organisms while the lizard searches for it. Prey vulnerability typically varies temporally and within as well as among generations. Thus, both immediate and lagged prey responses need to be considered when incorporating prey interactions into trophic ecology. We will introduce plastic prey responses that typically suppress predation called *trait-mediated indirect interactions* (TMII; Werner and Peacor 2003). The chapter will finish by briefly examining how humans alter ecosystems by changing prey responses in unintended ways.

6.3   ONTOGENY OF RISK

Predators make the world a dangerous place for most organisms, especially when they are young. Early life stages of species lie along a continuum from looking and behaving nothing like adults to being smaller, morphologically identical versions of older stages (Werner and Gilliam 1984; Figure 6.1). How prey animals contribute to energy transfer through their life is context dependent. Larval insects that live in aquatic systems or under the soil experience a very different predatory landscape than when they emerge onto the terrestrial surface as adults. To illustrate, dragonfly larvae are important prey in aquatic systems for fish and other aquatic vertebrates. When larval dragonflies emerge to defend terrestrial territories and mate, they fall prey to a number of larger species in a different ecosystem. To make things more complicated, dragonflies are important predators as larvae and adults, causing their responses as prey to affect their own prey (e.g., as trophic cascades; Chapter 10). Prey with less complex ontogenies may face a more predictable array of predators that may become less threatening as the prey age, grow larger, develop physical defenses, and learn how to avoid being eaten.

The growth and maturation trajectories of species influence life as prey and the transfer of biomass to predators. Many species are produced from externally fertilized

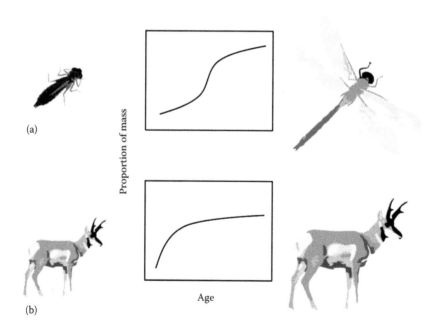

FIGURE 6.1   Patterns of growth that occur among taxa. Species with a high mortality rate during early life tend to produce a large number of tiny offspring, often with characteristics, locations, and behaviors that differ largely from adults (a). Biomass accumulation in these species lags early and then increases rapidly. Many of these species continue to grow through life (b). Species with low mortality during early life like antelope typically have few, large offspring that grow rapidly to adult sizes. Many of these species stop growing once they reach adulthood.

eggs and spend time exposed in the environment as tiny, vulnerable prey (Figure 6.1). The pattern of biomass accumulation and associated vulnerability in these species is typically low and then increases exponentially (Figure 6.1a). Offspring that are produced from large eggs or internal gestation often grow rapidly during early life, reaching adult sizes quickly (Figure 6.1b). These patterns of growth are linked with patterns of survival as we will demonstrate later.

The survival trajectory of species can be placed into at least three general categories (Pearl 1927), although the true *survivorship* patterns are likely far more complex (Figure 6.2). Predation contributes to the shapes of these survivorship curves, although other sources of mortality such as disease, starvation, and environmental insults also affect mortality and interact with predation. Species with high parental care and relatively low early mortality are categorized as Type I. Even with this life history pattern, there is usually relatively high mortality during very early life (i.e., pre-recruitment) that is not considered as part of the *traditional* hypothetical curve shown in most ecology textbooks (Figure 6.2). A nesting bird that cares for its young may incur high initial losses of eggs from the nest, leading to an initial dip in survival, with survival then becoming stable until late in life past reproductive age when survival declines rapidly with senescence (Figure 6.2). On the opposite end of the continuum, species with no parental care often experience an exponential decline in mortality called a Type III curve (Figure 6.2). Mortality before *recruitment* is usually even higher than that predicted for the remainder of life (Figure 6.2). This survivorship pattern is very common. For example, insects may grow through instars where they are continually consumed through life, but mortality risk declines exponentially as they grow larger and become less vulnerable. Many fish show this pattern.

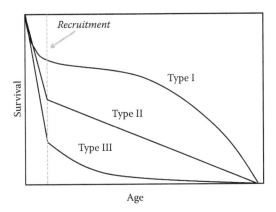

FIGURE 6.2 *Traditional* survivorship curves found in organisms as a function of their life histories. In all three models, early mortality before the point (dashed line) where *recruitment* occurs is typically very high and not incorporated into these models. Recruitment is the point in life where mortality rates along the curve can be predicted at any life stage. A Type I survivorship expects high survival of recruited offspring, with high mortality only occurring during old age. A Type II survivorship curve assumes linear mortality through life. Species with a Type III survivorship experience decelerating mortality through life. Most mortality occurs during early life stages.

A Type II curve assumes constant, linear mortality through life. In this case, mortality due to predation may be similar regardless of life stage. No one pattern fits all. Survivorship patterns vary among species and among populations within species.

The shape of the survivorship curve (Figure 6.2) and the distribution of biomass tied up in each life stage (Figure 6.3) dramatically affect the amount of energy available to higher trophic levels. The production of biomass by life stages depends on the summed biomass across individuals, their growth rates, and their reproductive output. We hypothesize that energy transfer to higher trophic levels should be minimized by natural selection during the most productive life stages, which will vary with life span, life history strategy, and other species-specific attributes (Figure 6.3). Species with a Type I survivorship are usually large-bodied with a high investment in few, large young. Because of the high energetic investment in young, relatively few predation events on young individuals can lead to a flux of energy to higher trophic levels (Figure 6.3a). Similarly, older, large-bodied individuals that have already reproduced can contribute a tremendous amount of biomass to their predators or to detrital and scavenging trophic levels (Figure 6.3a). An example of this would be an elephant seal. The mortality of young elephant seals is low due to parental care. But in the event of a successful predation event by a large predator like a killer whale, the whale scores a huge energy boost. Adult elephant seals dying of old age after a life of successful reproduction provide a huge amount of stored biomass to multiple trophic levels.

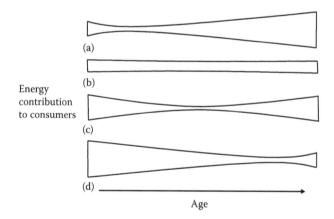

FIGURE 6.3 Relative contribution of prey biomass to predators (and other trophic levels) as a function of age, growth, and age-specific defenses in prey. The thickness of each bar represents the relative prey biomass available in the population to consumers. If prey are relatively rare and defended during early life but grow to vulnerable sizes as adults, then significant biomass may not be transferred until late in life (a). If prey remain small and defenseless through life, the biomass transfer may be constant (b). If prey have a specific defended size or age during life, then a bowtie pattern may occur with young and old individuals providing most energy to consumers (c). Prey may be produced in high densities during early life and contribute most to predation at this life stage (d).

Some short-lived species may contribute energy to consumers relatively similarly throughout life (Figure 6.3b). Cladocerans in lakes typically reproduce clonally, with their offspring being an identical, albeit smaller, copy of the parent. Young grow rapidly and thus may experience similar predatory-induced mortality to their parents (Figure 6.3b). Predators can count on a steady supply of relatively predictable prey, unless the prey change some aspect of their physical characteristics or behavior to reduce risk.

Although the shape of a Type III survivorship curve predicts decelerating mortality after recruitment, the pattern of energy flux to higher trophic levels will vary depending on the slope of the curve, the growth of the organisms, and reproductive patterns (Figure 6.3c,d). The pattern of adult versus young contributions to consumers will depend on life history strategy. A migratory species like Pacific salmon suffer their highest mortality during outmigration to the ocean and during their return to spawning streams (Figure 6.3c). Other fish species that live their entire lives in the ocean also have a Type III survivorship pattern, where fertilized eggs and larvae are distributed into the open water, with little chance for survival and thus a high contribution to consumers such as sea jellies. These fish grow exponentially and should be least vulnerable to predation at sizes of reproduction (Figure 6.3d). After reproducing, adult fish senesce and fall prey to various trophic levels. Periodical cicadas have a similar strategy, emerging as adults from the ground simultaneously, swamping predators, and dying after reproducing (Karban 1997).

Predictions relative to the relationship between survival patterns and trophic transfer at specific life stages are largely untested. Our conceptual model folds in the influence of natural selection on key life stages and their implications for ecosystem processes. The shape of the *bow-tie* pattern of life-stage-specific energy transfer in Figure 6.3 likely varies among species and requires a closer look. This pattern is a function of individual selection at specific life stages scaling up to the level of ecosystem processes. These predictions are similar to the ideas of MacArthur and Wilson (1967) about *k*- versus *r*-selected species. Species that have high survivorship during early life (*k*-selected) should follow a Type I pattern and have very different impacts on trophic transfer than *r*-selected species with Type III patterns (Box 6.1). If age/stage-specific selection for prey changes within communities, then ecosystem responses will change as well. Understanding how traits of individuals vary is critical for trophic ecology. Species with lower reproductive rates and long life spans (*k*-selected, Type I survivorship) should be more susceptible to predation and human consumption. One group of organisms that fit into this pattern is the sharks and other elasmobranchs throughout the world (Field et al. 2009). Sharks are long-lived and have low reproductive rates with high investment in young. Adults are harvested at high rates, which is responsible for their imperiled status at a global scale. Ironically, humans have again and again developed ways to strongly impact life stages of species (forest trees and harvested fishes) that are the most productive, likely the least vulnerable to other predators, and most likely to contribute to population and food web stability through top-down and bottom-up effects (Dulvy et al. 2004; Tscharntke et al. 2005).

BOX 6.1

MacArthur and Wilson (1967) developed the theory of island biogeography to understand processes that influence species richness as a function of the size of islands, with application to any patch in a landscape. They suggested that life histories of organisms are categorized as either r- or k-selected, relative to the terms in the logistic equation (Chapter 5). Species with an r-selected strategy are small, fecund, and rapidly growing—able to rapidly respond to a reduction in population density by reproducing quickly. In contrast, k-selected species have a strategy to maintain density near their carrying capacity by increasing survival of offspring and resisting perturbations that reduce adult densities. Species fall within a continuum between these two end points. A tremendous amount of predator–prey theory has been developed based on these concepts (see Chapter 5). Prey species often have r-selected characteristics. Similarly, vulnerable life stages within species (even if they are considered k-strategists) should behave more like r-strategists, with rapid growth and high densities to compensate for high mortality caused by predators.

The relative dominance of a system by r- versus k-selected species should influence patterns of trophic transfer and ultimately the stability of ecosystems. Because r-strategists time their production of biomass to be short, rapid, and usually in a confined area, these species will be resilient to predation but also will lead to short, flashy dynamics in ecosystems. Prey with k-selected strategies take a long time to transfer energy to higher trophic levels because they have evolved ways to avoid being eaten rapidly. If predation overcomes these defenses (e.g., human harvest), then these species will have difficulty persisting.

Relationship between Life History Strategy and the Potential Temporal and Spatial Patterns of Energy Transfer between Prey and Predator

Trophic Transfer	r-Selected	Intermediate	k-Selected
Timing	Short	Variable	Long
Rate	Rapid	Variable	Long
Spatial	Narrow	Variable	Broad
Pattern	All	Variable	Old
Age/stage			
Susceptibility to collapse	Low	Intermediate	High

6.4   PHYSICAL DETERRENTS

Organisms have evolved a vast array of ways to avoid being consumed during vulnerable life stages. Of course, all of these tactics involve trade-offs that ultimately reduce energy that can be contributed to reproduction. Any defensive or offensive tactic that works certainly is less costly than the total loss of fitness incurred by being injured or consumed.

Many organisms taste bad or have a memorable flavor. Taste preference of consumers is relative. Predators may not care much about how their prey taste, thus making bad taste a weak deterrent. Most ecologists suggest that bad taste is likely a signal indicating toxicity, although the circumstances have to be specific (Skelhorn and Rowe 2009). Presuming a predator survives consuming a toxic prey, it will learn to associate the taste with sickness and avoid the prey in the future. Taste and other sorts of warning signals will be addressed later.

Strong smells may be a way that prey organisms defend themselves from predators. Like taste, odor may be a sense that varies widely as a deterrent. However, olfaction is an important component of signaling in both terrestrial and aquatic ecosystems, and many organisms have well-developed chemosensory abilities. Emitting a strong, persistent smell such as that produced by skunks may swamp the ability for predators to use their sense of olfaction for detecting their own predators, finding mates, defining territories, and seeking food (Stankowich 2012). The strong smell of the predator can make it more conspicuous to prey, reducing its predatory success. If the prey is dangerous, smell also could be used as a learned warning signal, like taste.

Similar to many plants, toxicity is a very common tactic in prey animals. Toxins may either be accumulated exogenously from a food source or produced endogenously. For example, colorful neotropical poison frogs (Dendrobatidae) likely derive much of their toxins from ants and other prey (Darst et al. 2005; Figure 6.4). Monarch butterflies gain their toxicity from a class of compounds called *cardenolides* in the milkweed that they consume (Agrawal et al. 2012). Other taxa such as some marine sponges can synthesize their own toxins (Leong and Pawlik 2010). Ultimately, both external and internal sources have costs. Prey accumulating exogenous toxins must render the compounds nontoxic to themselves, which requires an energy investment, usually in enzyme systems that reduce toxicity (Agrawal et al. 2009). If organisms are capable of synthesizing their own poisons, producing toxic secondary metabolites wrests internal energy from growth and reproduction. Adult fruit flies "medicate" young to protect them from predators. In the presence of parasitoid wasps, adults lay eggs in fruit

FIGURE 6.4 Poison dart frog showing clear aposematic coloring to alert predators of danger. (Courtesy of Shutterstock.)

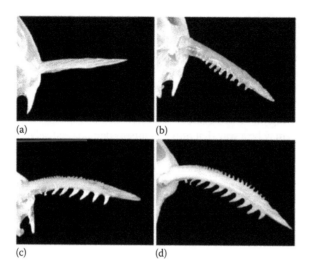

FIGURE 6.5 Pectoral fin spines of four species of North American madtoms used to inject venom into potential predators. (From Wright, J. J. 2012. *Journal of Experimental Biology* 215 (11):1816–1823.)

food of high alcohol concentrations to deter wasps from using drunken hatched larvae. Alcohol in this regard is a protective toxin (Kacsoh et al. 2013).

Many prey use a delivery system to inject venom into their predators. Small North American stream fish called madtoms have pectoral spines that inject venom to deter predation (Figure 6.5). Bees, wasps, and many other hymenopterans are well known for their ability to sting their opponents. Venom in many of these organisms can serve as a tool for predation or territoriality as well as defense. Baracchi et al. (2011) suggest that venom also may have antimicrobial properties that are beneficial to social insects such as bees. A similar mechanism occurs for ants (Tragust et al. 2013).

Venom exists in many forms. Formic acid is a common venom and can be sprayed as well as injected by many insects. Other venoms are far more complex, with hemolytic and neurotoxic properties. Venom can be found in just about every taxon from invertebrates to mammals. Wong and Belov (2012) argue that venom evolution is prevalent due to the duplication of genes for venom production. Gene duplication is a process by which a gene is copied and no longer subject to selection. Duplicated genes for venom can rapidly accumulate mutations that give them diverse and often deadly characteristics. In high doses, venom can be toxic to their host, as a classic paper on rattlesnakes showed (Nichol et al. 1933).

Armor is another deterrent that is common among animals. The calcium carbonate shells of mollusks are an obvious defense. Bony scales or scutes on fish and reptiles provide protection. An important, flexible proteinaceous molecule called chitin is found in many organisms, especially in the exoskeletons of arthropods. Chitin can be combined with minerals to make it particularly resilient to external attack by predators. A testament to the stability and strength of this molecule is that it has been found in beetle fossils dating back 25 MYA (Briggs 1999). Thick skin or fur on mammals can be an effective if not cumbersome deterrent. In addition to armor, animals can develop all sorts of structures such as spines and pinchers from thick materials like bone, skin, and chitin to make them difficult to consume.

6.5 ADVERTISING

Aposematic warning signals are very common in dangerous prey (Figure 6.6). The root *apo* means away and *sematic* refers to meaning. Bright colors, unusual physical patterns, distinct sounds, strange odors, and specific behaviors are among the many aposematic traits that allow vulnerable but harmful life stages to avoid being eaten. The general assumption is that predators encounter or consume a dangerous prey item and are harmed. If the predators live to survive the encounter, the warning signal serves as a deterrent to future encounters. Presumably, the offspring of some predators may learn to identify warning signals from adults. An avoidance of novel signals in prey such as unusual coloration or odors also may be innate in some predators (Jetz et al. 2001; Exnerova et al. 2007). Anyone trying to protect edible garden plants from herbivores knows that spraying distasteful substances is only a temporary fix. Once the herbivores become accustomed to the substance, the plants are eaten, especially when wild food resources are limited and herbivores are hungry.

Aposematic signaling may require that at least one prey individual lose its life in order to alert predators of the species' danger. From the perspective of natural selection, which operates at the scale of the individual, the evolution of aposematism seems unlikely because the signaling prey is the one being harmed or consumed and not passing its genes to future generations. Alatalo and Mappes (1996) suggest

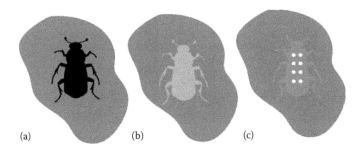

	(a)	(b)	(c)
Tactic:	Visible, conspicuous	Crypsis	Warning signals
Mechanism:	Compensate	Hide	Toxic/mimicry
Tropic outcome:	High energy transfer	Reduced energy transfer	Reduced energy transfer

FIGURE 6.6 Different methods by which prey respond to predation. Prey may not alter appearance or behavior but compensate for increased mortality by producing high densities as groups over a short period of time (a). Prey may adopt cryptic behavior and characteristics (b). A third alternative of prey is to be conspicuous, which either advertises danger or reduces consumption by wary predators (c).

that the tactic likely evolved in prey aggregates, where some conspicuous individuals cued predators to avoid closely related conspecifics. Thus, this is an example of selection by association. Interestingly, the strength of the warning signal and the toxicity of the organism are not always strongly linked, suggesting that other processes are influencing the evolution of these patterns (Wang 2012).

With the evolution of warning signals in harmful prey species, nontoxic mimics arose (Figure 6.6). *Batesian mimicry* occurs when nonharmful species evolve signaling traits that are similar to sympatric harmful species. Examples of this abound, such as harmless syrphid flies (e.g., sweat bees) that have similar color patterns to bees. Some harmless mimics exist in areas without the harmful species (Pfennig and Mullen 2010), supporting the idea that predator avoidance of unusual, potentially harmful prey may be innate. *Mullerian mimicry* is the condition where more than one harmful prey species advertises using similar patterns. The monarch and viceroy butterflies, once thought to be a classic Batesian model, were reclassified as Mullerian mimicry when viceroys were found to be toxic as well (Ritland and Brower 1991). The familiar yellow-black stripe pattern in so many venomous hymenopterans is a typical example of Mullerian mimicry.

Ecologists need to be careful about inferring that certain unusual characteristics are favored by natural selection to deter predators. Large claws in decapod crustaceans, pinchers in beetles, bright plumage in birds, and aggression in ungulates may be inferred as deterrents and may in fact work to reduce predation. However, these characteristics may have evolved for other reasons, most likely through sexual selection that favors characters that signal mate quality or the ability to access mates. Careful research is necessary to tease apart the differences as was done for Mullerian mimicry within a group of cyanide-producing millipedes of the Appalachian region of the United States (Marek and Bond 2009). These organisms are blind and their similar coloration patterns can only be intended for predators (Figure 6.7).

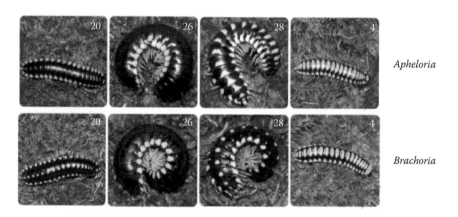

FIGURE 6.7 Two millipede genera (*Apheloria* and *Brachoria*) from four sites (sites 20, 26, 28, and 4) within the Appalachian region of the United States with similar coloration patterns. At each site, both species look alike and are more similar in appearance than conspecifics at other sites. These two genera are part of a group of millipedes that are toxic and a clear example of Mullerian mimicry. (From Marek, P. E., and J. E. Bond. 2009. *Proceedings of the National Academy of Sciences of the United States of America* 106 (24):9755–9760.)

6.6 SIZE MATTERS

Both prey and their predators grow through time (Wilbur 1988). Growing large should provide advantages to prey. Many predators, especially those in aquatic ecosystems, are gape-limited because they consume their prey whole (Hambright et al. 1991). At the simplest, prey can grow to sizes that are too big to cram down a predator's esophagus. Big prey can run or swim faster than small counterparts and presumably are stronger and more dangerous to predators (e.g., moose; White et al. 2001). Thus, the idea that *bigger is better* is common in trophic ecology, with some suggestion that natural selection should favor rapid growth and large body size, allowing prey to outgrow their predators (Mittelbach and Persson 1998). Gizzard shad is a fish species in North America that grows rapidly early in life and evades most gape-limited predators, suggesting that this tactic has a selective benefit (Box 6.2). Many mammals and birds have young that rapidly reach adult sizes, again providing protection. Although the idea that rapid growth is a strongly selected trait is compelling, the reality is that several factors likely constrain selection for rapid growth in many species (Sogard 1997).

Being big and growing fast may be a disadvantage. During early life, prey development and prey growth may be decoupled (Werner 1986; Bertram and Leggett 1994). This is particularly prevalent in species like amphibians and insects that undergo metamorphosis, where there may be trade-offs between growing fast and developing into an adult-like form. A tadpole may face selective pressures to grow rapidly to avoid predation in a pond (Skelly 1996), but this tactic may lead to it developing early as a small juvenile that is susceptible to predation on land (Skelly and Werner 1990). Many predators detect their prey visually. Thus, the larger a prey organism is, the more conspicuous it becomes (Figure 6.8; Litvak and Leggett 1992). In this model, growing confers a defensive advantage (Figure 6.8). The intersection between the two curves in Figure 6.8 is the theoretical point where prey should be most vulnerable and contribute most to higher trophic levels. The intersection will depend on the visual acuity, size, and other characteristics of the predators. Small prey may be more susceptible to invertebrate predators, which use tactile and chemosensory information to locate prey while being limited by handling of large prey. The opposite may be true of larger, visual-oriented vertebrate predators (Litvak and Leggett 1992).

BOX 6.2

Gizzard shad is a member of herring family that is common in rivers and lakes in North America. This species is an important prey for piscivorous fishes and waterbirds. However, it grows swiftly during early life by rapidly increasing its body depth, allowing individuals to exceed the gape limits of many of their predators. We found that early growth of a piscivorous fish, largemouth bass, depended on the first-year growth rates of gizzard shad in Midwestern US lakes (Garvey et al. 2000). During years when gizzard shad grew slowly, young largemouth bass grew rapidly. Conversely, when gizzard shad grew rapidly, largemouth bass suffered poor growth.

Gizzard shad dominate many of these lakes, comprising more than 50% of the fish biomass (Vanni et al. 2005). Thus, their population dynamics drive ecosystem structure and function.

A typical pattern for largemouth bass and other gape-limited species is that there is an upper limit to the size of prey that can be consumed. In the case of largemouth bass and gizzard shad, largemouth bass can consume gizzard shad up to 50% of their gape width.

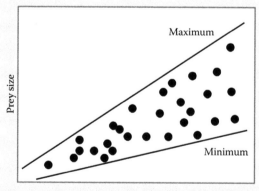

Relationship between a gape-limited predator and the prey that it can consume. Note that small prey are still consumed. However, there is an upper limit to the size of prey that can be consumed. (From Scharf, F. S. et al. 1998a. *Ecology* 79:448–460.)

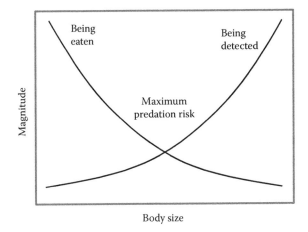

FIGURE 6.8 Hypothetical curves of risk of being eaten and risk of being detected by predators as a function of body size. The intersection between the curves should be the size where prey are most vulnerable to predation. (From Litvak, M. K., and W. C. Leggett. 1992. *Marine Ecology Progress Series* 81:13–24.)

6.7 HIDING

Another way to avoid being consumed is to avoid detection by predators. One way to accomplish this is to hide in plain sight by blending in with the surrounding environment (*crypsis*; Figure 6.6). Cryptic coloration in terrestrial systems involves adopting colors similar to the vegetation or ground. In aquatic systems, many pelagic fish have light-colored ventral surfaces that blend with the sky above to make them less conspicuous to predators below them (i.e., countershading; Ruxton et al. 2004). Other organisms like walking stick insects physically mimic their surroundings. Examples are countless. The energetic costs of these tactics likely vary markedly across taxa. Modifying coloration is likely far less energetically costly than generating elaborate external structures. However, for animals that must bask to thermoregulate, coloration can be a problem. Many reptiles must bask in the sunshine to raise their body temperatures. Darker colors may speed the warming process but make them more conspicuous. Small birds that need to warm in the sunlight during winter may face similar problems. Coloration and conspicuousness may need to be balanced with the need to heat their bodies (Carrascal et al. 2001).

Reducing activity is another way to reduce predation. Stein and Magnuson (1976) were among the first researchers to show that prey (crayfish) reduced the rate of predation by modifying their behavior. When confronted by fish, vulnerable sizes and life stages of crayfish reduced activity and burrowed. This led to disciplines exploring inducible defenses in prey and eventually the trophic-evolutionary concept of TMII. As we will describe, reduced activity by prey results in reduced feeding and growth, which may be detrimental by suppressing fitness. Migratory patterns of prey such as fish also may be affected by predators (Skov et al. 2013) affecting seasonal patterns of prey abundance in a spatial context.

6.8 INDUCED DEFENSES

Many prey species only respond to predation when necessary, leading to the concept of inducible defenses (Harvell 1990; Figure 6.9). Prey only need to invest energy when risk is present, thereby having the ability to invest more energy into growth and reproduction when predators are scarce. This strategy should only emerge if the lifetime fitness cost of a permanent defense is greater than that of an induced defense. In other words, the *cost* area under the solid cost curve in Figure 6.9 needs to be greater than the area under the dashed curve. For defenses to be effective, prey need a reliable, *honest* signal and the ability to respond quickly. The sight of a predator or a physical sign of it (e.g., its shadow or silhouette) will serve to warn prey with vision. The landscape is full of potential threats, and visual cues are often learned (Kelley and Magurran 2003). If prey cannot see, then they must rely on alternative cues.

Olfaction or related forms of chemosensory signaling are probably one of the most common and reliable ways that prey detect predators (Peckarsky 1980). Many prey organisms also may detect the chemicals released by a conspecific (Mathis and Smith 1992) or another prey species when it is consumed or injured (Gonthier 2012). This chemical substance was described initially by Karl von

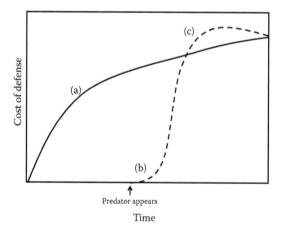

FIGURE 6.9 Hypothetical cost of prey defense through life for two different strategies. Prey may generate a defense through life (a) incurring a lifetime energetic and fitness cost. Prey with inducible defenses may mobilize the defense only when predators are present (b). The cost of inducing the defense may initially be higher than the average experienced by the prey that has a lifetime defense (c). Ultimately, the area under the curves should determine which of the life history strategies is favored.

Frisch as *schreckstoff*. This response to the smell of death likely evolved second-arily (Ferrari et al. 2010). The alarm substance within cells had some other pur-pose (e.g., protection against UV radiation, immune function) and later became associated with risk. Sound is another form of communication about predators that can work effectively in either water or air. Birds can produce calls that warn conspecifics of predators (Evans et al. 1993). Small, vulnerable birds respond to the calls of their predators by reducing reproductive output (Zanette et al. 2011). Insects are quite adept at hearing the ultrasonic chirps of bat predators (Conner and Corcoran 2012).

Prey respond to predator signals in many ways. Inducible defenses can be cat-egorized as behavioral and physical. Aggression is one potential defense. Garvey et al. (1994) showed experimentally that aggressive rusty crayfish were less vulner-able to fish predators than passive congeners, the virile crayfish. Passivity is another response, where prey become inactive and thereby less conspicuous. Spring peeper larvae have evolved to be stealthy around fish predators, allowing them to coexist with them, whereas their more active congener, the chorus frog, suffers high preda-tion (Skelly 1996). In this case, the two species separate out in space between ponds with and without fish predators.

Schooling or flocking is another immediate response that organisms may use to reduce predation risk (Roberts 1996; Bednekoff and Lima 1998). One argument for this behavior is that it increases vigilance within the group because multiple eyes are searching for predators. Thus, each individual can spend more time foraging and respond more rapidly to an attack. Ultimately, the vigilant group lowers its indi-vidual risk of being eaten. Although valid in some species, Lima (1995) did not find strong support for this in birds. Another reason for grouping could be *predator swamping*, whereby the individual risk of being eaten is reduced by the sheer number of conspecifics present. Flocks, schools, or swarms may confuse predators, prevent-ing them from accurately targeting any one individual and reducing the predator's individual capture success.

Inducible physical responses vary markedly among species (Harvell 1990). The lag time for these responses can range from rapid (e.g., a squid producing a cloud of ink) to gradual (e.g., generational changes in morphology). Zooplanktons have long been known to change body morphology to make them more difficult to con-sume. These small invertebrates will develop spines and exaggerated heads, which may make it hard for small fish to eat or for invertebrate predators to grasp (Zaret 1980). Morphological change to avoid predators is not restricted to invertebrate prey. Bronmark and Miner (1992) used experiments to show that crucian carp, a cyprinid, increased their body depth in the presence of pike predators within the same generation (Figure 6.10). This served to make it impossible for pike to get their mouths around the carp. Although these sorts of rapid changes are common in invertebrates (e.g., metamorphosis), this energetically costly response is particu-larly intriguing for a vertebrate. The timing and extent of the changes in physical characteristics have implications for prey production and ultimately trophic transfer of energy.

(a)

(b)

FIGURE 6.10 Crucian carp is a fish species that has a relatively shallow body morphology when alone (a). When overlapping with pike predators, individual carp will increase its body depth to exceed the pike's gape limit (b). (From Bronmark, C., and J. G. Miner. 1992. *Science* 258:1348–1350.)

6.9 TEMPORAL CHANGES

If predation is temporally predictable, it might be beneficial for prey to time periods of reproduction or activity to avoid consumption (see Blouin-Demers and Weatherhead 2000 for rat snakes). Cushing (1974) posited the *match–mismatch* hypothesis that suggests that predators time their reproduction with seasonal peaks in their prey to enhance their cohort success. The better the predator is at producing young when prey are abundant, the faster the predator population will grow. In turn, evolution should favor prey that reproduce or are most active before predator reproduction peaks. Although a compelling concept, this idea has mixed support in the literature (Leggett and DeBlois 1994).

The reality is that prey can respond to predation during reproduction in many ways including providing parental protection of young, swamping predators with huge numbers of dispersing offspring, or reproducing asynchronously (Ims 1990; Winemiller and Rose 1992; Figure 6.11). Using an extensive data set for fishes, Winemiller and Rose (1992) expanded upon MacArthur and Wilson's *r–k* continuum idea described in Box 6.1

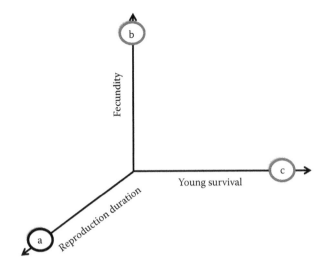

FIGURE 6.11 Three potential reproductive strategies that affect how young prey contribute to predator diets. Prey can produce a small number of offspring many times during their life, potentially missing periods of overlap with predators (a). Swamping predators with high numbers of offspring during a short period is another option for avoiding predation (b). Prey may invest high energy in a few young and use parental protection to ensure that predation on offspring does not occur (c). Species likely fall within this continuum of strategies as a function of the predatory impact on offspring. (Adapted from Winemiller, K. O., and K. A. Rose. 1992. *Canadian Journal of Fisheries and Aquatic Sciences* 49:2196–2218.)

to suggest that species have evolved reproductive strategies to deal with predation (and other mortality sources) along a continuum of at least three factors. If predation or other environmental variables are unpredictable, prey species can be small-bodied and reproduce in short spates throughout the year (Figure 6.11a). If predation is predictable, then big adults swamping the predator with very high densities (i.e., millions) of small offspring over a short duration is the option (Figure 6.11b). The third option is for large-bodied organisms with parental care and large offspring with high survival (Figure 6.11c). Knowing how prey will respond to predation through reproduction along this continuum will inform predictions for trophic flux during early life.

Some prey populations can opt to *check out* when predators are abundant by completely ceasing their activity. Bryozoans initiate dormancy during summer to avoid predation (Callaghan and Karlson 2002). Many species of zooplankton produce resting eggs through diapause when predation is high (Cáceres and Tessier 2004). This phenomenon also occurs in terrestrial species like mites (Kroon et al. 2008). Of course, the downside to the prey is delayed reproduction and growth. To the predator and ecosystem, this means lost trophic transfer.

6.10 PREDATOR–PREY COEVOLUTION

Many prey and predator species have coexisted for millions of years. The unusual and often bizarre characteristics and inducible defenses of prey are testament to the fact that prey evolve many ways to avoid being eaten. Predators should evolve ways to circumvent these defenses. This indeed occurs. As shelled organisms evolved, species such as whelks that could drill through the shells quickly appeared (Dietl 2003). Sea otters are well known for using rocks on their bellies to crack through the spiny bodies of sea urchins. For every defense, there is usually a plausible offense. Marrow et al. (1992) used modeling games to determine the stability of predator–prey interactions as they evolve through time. If either the prey or the predator evolves a way to overcome the other, prey may strip their resources while predators starve to death. Although extinction is a possibility, the interactions can lead to evolutionary stable strategies where the predator and prey maintain their characteristics and evolution does not proceed further (Marrow et al. 1992). The most likely outcome is a weakening of the predator–prey interaction. The implications are that the transfer of energy between the two species will decline through time, but the ecosystem will become more stable because extinctions are unlikely. This is an important ecological premise that is critical for food web theory (Chapter 10).

6.11 POPULATION CONSEQUENCES

To this point, we have focused on individual prey responses and how they scale up to energy transfer in ecosystems. Predator interactions with prey populations elicit many responses that are important to the stability of both populations as shown in Chapter 5. Assuming that prey populations grow in a logistic fashion, then the impact of predation may lead to a per capita increase in the birth rates and decline in the death rates of the surviving prey (Figure 6.12). This *thinning* of the prey population leads to the ability for the individual survivors to *compensate* for predation by

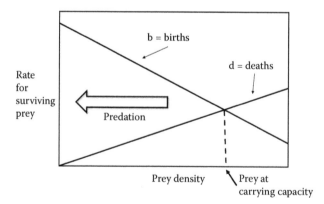

FIGURE 6.12 Vital rates for surviving prey in populations if compensation is occurring. As predation drives prey densities lower (arrow), individual birth rates increase and death rates (of prey not being consumed) decline. This response serves to allow the survivors to produce offspring and biomass that compensate for predation. This should occur because *thinning* by predators reduces intraspecific competition for limited resources in the surviving prey.

increasing reproduction and experiencing higher condition (Cole 1954). This compensatory response should stabilize predator–prey interactions, although the reality is much more complex (Arditi and Ginzburg 2011; Chapter 5).

Prey populations may be particularly susceptible to predators if predator populations grow dense or evolve ways to circumvent prey defenses. Density-independent forces also may reduce prey populations to low densities (Davidson and Andrewartha 1948), making the survivors susceptible to the impact of predation. Under these circumstances, the prey population is unable to compensate for further mortality and suffers an increase in death rates and a decline in birth rates, termed population *depensation* (Post et al. 2002). Allee effects where prey are unable to find mates are one common depensatory problem (Gascoigne and Lipcius 2004). A species with a complex mating system that requires several adults like nesting seabirds may be negatively affected by a reduction in the number of nesters. Predation may contribute to the depensatory collapse of the prey population. The few remaining prey may seek refuge by flocking or schooling, making them more conspicuous and likely to be eliminated by predators. An example of this is passenger pigeons that went extinct in North America in the early twentieth century. This species may have had as many as 3 billion individuals but was hunted to extinction, because the remaining individuals kept contracting into flocks that were easily hunted and preyed on by predators.

Prey populations may respond spatially to predation, where individuals occupy habitats that reduce their risk of being eaten. As with all tactics, finding a refuge may involve trade-offs, especially with food intake and growth. Given that within species, different life stages experience different predators and levels of risk, it makes sense to look at stage-specific habitat use. Werner and Gilliam (1984) formalized this by hypothesizing that habitat occupancy should be a function of minimizing the

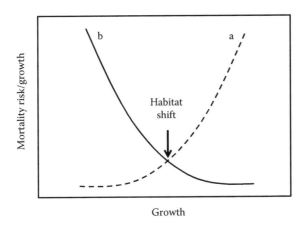

FIGURE 6.13 Ratio of mortality and growth within habitat patches as a function of growth. Prey organisms should shift habitats from patch a to patch b when the ratio in a exceeds that in b. Mortality risk in patch a increases with increased body size because of lack of food resources. Mortality risk in b declines because of increased size (which reduces predation risk) and more food.

ratio of the mortality to growth it provides (Figure 6.13). A habitat shift within a prey species can be predicted when the mortality/growth ratio of one habitat exceeds that in the other (Figure 6.13). Mittelbach (2002) reviewed this, showing that this was common in species. Young bluegill sunfish occupy vegetated, shallow-water areas of lakes until they grow to sizes where the food intake in these zones is low relative to their mortality risk. As they grow to invulnerable body sizes, the fish then shift to open-water areas to feed on zooplankton prey that provide a high energetic return (Mittelbach 1981).

6.12 TRAIT-MEDIATED INDIRECT INTERACTIONS

In recognition that both permanent and induced traits of prey influence food webs, the term trait-mediated indirect interactions (TMII) arose in the late 1990s (Werner and Peacor 2003), although the process had been recognized for decades under a general food web term: *indirect effect*. The need for a more specific term is that TMIIs are a plastic, evolutionarily relevant component within species and populations, whereas the category of indirect effects (of which TMIIs fall) includes other factors associated with prey such as nutrient excretion or physical modification of habitat.

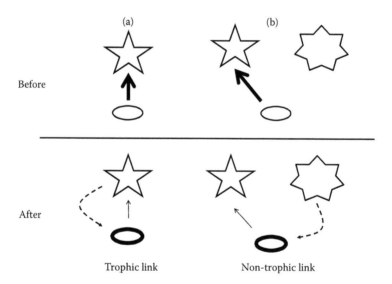

FIGURE 6.14   Two potential trait-mediated indirect food web interactions occurring between a prey organism (oval) and predators (stars) before and after traits are induced. Solid arrows and their widths depict the direction of energy flow. Dashed arrows depict a predator signal that induces an antipredatory trait. (a) A predator begins consuming the prey. The signal induces the prey to increase its defense, reducing its susceptibility to the predator. (b) A signal from another predator (7-pointed star) that is not currently consuming the prey induces a defense that reduces predation success of the other predator (5-pointed star). There are many possible complex scenarios and interactions that might occur.

The impacts of TMIIs on food webs are myriad and still not well understood (Figure 6.14a). Most simply, a density-dependent, compensatory response can be considered a TMII in a population. As prey individuals decline, resources may become more available to them, allowing them to improve the physical traits of body condition and reproductive capacity. The impact of predators on prey, however, is more than just density related. Although prey may have increased resource income, their ability to obtain food and other resources and use them may be limited by the *landscape of fear* created by the predators (Preisser et al. 2005; Matassa and Trussell 2011; Zanette et al. 2011). This is analogous to the predators turning off the spigot of energy flux and may account for up to greater than a 50% reduction in energy return to higher trophic levels (Preisser et al. 2005). Prey suffer reduced growth and reproduction, further reducing the productivity within an ecosystem (Zanette et al. 2011).

Predator-induced TMIIs may have effects that extend beyond their prey (Figure 6.14b). Ladybeetles prey on scale insects, which are protected by ants. When ant activity is reduced by parasite infections, ladybeetle predation on scale insects increases (Liere and Larsen 2012). Thus, changes in behavior induced by one type of predator affect other predatory interactions. Montipora corals respond to amphipod infestations by changing their morphology into finger-like projections (Bergsma 2012). These morphological changes reduce starfish predation, shifting predation on other species. Responses in traits to deal with one predator had implications for other prey species. Understanding phenotypic variability and its impact on trophic interactions is a necessary consideration in food web modeling (Kishida et al. 2012). Without predicting trait-specific responses and their impacts on interacting species, food web models may grossly overestimate or underestimate energy transfer depending on the complex responses of component species.

6.13 HUMAN EFFECTS

Humans have long understood that species have plastic traits and have used this information to domesticate organisms. However, we are only beginning to understand how human activities modify the physical traits of wild organisms and how these changes influence trophic interactions. Harvesting fish populations often modifies growth and maturation patterns, allowing some species to persist with increased human predation. Conover and Munch (2002) showed this experimentally with Atlantic silversides undergoing experimental, size-dependent harvest. Changes in the structure of harvested populations have dramatic consequences for trophic interactions (Biro et al. 2003). For example, small survivors within harvested species may have less impact on lower trophic levels and provide less energy to apex consumers.

Humans also induce effects on the traits of prey through other activities, such as by releasing environmental contaminants. Predator or prey behavior may be modified by toxins that affect their performance. Dragonfly foraging efficiency was negatively affected by exposure to a pesticide, which indirectly protected their tadpole prey (Cothran et al. 2012). Some contaminants are endocrine disruptors or mimics, potentially reducing the reproductive output of species (Vos et al. 2000; Hillis et al. 2015). If prey populations are negatively affected, this will alter food web interactions in complex ways. Reduced prey availability may cause predators to shift to other species or cause the predators to decline in growth and density.

6.14 SUMMARY AND FUTURE DIRECTIONS

All organisms transmit most of their biomass to other trophic levels during some point of their life. Predators seek to consume this biomass sooner in a prey's life than later (i.e., death). Luckily, natural selection works to ensure that some individuals do not fall prey until after they have reproduced and successfully transmitted offspring to the next generation. From an autecological perspective, this has led to interesting traits in organisms that extend from birth through reproduction. Of course, other selective factors are at work. Identifying how traits, whether they are permanent or inducible, are associated with predation versus other selective forces will continue to be an important research challenge in trophic ecology.

A goal of trophic ecology is to predict the flux of materials and energy through ecosystems. As we noted, the biomass of prey accumulating in the ecosystem relative to its vulnerability is an interesting topic of research. In fact, Kerr and Dickie (2001) predicted that general patterns exist in aquatic ecosystems where the biomass tied into individual size spectra (not trophic levels) is constant. Thus, they suggested that the production of aquatic ecosystems can be set based on these broad, size-dependent patterns (Box 6.3). Changes in ecosystem health may be delineated by sudden trait-mediated shifts in these size spectra and are likely linked to changes in energy transfer among organisms.

Regardless of whether size-spectrum analysis works to generate broad predictions for ecosystems, mechanistic information will still be needed. The reaction norms of morphological and behavioral traits of prey relative to the field of predators both familiar and novel will need to be evaluated to predict the stability of interactions and their consequences for ecosystems. A coevolved predator–prey system in which prey have developed a keen sense of their predator will likely be more stable than one in which a novel (i.e., invasive) predator appears. With the novel predator, the ability for the prey to respond to the predator's signal via natural selection will be critical for whether the system will compensate, diminish, or collapse (Figure 6.15). This is a strong argument for maintaining diversity within prey populations to avoid negative ecosystem responses to consumption by invasive species or other unanticipated effects.

Humans are the ultimate predator, exerting important effects on prey populations through various pathways. Although prey may temporarily respond to human effects, humans adapt well to changes in prey populations. Conservation and management of species and their ecosystems need to recognize plastic prey responses and preserve their diversity to be protective. Humans typically overcome any defense a species can muster until the species is either ecologically insignificant or extinct. Most management of fish populations, forests, wildlife, and other natural resources still does not account for the evolutionary consequences of exploitation. Ecosystem consequences are even less understood. We ignore these at our peril.

BOX 6.3

Kerr and Dickie (2001) summarized the biomass spectra found in many fresh-water and marine systems. Sampling all the prey available at each trophic level is an impossible task (see Chapter 10). Thus, the authors suggested that ecologists quantify the distribution of sizes of organisms and assess the flow of energy in this manner. In water, sizes of organisms are relatively simple to quantify cheaply and quickly using various remote-sensing (e.g., hydroacoustics) and sampling (e.g., Coulter counter) techniques. When done, the body size–biomass patterns are fairly uniform with the intercepts differing between ecosystems but the slopes being surprisingly similar.

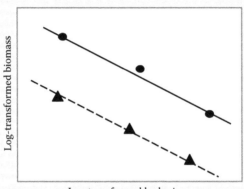

Distributions of biomass in phytoplankton, zooplankton, and fish from two lakes (circle = lake 1 and triangle = lake 2). (Adapted from actual patterns in Boudreau and Dickie 1992.)

Energy flow can then be modeled as a function of the slope of the relationship. The underlying mechanisms influencing the transfer among prey and consumer are associated with the allometric scaling of growth within the interacting organisms (Thiebaux and Dickie 1993). In other words, as each prey level grows larger, significantly less biomass is available for consumption. This acts to *stabilize* the trophic relationships between size spectra in the ecosystem.

Data sets of size spectra are scarce in most systems and certainly not available in terrestrial systems, because obtaining these sorts of size data is much more difficult. Although this approach is enticing, suggesting that emergent properties of aquatic ecosystems arise as a function of complex interactions among prey and predators at mind-boggling differences in spatial and temporal scales, the danger is that the patterns are plotted logarithmically. This approach tends to compress variation and make large changes seem small. Thus, it is important to evaluate patterns of biomass production at realistic (i.e., nontransformed), biologically relevant scales for conservation and management. In both aquatic and terrestrial ecosystems, the underlying mechanisms and the ecosystem output can be explored using mass-balance modeling described in Chapter 8.

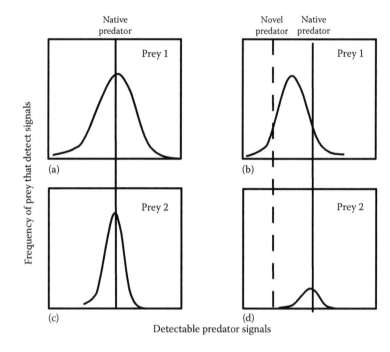

FIGURE 6.15 Reaction norms for predator detection for two prey species (prey 1, panels a, b; prey 2, panels c, d). The further the prey is from the predator signal (vertical bars), the less able it is to detect the signal. Individuals with perfect overlap with the bars are the best detectors. Species 1 has a broad reaction norm and is able to detect the signal of its native predator (a, solid bar) as well as a new predator when it appears (b, dashed bar). Selection within species 1 should shift the frequency of reaction norms and possibly the underlying genotypes to individuals that are able to detect both predators. Species 2 has a much narrower reaction norm for detecting predators other than the native (c). When the new predator appears (c), density of species 2 should be reduced by it (d), because it is unable to detect the invader in the environment.

QUESTIONS AND ASSIGNMENTS

1. How common are prey in the environment? Are all species prey? Or just a subset of organisms?
2. What is ontogeny of prey and why is it important for understanding vulnerability to predation?
3. How are patterns of survivorship and susceptibility to predation related?
4. Are prey with r- or k-selected life histories more susceptible to predation by humans? Why?
5. What are the potential fitness trade-offs between permanent and induced defenses in prey?
6. What is the difference between Batesian and Mullerian mimicry? Are these simple concepts or are there more complex reasons for how deterrent traits arise?

7. How do prey interactions with predators differ between aquatic and terrestrial ecosystems?
8. How do prey hide in plain sight?
9. Many characters that make prey invulnerable were not evolved to reduce predation. Explain.
10. What are the differences between compensatory and depensatory responses of prey to predation? How do these affect the stability of prey populations and ultimately food webs?
11. Explain how mortality and growth trade-offs can be used to predict the distribution of prey in the environment.
12. What is a trait-mediated interaction in a food web?

Section III

---

*Diet Data, Modeling,
and Energetics Approaches*

7 Analyzing Diets

7.1 APPROACH

Two primary reasons for collecting diet data from consumers are assessing the ecology and evolutionary decisions that organisms make when foraging and quantifying the relative contribution of prey to energy flux in ecosystems. We will find that diets provide useful information, with many caveats. Technological ways to assess consumer behavior and diet intake are increasing. Similarly, tools for analyzing data are becoming more sophisticated and useful.

7.2 INTRODUCTION

Collecting accurate information about the diets of consumers is critical for trophic ecology. Evaluating diets often falls in the realm of natural history, where diet contents are described but not treated quantitatively. In the 1960s and 1970s, two investigators were critical in moving diets toward quantitative trophic ecology. A Russian ecologist, Viktor Sergeevich Ivlev (1961), focused on the foraging of fishes to quantitatively assess trophic relationships, writing a book that led to an important index described later. His approaches were inspired by the work of Charles S. Elton (1927). Thomas W. Schoener (1971), drawing on techniques developed earlier (Renkonen 1938; MacArthur and Pianka 1966; Emlen 1967), published a paper that placed diet choice of organisms in both an ecological and evolutionary context, creating a framework for identifying and evaluating the niches of organisms to test the hypothesis that no organisms use the same niche space (Hutchinson 1959).

Although Ivlev attempted to link diet choice to energy flow in fish communities (Figure 7.1), Schoener's further developments were not specifically geared toward understanding patterns of energy flow in ecosystems. Rather, his goal was to test ecological and evolutionary predictions about the composition of communities and patterns of diversity. The behavior of the consumer was his focus: the hypothesis of optimal foraging arose from this conceptual and quantitative development (Chapter 5). At this point, it seems that the division between community-level foraging concepts envisioned by Schoener (1971) and the ecosystem-level consequences of foraging decisions alluded to by Ivlev (1961) was fortified and continues to this day (Figure 7.1).

The goal of the previous chapters was to link individual predator and prey characteristics to their ecosystem-level interactions, bridging the gap between the approaches developed by Ivlev and Schoener. In this chapter, we provide a menu of ways to collect and analyze foraging data to test inferences about trophic relationships. Data from foragers can include or exclude information about the prey they are encountering. If prey organisms in the environment are included in the analysis, a variety of indices of electivity, selectivity, or preference are available. These three

	Ivlev	Schoener
Goal	Diet choice for growth	Diet choice as an evolutionary decision
Result	Indices of electivity, eventually bioenergetics models	Optimal foraging models
Users	Managers/ecologists	Ecologists/evolutionary biologists

FIGURE 7.1 Historical sources of contemporary approaches in trophic ecology. Ivlev developed an index in the 1960s that is used today by ecologists and managers to interpret species-specific diet electivity. Schoener developed an evolution-based modeling approach in the 1970s that attempted to predict diet patterns in the field. Both are useful but have different goals.

terms are often used interchangeably in the literature, although each has a slightly different meaning. *Electivity*, as coined by Ivlev (1961), is probably the most correct term for any diet index because it suggests that a forager elects to eat a prey item but that the forager is not necessarily being selective or preferential about its choice. As we will see later, several indices that use the term *selectivity* or *preference* include the term *neutral* when the forager is not electing to consume rare prey. Saying that an index value is showing neutral selection or preference is an oxymoron, although the statement is quite common in the literature.

Putting semantics aside, we will first address how foraging data can be collected. Then we will explore how to use these data to assess how broad the dietary niche of consumers can be. For many foragers, diet data are complex, and our conclusions about their choices depend greatly on how the data are collected and treated. We fold in information about the prey that are encountered by consumers and use this to determine whether consumers are electing to eat prey in proportion to their availability or actually selecting or avoiding food items. We conclude by determining ways that diet data may be used to infer ecosystem patterns of energy and material flux.

7.3   COLLECTING DATA

Collecting foraging information is tricky business. Catching an organism in the process of eating or removing its gut contents may seem an easy way to evaluate diet choice. However, this information represents a mere snapshot in the organism's life as it accumulates energy to grow, mature, and reproduce. The true foraging habits of an organism may be very complex, depending on time, physical state, ontogeny, and many other factors. Inferences about foraging may be influenced by the collection methods used. In this section, we will weigh the options of determining foraging information from both live and dead consumers.

We are all familiar with media footage of organisms foraging like a crocodile attacking a wildebeest at a watering hole in Africa or a harpy eagle swooping down to fetch a monkey from the trees in South America. However, the foraging behavior of most organisms is difficult to capture and often far less exciting to observe (i.e., not posted on social media). Obviously, foraging occurs in the field, and the presence of the ecological investigator may alter the behavior of the organisms perhaps making the consumer less efficient or increasing the susceptibility of the prey. Foraging behavior of organisms can be cryptic or impossible to observe. Sperm whales forage hundreds of meters below the ocean's surface in the lightless abyss. The evidence of their battles with giant squid prey is only betrayed by sucker scars on their skin (Figure 7.2). Past foraging interactions are also difficult to quantify. D'Amore and Blumenschine (2012) used Komodo dragon teeth marks in prey to assess the foraging behavior of theropod dinosaurs in fossils. Their approach was that Komodo dragons likely have similar teeth and jaw morphology to this group of dinosaurs and can be used as a surrogate.

Field ecologists have a growing menu of technology available for remotely quantifying the foraging behavior of organisms. In the past, a running joke was that the only equipment field ecologists needed for foraging research was a pair of binoculars and a notepad. This is no longer the case. One recent advance is the miniaturization and ease of use of cameras. Cameras can be placed in the field and activated by wildlife (i.e., camera traps). Cameras also can be placed on the organism directly, allowing researchers to observe foraging from the animal's perspective. Cameras placed on the collars of feral cats recently showed how damaging they are to local wildlife (http://www.kittycams.uga.edu/index.html). Aerial and satellite photography, once beyond the reach of an ecologist's budget, may soon be available for remotely assessing foraging of some organisms (Figure 7.3).

Foraging and location are typically linked in organisms. Global Positioning Systems (GPS) are nearly ubiquitous around the globe and can be used to assess

FIGURE 7.2   Piece of sperm whale skin with sucker scars from a giant squid. (From Hjort, J. 1914. *Rapports et Proces-Verbaux des Reunions Conseil International pour l'Exploration de la Mer* 20:1–228.)

FIGURE 7.3 Remote sensing of a tortoise foraging in a forest. Data about the forest surrounding the tortoise is collected by a remote sensing satellite that can image the vegetation and soil characteristics. A tag attached to the tortoise's shell transmits location information to a cell phone tower, which is relayed back to the laboratory.

the movement patterns of mobile foragers. A related technology, which can be integrated with GPS, is biotelemetry where transmitters are either affixed externally or surgically implanted in foragers (Box 7.1). Keefer et al. (2004) quantified the consumption of radiotagged endangered salmon by sea lions. Once relegated to large organisms, small animals including dragonflies (Wikelski et al. 2006) can now be tagged and stream position data to a receiver (Figure 7.3). A particularly clever use of this technology involves tracking the prey rather than the predator (Hirsch 2012). Hirsch (2012) quantified how scatter-hoarding animals moved palm seeds around a neotropical forest by attaching radio transmitters to the seeds and then tracking their dispersal through the landscape. Passive integrated transponder (PIT) tags, otherwise known as radio frequency identification transmitters, also can be used to quantify the movement of foragers and their prey (Bonter and Bridge 2011; Figure 7.4). These transmitters do not provide location data until they are activated by an external antenna that records the proximity event. This technology might be used to quantify when a forager enters a fixed area to consume its prey or when a prey uses a refuge. An array of PIT antennas could be placed in these locations and determine how frequently they are used. Mariette et al. (2011) used PIT-tagged zebra finches to quantify their visits to feeding boxes and their effort in provisioning their offspring.

The ability to quantify foraging data in areas without light no longer is in the realm of science fiction. Nocturnal foraging can be tracked using night-vision technology, where infrared light is used to illuminate and detect organisms. In murky or deep water where light is scarce or absent, acoustic technology is available to detect organisms, much like a bat or dolphin uses echolocation to gather information.

BOX 7.1

Using telemetry to assess foraging behavior. Calkins et al. (2012) were interested in how food influences the distribution of silver carp, an invasive fish, in the Mississippi River. This fish primarily eats phytoplankton, which is suspended algae. Fish were captured, anesthetized, and surgically implanted with acoustic transmitters. Because sound travels well in water, the sounds these tags produce are easy to detect with a hydrophone. Each tag generates a specific code that allows individual fish to be identified. The investigators followed each fish in the river and quantified phytoplankton both where silver carp were located and at random locations.

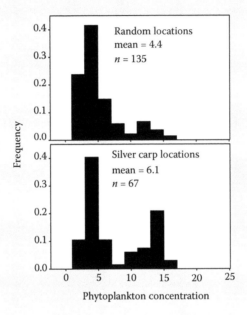

On average, silver carp were located in areas in the river with higher phytoplankton concentrations than those occurring randomly in the river. From this, the authors concluded that silver carp distribute themselves in part as a function of food availability in the river.

High-frequency sound can be used to detect and image organisms underwater similar to the ultrasound techniques used for anatomical imaging (Figure 7.5).

Technology is a helpful tool. But it needs to be used in conjunction with sound ecological techniques. The use of certain technologies may alter the behavior of predators and prey, much like the presence of a human observer. Careful consideration needs to be taken about how cameras, transmitters, and other gear influence feeding behavior. For example, the presence of an internally implanted transmitter may reduce the feeding activity of a forager as it recuperates from surgery.

Implanted tag

Birdhouse with
RFID receiver

FIGURE 7.4   Using radio frequency identification tags (i.e., RFID or PITs) to quantify the number of foraging trips that a bird makes to feed its young in a box. The tag is implanted in the bird and activates an antenna each time it enters the hole. The visit is logged on a computer.

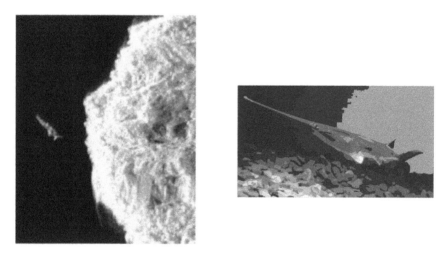

FIGURE 7.5   In water, it often is difficult to see foragers. A high-frequency side scan SONAR is used to image a paddlefish by a rock in a lake.

A conspicuous camera or receiver may alarm prey and alter their susceptibility to predators. Experiments or careful field observations including control animals need to be used to assess the impact of the gear on foraging data.

Some foraging data may be collected experimentally, although the investigator needs to be cautious about the design. Before commencing, thoughtful, well-developed, and realistic hypotheses need to be developed (Box 7.2). Probably most

BOX 7.2

Designing experiments for foraging can be difficult. Hypotheses need to be clearly defined at the outset. Each experimental treatment should only vary one factor in a consistent way that addresses the foraging hypothesis. A researcher interested in the impact of temperature on foraging of a skink should keep all other environmental conditions as similar as possible and only vary temperature. Replication is important. Observing a single organism consuming prey will not provide useful information that can be applied to the population or community. Rather, replicate experiments with multiple foragers need to be conducted where the conditions are kept similar. This can lead to a great deal of work, time commitment, and data analysis.

To minimize effort, the statistical power of the experiment should be assessed. Power is roughly defined as the number of replicates required to detect an effect if one indeed exists. In the case of foraging responses, minimizing the variance among individuals is important for generating precise estimates. A way to know when sufficient replicates have been collected is to plot the variance as replicates increase.

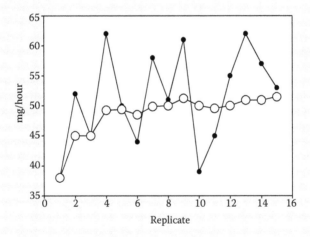

In this example, a researcher is quantifying the consumption rate of a forager at different temperatures. The true mean foraging rate is 50 mg/h, which is unknown. As she is conducting the treatment, she plots each individual observation (black circles). Each forager's consumption rate varies around the mean. The cumulative mean (open circles) across foragers begins to stabilize near the true mean at four replicates. From this graphical analysis, the researcher can conclude that about four replicates are sufficient to capture the true average for that treatment.

important to the success or failure of a foraging experiment is the spatial scale by which it is conducted (Figure 7.6). Large predators or prey confined to small experimental systems will most likely not behave as they would in the field. These are often called *cage effects* and lead to skepticism about the results (Winkler and Van Buskirk 2012). Another factor to consider when conducting foraging experiments is the physical state of the organisms. If the experiment influences the hunger level, stress, aggression, or activity of the organisms or their prey, then it will produce biased foraging results. The presence of other organisms in the experiment such as conspecifics may alter behavior as feeding hierarchies develop among individuals or interference competition occurs between feeders.

Given so many potential confounding variables, the use of experiments to assess foraging seems suspect under any condition. Even given the loss of realism, important mechanistic information can arise from experiments that can then be tested at larger scales with less control (Figure 7.6; Thomson et al. 1996). The best experiments are conducted at spatial scales that are large relative to the organisms. Gause (1934) conducted his famous experiments on protozoa to test ecological foraging theory. The small size of these organisms lent itself quite nicely to the realm of the laboratory bench. Clearly, ecological information about foraging is better collected experimentally with small organisms like birds or rodents where conditions can be kept somewhat realistic. Kotler et al.'s (1997) research is a great example of part of a long-term effort to understand factors influencing foraging. In this experiment, the behavior of gerbils foraging under the threat of red fox predation at night was quantified in a 34 m × 17 m enclosure divided into areas with different levels of cover. Here, the authors could accurately test foraging models at a relevant, realistic scale.

Conducting experiments at large spatial scales or with large organisms is more difficult. Obviously, manipulating the conditions that whales or elephants experience is impossible to do, and thus observation is the only possible way of collecting data. Under some circumstances, conducting experiments on whole ecosystems is possible and can lend itself to learning about responses of component foragers (Carpenter et al. 2001; Chapter 10). Forest plots, small lakes, streams, and even sections of ocean

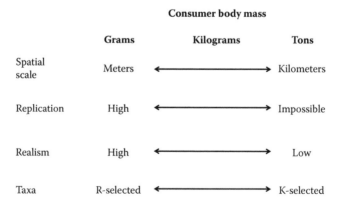

FIGURE 7.6 Impact of body size of a forager on the ability to make inferences about foraging behavior.

coast may be changed in a controlled manner to assess how foraging behavior and ultimately trophic flux occurs.

Both in the field and experimentally, foraging decisions may be determined visually or indirectly (e.g., counting the number of prey removed). Other information may be obtained from live organisms through various techniques. Lavage is often used, where the organism is sedated and the gut contents removed mechanically through the mouth (Foster 1977; Box 7.3). This method is used frequently for fish and other organisms that consume their prey whole, making identification of the diet items possible. Although even less appetizing than sorting through regurgitated food, the scat or feces of organisms may also contain clues to their diet by including indigestible materials like hair, bones, teeth, and exoskeletons (Lee and Severinghaus 2004). Many raptorial birds including hawks and owls cough up pellets of indigestible materials, again allowing ecologists to assess their diets (Leveau et al. 2004; Heisler et al. 2016). Animals that capture their food in webs (Pekar et al. 2012) or cache or hoard their prey provide a convenient record of their foraging habits. All of these approaches need to be evaluated cautiously, realizing that the remaining or discarded diet items likely overestimate indigestible items that may contribute little to energetics.

Under some circumstances, it is possible to collect diets from organisms that are killed for research. This depends on taxa, availability, and permission from institutional animal care groups. Obviously collecting this information from large mammals or rare and endangered species is impossible. However, for small, common taxa, especially invertebrates, this methodology is still generally acceptable. As with lavage from live animals, the feeding mode of the organism needs to be considered. Diets of foragers that eat their prey whole will be easier to process and analyze than those that crush, shred, or chew their food. Once the food enters the alimentary canal, it is further processed by organs such as gizzards and stomachs as well as through enzymatic and microbial activity. Diet items digest at different rates, and warmer temperatures should increase this process. Thus, it is important that researchers consider these factors when evaluating diet contents.

Many decisions need to be made before killing organisms to collect diet data. The alimentary canal of organisms can vary both within species depending on life stage and among organisms as a function of morphology. Organisms with well-developed stomachs can lead to a simpler sampling protocol than those with long, undifferentiated guts. A stomach is a distinct, well-defined organ that can be removed consistently among foragers. In contrast, if the gut is a tube with no discernible characters, then the

BOX 7.3

Gastric lavage is a technique in which the gut contents of a live organism are removed manually. Anatomy and foraging behavior limit the organisms that are candidates for this technique. Animals with complex digestive tracts, rapid digestion, and physiology that makes sedation difficult are likely not good choices. Data may be unreliable and the technique likely will cause harm to these organisms. Conversely, diets from animals like fish and amphibians with relatively low metabolic rates and simple guts are often sampled in this fashion. The technique is relatively straightforward. The organism is sedated with an anesthetic and held in a secure harness, often a sling or curved board. For aquatic organisms, freshwater is pumped onto the board to keep the animal from drying. A tube with flowing water is inserted into the stomach. The water displaces the diet contents, which exit the mouth and are captured with a cup. The contents are then filtered, washed, and preserved for analysis. After the lavage is finished, the organism is allowed to recover before being released back in the field.

Lavage is not foolproof. As with all diet content data, the collected diets depend on the time of day, food availability, and the organism's preference. The diet may not represent the long-term food intake of the organism. Not all items are easily removed from the stomach. To assess the efficiency of the technique, researchers should evaluate the relative amount of contents removed versus that which remains. Lavage is obviously stressful and can have extended effects on the behavior, health, and perhaps survivorship of the organism. Investigators need to be careful to not damage the lining of the alimentary canal. For animals with lungs or in the case of fish, open swim bladders, the water and contents should not be allowed to enter these organs. If so, infections can occur.

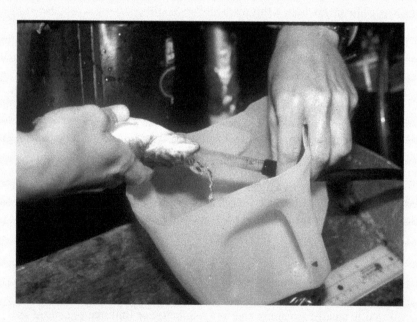

In this picture, a sedated smallmouth bass was lavaged using a small pump and hose. The displaced diet contents were collected in the filter (i.e., half a milk jug) and then preserved. The fish was then revived and released back into the lake.

whole organ needs to be processed, understanding that the level of digestion advances from the anterior to the posterior. The choice of preservation also is critical. For small organisms, the entire body may be preserved. However, for larger individuals, the guts may have to be removed to arrest digestion and decomposition. Many preservation methods are available including freezing, ethanol, and formalin. As with data from live organisms, differential digestion will continue to be an issue. In an experiment with large fish predators, Kim and DeVries (2001) found that tiny fish larvae rapidly disappeared in enclosures due to consumption. However, these larvae rarely appeared in predator diets collected from the enclosures because of rapid digestion. Investigators quantifying diets without knowledge of prey in the environment would wrongly deduce that larvae were unimportant. Experiments should always be conducted to assess the differential rates of loss of prey to assess potential biases in diet data.

In addition to identifying prey in diets morphologically, molecular techniques can be used. Overlap between prey fatty acid or stable isotope composition in the field and the items in the diet can be compared to determine the relative contribution of prey (Chapters 13 and 14). Another technique that has become widely used is to compare DNA in the diets to that of the prey (Tiede et al. 2016). Because DNA are large, stable molecules, some of this material typically survives the digestion process and can be found either in the alimentary canal or the feces. With the increasing availability of high-throughput DNA sequencers and low costs per sample, it is possible to compare the DNA in diets to that of the prey in the environment. Carreon-Martinez et al. (2011) is among the many researchers to have found a strong match between the prey selected by predators and the DNA composition in the diets.

7.4   PREY DATA

The world is full of potential prey. However, as the previous chapters show, the choices foragers make about these prey are complex. This poses some particular problems for developing quantitative ways to assess electivity of predators. A consumer's knowledge of the prey environment depends on its combination of senses as well as local conditions. The gear ecologists use to sample prey may greatly misrepresent the prey available to the foragers. Ecologists may underestimate prey. Foragers are well known for detecting rare organisms that sampling cannot find. An invasive zooplankton species was found in the diet of a native fish in a Midwestern US lake, although intensive zooplankton sampling never revealed the invader (Ferry and Wright 2002). Clearly, the fish was better at finding the organism than the ecologists. As we will see, developing an index of electivity in this case would be impossible.

Ecologists may sample food items that foragers never encounter. The sampling gear and location of sampling affect the estimate of prey for quantitative analysis. Ecologists with little knowledge about the behavior of their forager may find themselves grossly overestimating prey. Classic research by MacArthur (1958) showed that different species of warblers occupied different locations within a tree. Without proper knowledge about the distribution of these birds, collecting prey in the wrong portion of trees would bias the estimate of availability. If four species of warbler were expected to forage throughout a tree (Figure 7.7a) but in fact restrict foraging to one of four locations (Figure 7.7b), we would generate very different conclusions. The

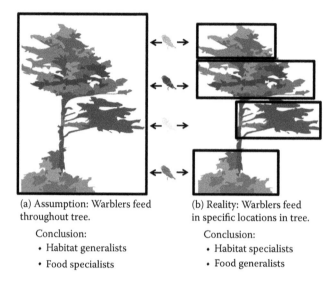

(a) Assumption: Warblers feed   (b) Reality: Warblers feed
throughout tree.   in specific locations in tree.

    Conclusion: Conclusion:
     • Habitat generalists   • Habitat specialists
     • Food specialists   • Food generalists

FIGURE 7.7 Four species of warblers occupy different portions of tree (top, upper middle, lower middle, and base). (a) The warblers are assumed to occupy the entire tree (box), although this incorrect. (b) The correct habitats for each warbler (boxes) are delineated.

warblers assumed to feed anywhere in the tree would be considered specialists on prey found in specific locations (Figure 7.7a). In reality, they are generalists on prey restricted to their preferred location (Figure 7.7b). This distinction may seem subtle, but it may affect decisions about management and conservation of these bird species.

Surprisingly little is known about how to develop accurate null expectations for prey in the environment. Still, there are hundreds of studies in the literature that incorporate an estimate of prey availability in their quantitative analyses of electivity, whereby the researchers proclaim that preference or avoidance of prey occurs. In our view, these analyses require strong natural history and behavioral information about the forager in its environment to assess whether assumptions of availability are supported.

7.5 FORAGING INDICES

7.5.1 Choice or Diets Only

Several options for analyzing prey choice data exist. The simplest analysis involves combining all the diet information for each forager and then averaging these data among individuals. Information can be the total mass, energy content (kJ or calories), or volume of all the prey consumed and found in the gut. These totals can then be compared among foragers using standard statistics (Box 7.4). If the foragers vary in size, then the data can be standardized by dividing the values by the mass of each individual. A 1 kg predator and a 2 kg predator may have 100 and 200 g of prey in their diets, respectively. Thus, they would both have 100 g/kg prey in diets, consuming proportionally comparable amounts of food. It is important to note that gut capacity may change with body size or age as a function of allometric

BOX 7.4

Foraging data are surprisingly complex because most organisms consume more than one thing. Further, individual foragers differ in their behavior and food intake, making generalities difficult. We provide data on a bird species that forages on four general items. Within each group, there were multiple species of prey, but we lumped them for simplicity. Prey items were counted and their total weight in each bird was quantified. The simplest analysis was frequency of occurrence (O). For both beetles and ants, $O = 1$, because these items were present in the stomachs of all six birds. Grasshoppers were present in 83% of birds ($O = 0.83$); crickets were rarer, $O = 0.5$.

Bird	Grasshopper		Beetle		Ant		Cricket		Totals	
ID g	#	mg	#	mg	#	mg	#	mg	#	mg
1 50	12	40	3	100	300	30	1	2	316	172
2 45	15	45	5	123	700	55	0	0	720	223
3 43	20	60	7	144	400	40	4	5	431	249
4 48	0	0	2	80	330	27	3	6	335	113
5 50	16	44	4	90	810	51	0	0	830	185
6 46	11	39	3	102	675	54	0	0	689	195
Avg.	12	38	4	107	536	43	1	2	554	190

Because the birds were all similarly sized, there was no need to standardize the amount of food within each diet (e.g., number/g or mg/g grasshopper eaten). The proportion of prey eaten per bird was calculated by dividing the number of each item in each stomach by the total number of prey items in each stomach. These proportions were then averaged across the birds to get N for each item. Ants were numerically dominant in the diets, generating $N = 0.965$. Thus, the birds expended much time and energy consuming many ants.

Bird	N			
ID	Grasshopper	Beetle	Ant	Cricket
1	0.038	0.009	0.949	0.003
2	0.021	0.007	0.972	0.000
3	0.046	0.016	0.928	0.009
4	0.000	0.006	0.985	0.009
5	0.019	0.005	0.976	0.000
6	0.016	0.004	0.980	0.000
Means	0.023	0.008	0.965	0.004

The limitation of N is that it does not account for the mass and therefore the energy contribution of the prey to the diet. Ants were frequent in the diets but also small relative to other diet items. Proportion by mass (W) was calculated by first generating the proportion mass of each item in each bird relative to the total mass of the stomach content. These values were then averaged across all birds to get W.

Bird	Mass			
ID	Grasshopper	Beetle	Ant	Cricket
1	0.233	0.581	0.174	0.012
2	0.202	0.552	0.247	0.000
3	0.241	0.578	0.161	0.020
4	0.000	0.708	0.239	0.053
5	0.238	0.486	0.276	0.000
6	0.200	0.523	0.277	0.000
Means	0.186	0.571	0.229	0.014

In contrast to N, the proportion by mass was highest for beetles at $W = 0.571$. Ants were still the second most significant diet item at $W = 0.229$. Thus, beetles likely comprised the most energetically valuable component of the birds' diets.

These values show that conclusions about foraging behavior depend on the analysis of the data. All are biologically relevant. Frequency of occurrence shows how consistently prey items were consumed among foragers. Proportion by number (N) reflects the effort expended by the consumer, whereas W provides insight into the energetic benefit of consuming the prey. Likely, ants were easy for birds to find and eat. Beetles were harder to locate and consume but resulted in a greater payoff.

growth in some organisms (Figure 7.8), making comparisons difficult to interpret. The digestive tracts of many organisms start out small and simple and do not develop until later in life (Figure 7.8a,c). Thus, a young organism may be limited in the total amount of prey it can consume compared to adults.

If the prey items within each forager are treated separately rather than lumped into a single group, then analysis and interpretation become more complex. One of the simplest analyses is frequency of occurrence (O_i), where the presence of a diet item, i, is evaluated for each forager, and the proportion of foragers consuming the item is calculated (Box 7.4). Across the foragers, then, there are as many O values as there are prey items, ranging from 0 to 1. In our view, O might be one of the most useful and realistic indices available, depending only on presence or absence. Unfortunately, this index provides little insight into the relative contribution of each prey item to individual energy intake. Slightly more complex indices are the proportion by number N_i and the proportion by mass W_i. Here the total number or mass of

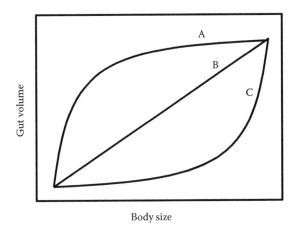

FIGURE 7.8   Hypothetical relationships between a forager and the capacity of its gut.

prey items within each forager is calculated, and the proportion of each relative to all prey is determined. These values are then averaged across all foragers (Box 7.4).

Issues or concerns arise when using O, N, and W for evaluating the foraging behavior of organisms. The values of these for each prey item are not independent of each other. In a snake, if more frogs are found in the diets, then fewer rodents would be present, because the snake's gut is a finite space. Thus, traditional parametric statistics comparing frogs and rodents in snake diets such as a t-test or analysis of variance (ANOVA) are violated, because the amount of one diet item depends on the other. Rather, techniques that incorporate the non-independence of the items need to be used if the diet items are to be evaluated simultaneously and compared between populations or treatments in an experiment. A multivariate ANOVA (MANOVA) is an option for analysis because it allows for multiple dependent variables in the model to be tested (Box 7.5).

Interpreting the index results requires an appreciation for how the data are compiled. Frequency of occurrence, O, does not provide information about the relative abundance of prey, just whether they appeared in the diets of individuals. A particular item may be eaten by a large proportion of foragers, having a high O, but make up a relatively small proportion of the diet of each. A large mean N for a prey item shows that foragers consumed a large number of that food. However, the contribution of the prey to the diet may be relatively meager if the prey items are small-bodied and low in energy relative to other diet components. A bird may consume several ants and one large, energy-rich earthworm. Although the ants produce a greater N, their combined biomass is lower than that of the single worm. The W index addresses this issue by accounting for the mass contribution of the items. In this case, the worm would have a greater W value than the ants. We will see later that combining O, N, and W provides interesting insight into the diet habits of organisms and how this translates to their ecosystem effects.

None of the indices mentioned up to now address the true energetic contribution of the prey to the forager. Small prey items may have a higher energy density than large ones making them more valuable to consume. The *prey importance index* (PII_i)

BOX 7.5

Diet index values like proportion by number (N) are useful for understanding how diets vary within a population of foragers. However, comparing diet patterns with these data is tricky because the indices are composed of multiple values for different diet items. In this case, a MANOVA may be appropriate for comparing diet patterns between populations. Diet selection by ten mongooses was compared between two locations, and N was calculated for each individual. We wanted to know whether the average N for snakes, frogs, rodents, and insects differed between the two patches.

	Mongoose	Snakes	Frogs	Rodents	Insects
Patch 1	1	0.167	0.500	0.000	0.333
	2	0.250	0.500	0.125	0.125
	3	0.286	0.429	0.000	0.286
	4	0.091	0.545	0.091	0.273
	5	0.000	1.000	0.000	0.000
	Mean	0.159	0.595	0.043	0.203
Patch 2	1	0.000	0.154	0.769	0.077
	2	0.000	0.250	0.583	0.167
	3	0.077	0.231	0.615	0.077
	4	0.071	0.143	0.643	0.143
	5	0.111	0.222	0.444	0.222
	Mean	0.052	0.200	0.611	0.137

Most statistical packages will easily calculate a MANOVA for these data. In traditional ANOVA models, only one dependent variable is tested as a response to the factor we call patch. In the MANOVA model, the effect of patches on all four dependent variables was tested using a test statistic called Wilk's lambda. When we did this for these data, we found that the diets differed between the two patches (Wilk's lambda: $F_{3,6} = 26$, $p < 0.0008$). Looking at the data, it is clear that mongooses in patch 1 fed primarily on frogs. Counterparts in patch 2 consumed rodents. Analysis can proceed further, contrasting the means within prey categories. Other MANOVA test statistics are available.

incorporates the energy density of prey and can be used to include the actual assimilated energy of each item (Probst et al. 1984; Box 7.6). The index thus is assessing the proportional contribution of each prey type to the total energy available in the dietary pool. The PII is the most accurate assessment technique for diet contents, but it requires information about prey items that may not be available or easy to quantify such as the energy density and the energy being assimilated into the forager.

Rather than the energetic contribution to the diet, ecologists may be interested in quantifying diet choice to determine the niche breadth of the forager. This

BOX 7.6

The PII is a way to determine the proportional energy contribution of each prey item to the forager. Both the mass and the energy density (in kJ/g) of the prey are quantified. Energy density of each item is determined using calorimetry. Both the total mass and the pooled energy density (both by wet mass) are quantified for each for diet item within each forager. Energy density of each item is multiplied by its mass to determine its total energy contribution in kJ to the diet. This is assumed to be the energy that can be assimilated by the forager. The total energy content of each gut is calculated, and the proportion contribution of each item is calculated, which is the PII for each forager.

g			kJ/g			kJ			Total		PII		
Fish	Insect	Crayfish	Fish	Insect	Crayfish	Fish	Insect	Crayfish	Fish	kJ	Insect	Crayfish	Fish
1	0.10	0.30	1.00	3.64	2.80	4.60	0.36	0.84	4.60	5.80	0.06	0.14	0.79
2	0.30	0.50	1.70	3.22	2.51	5.02	0.97	1.25	8.53	10.75	0.09	0.12	0.79
3	0.23	0.33	2.00	3.55	3.01	4.81	0.82	0.99	9.61	11.42	0.07	0.09	0.84
4	0.05	1.20	0.75	4.05	2.47	4.18	0.20	2.96	3.14	6.30	0.03	0.47	0.50
5	0.20	0.00	3.00	2.80	3.64	4.14	0.56	0.00	12.41	12.97	0.04	0.00	0.96
6	0.11	1.20	0.00	3.34	2.76	4.77	0.37	3.31	0.00	3.68	0.10	0.90	0.00
7	0.40	2.10	2.00	3.64	2.63	4.22	1.45	5.53	8.44	15.43	0.09	0.36	0.55
										Mean	0.07	0.30	0.63

In this example, the diet contents of seven foraging fish were categorized into insects, crayfish, and fish. In most of the foragers, fish were the dominant prey item by mass. Also, fish consistently had the highest energy density. Crayfish had the lowest because much of their biomass is locked up in indigestible and energy-poor chitin. The resulting averaged PII for the three items showed that fish comprised 63% of the energy in the diets. Crayfish, although frequently abundant in diets by mass, made up only 30% of the energy available to the foragers. Insects generated a mean PII of only 7%.

information allows us to determine whether an organism is a specialist or generalist consumer. A *niche-breadth* index by Levins (1968) is widely used:

$$B = \frac{Y^2}{\sum N_j^2}$$

where N_j is the number of diet items of type j found in the diet of a forager. Y is the total number of diet items across all prey taxa. The index B can vary between 0 and j. If $B = j$, then all diet items are equally represented in diets (Box 7.7). The B values can be averaged across all foragers to be treated statistically.

BOX 7.7

Trophic ecologists often want to know how specialized a forager is. The more specialized, the stronger the trophic linkage between the predator and the prey. Levin's diet breadth index (B) allows us to calculate this in a simple and biologically interpretable way:

$$B = \frac{Y^2}{\sum N_j^2}$$

Prey (j)	Forager 1	Forager 2
1	50	20
2	50	20
3	0	20
4	0	20
5	0	20
Y	100	100

Both foragers had the same total number (Y) of items in their guts. However, the distribution of diet types differed. Forager 1 consumed only prey items 1 and 2, whereas forager 2 consumed all five prey types. Calculating B was straightforward. For both foragers, $Y^2 = 10,000$. For forager 1, the sum of N^2 for the prey items = 5000. Thus, $B = 10,000/5000 = 2$. For forager 2, $B = 10,000/2000 = 5$. In this case, the results were intuitive. The index value for both foragers was the same as the number of prey items represented in the diets. If prey items were more haphazardly distributed in diets, then indices of B would be fractions, reflecting the variation in the diet breadth.

Graphs can be used to assess the relative behavior of foragers relative to the abundance of their prey in diets. Costello (1990) developed a technique modified by Amundsen et al. (1996) to explore forager feeding strategies as a function of prey-specific abundance in diets in numbers, grams, volume, or calories,

$$P_i = \left(\frac{\sum S_i}{\sum S_{ii}} \right) 100$$

where S_i is the abundance of prey i in stomachs, and S_{ii} is the total amount of prey in the guts of foragers that consumed prey i. These values can be plotted against frequency of occurrence, O_i, to assess foraging behavior. Prey-specific abundance indicates the relative dominance of a prey item within individuals, whereas O_i indicates the relative abundance among individuals (Box 7.8).

BOX 7.8

Graphical analysis. Four foragers were used to determine the relative importance of prey to diets. The frequency of occurrence (O) and prey-specific abundance (P) were calculated for each prey item.

Prey Item	Forager				Index	
	1	2	3	4	O	P
A	50	44	33	22	1	0.06
B	100	45	43	99	1	0.12
D	0	0	600	0	0.25	0.24
E	14	0	13	0	0.5	0.01
F	500	400	440	90	1	0.57
Totals	664	489	1129	211		

To understand the relative contribution of prey to diets, O is plotted against P for the five prey types.

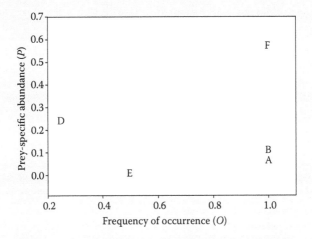

Prey F was found in all consumers and was numerically dominant (i.e., high P). Thus, graphically, it fell in the upper right quadrant, which suggested that it was the dominant prey. Items A and B occurred in all foragers as well. However, their contribution to the total diet varied among individuals, causing their values to cluster in the lower right quadrant, which indicates differing total energy contribution among foragers. Prey E was rare among individual foragers and contributed little to diets (lower left). Prey D was rare among foragers but comprised a high proportion of the diet of one individual. Hence, its value was near the upper left corner of the graph. From this analysis, investigators can further assess why certain prey such as D vary so much in importance to trophic interactions.

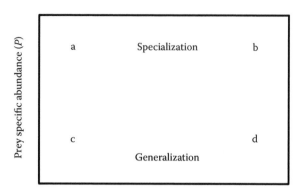

FIGURE 7.9 Graphical analysis of foraging data. A prey item appearing in quadrant a is rare among foragers but numerically abundant as a diet component in a few individuals. Prey in area b are both abundant among foragers and within their diets. Prey c are rare both among and within foragers. Prey in quadrant d are eaten by most foragers but do not comprise a large proportion of gut contents.

Unlike most indices that provide a single value that has limited interpretation, the graph that results by plotting O_i by P_i provides four possible conclusions (Figure 7.9). If P_i is high but O_i is low, then a few foragers are specializing on specific prey (Figure 7.9a). Conversely, if all the foragers are consuming a high proportion of the same prey, then data will fall in the quadrant with high P_i plus O_i (Figure 7.9b). Under both of these circumstances, all the foragers are specialists, although variability in diets among foragers varies from low (a) to high (b). Prey may be rare and uncommon in diets (Figure 7.9c), meaning that each unique item likely contributes little to energy intake. A large number of foragers may share a common prey item in their guts (high O_i), but it contributes little to the overall relative amount of food present (Figure 7.9d). Under both of these circumstances, foragers may be consuming prey items opportunistically but are not specializing on them.

Comparing diet contents among consumers may be important for assessing competition or responses to foraging conditions. Many indices for assessing dietary overlap among consumers have been developed (Schoener 1971; Hurlbert 1978; Wallace 1981). Smith and Zaret (1982) evaluated the potential biases of several of the conventional indices used by ecologists to make pairwise comparisons between consumers j and k for a common resource i. The Morisita index (1959) works best when the number of diet items is known (Box 7.9):

$$C = \frac{2\sum_{i}^{n} p_{ij} p_{ik}}{\sum_{i}^{n} p_{ij}\left[(n_{ij}-1)/(N_{j}-1)\right] + \sum_{i}^{n} p_{ik}\left[(n_{ik}-1)/(N_{k}-1)\right]}$$

where p_{ij} is the proportion of prey items in each category i used by consumer j, and p_{ik} is the proportion of prey items used by consumer k. n_{ij} is the number of

BOX 7.9

Overlap indices are useful for determining how similar diets are between individuals within a population as well as between closely related species. We used Morisita's index (C) to compare the total number of diet items consumed by two species of squirrel in a woodlot.

Diet Item	Gray Squirrel	Red Squirrel
Acorn	20	100
Walnut	20	100
Pine seed	100	30
Apple	100	20
Hickory	100	40

The value of C for this was 0.49, indicating moderate overlap between the two species. If the species consumed equal amounts of items, then $C = 1$, indicating complete overlap.

If the mass contribution of the diet items was known, then Horn's index (R) would be appropriate. The results for mass showed that overlap was greater with $R = 0.79$.

Diet Item	Gray Squirrel	Red Squirrel
Acorn	57.3	87.2
Walnut	44.3	94.3
Pine seed	127	13.3
Apple	130	22.2
Hickory	99.7	41

Note that Horn's index will not handle zero values for diet items, because the data are converted to a log scale. Under those circumstances, either the diet item needs to be removed from the analysis or the data need to be transformed by adding a small, equal amount to each item.

consumers, j, using prey resource I, and n_{ik} is the number of consumers k using i. N_j and N_k are the total numbers of consumers j and k in the sample, respectively (Box 7.10). The maximum value of C is 1, indicating complete overlap between consumers.

When volume, biomass, or energy contribution of prey to diets is available and numbers are not, then Krebs (2001) recommends using Horn's index (1966). Also, Morisita's index uses a measure of frequency of occurrence to compare diets, which

BOX 7.10

Many electivity indices are available for assessing the choice of prey relative to availability in the environment. In this analysis, r was the proportion of prey items consumed by an aardvark, whereas p represented the proportion of these prey in the field.

Diet Item	r	p	Fraction	Alpha
Fly	0.01	0.01	1.00	0.05
Worm	0.80	0.98	0.82	0.04
Ant	0.19	0.01	19.00	0.91

Strauss's index is the simplest to use where $E'' = r - p$. $E'' = 0$ for flies meant that the aardvark did not select or avoid this prey in the environment. This makes sense given that flies were rare in the field. Worms comprised a large proportion of the diets, but $E'' = -0.18$ showed that the aardvarks were not selecting them. Conversely, ants were quite rare in the area, but $E'' = 0.18$ showing that there was some preference for them.

Because Strauss's index has problems with the statistical properties of the test statistic, E'', the Manly–Chesson alpha was developed. In this example, the threshold for selection was $1/m$, where m = number of prey categories. Thus, if the calculated alpha was $>1/3 = 0.33$, then the aardvark was selecting that prey item. Both flies and worms generated alphas much less than 0.33. Conversely, the alpha for ants far exceeded 0.33, suggesting that ants were eaten at quantities much greater than their availability.

The indices provided similar information, with intuitive results. As the number of foragers and diet items increases, it becomes more difficult to tease apart the patterns by simply looking at the raw data. Under these circumstances, the indices are invaluable for identifying important trends in trophic relationships.

does not incorporate the relative energetic contribution of items to consumers. Horn's index uses proportions only to compare diets of consumers j and k:

$$R = \frac{\sum (p_{ij} + p_{ik}) \log(p_{ij} + p_{ik}) - \sum p_{ij} \log p_{ij} - \sum p_{ik} \log p_{ik}}{2 \log 2}$$

where p_{ij} and p_{ik} are the proportions of biomass, volume, or energy in prey items i in diets (Box 7.10). Any base of log can be used. As with Morisita's index, complete overlap would yield $R = 1$. Horn's index appears to be relatively unbiased, although there is some increase in imprecision with the number of prey categories added to the analysis (Krebs 2001).

All of these indices may be generated for a relatively small number of consumers, rendering variance estimates difficult. Bootstrapping may be used to generate confidence intervals around them. The procedure is relatively simple and can be done with a spreadsheet program. Values to calculate the bootstrap means are randomly selected from the pool of original values with replacement. Replacement means that the same value may be selected more than once or not at all from the pool of original values. Each mean will be slightly different from the original allowing for a distribution to be generated by repeating the procedure with the original data set at least 1000 times. The bootstrap mean should be adjusted to avoid bias:

$$\text{bootstrap mean}_{adj} = 2\bar{X}_s - \bar{X}_B$$

where $\bar{X}_s$ is the sample mean and $\bar{X}_B$ is the unadjusted bootstrap mean. The confidence intervals around the adjusted bootstrap mean are calculated in a conventional manner.

Assis (1996) introduced the *Geometric Index of Importance (GII)* as a way to combine multiple index values of relative prey importance in diets into one analysis. As we have pointed out, each index value differs in information and combining them should provide a way to standardize them. As many indexes can be used as deemed relevant. Each is included in the analysis as a standard, weighted as the deviation from an index value of rare or abundant (an uneven weight) relative to common (an even weight). For example, $O = 0$ or $O = 100$ percent would have large deviations and $O = 50$ percent would have 0 deviation. Values of N and W could be treated similarly. The *GII* treats these deviation values as a vector (a matrix with all the data in a column) and determines their contribution to prey choice. The equation used is:

$$GII_i = \frac{\left(\sum_{i=1}^{n} V_i\right)_i}{\sqrt{n}}$$

where the *GII* for each prey item i is the sum of all the standardized prey index values, V, across all the possible prey index values, n. Values range from 0 to 100 × the square root of n (Assis 1996).

7.5.2 Incorporating Prey

Incorporating prey that foragers encounter may allow ecologists to determine their electivity. The index has been called various terms, most commonly the foraging ratio:

$$E' = r_i/p_i$$

where r is the proportion of prey i in the diet, and p is the proportion in the field. The use of the symbols r and p as parameter values may seem a strange choice. The convention seems to come from Ivlev (1961), who used r for ration in diets and p for prey population in the field. For the foraging ratio, values of E' can be +1 for positive electivity (i.e., selection) to 0 for no electivity. If prey appear in the diet and not the field, then this value is invalid because 0 is in the denominator. Ivlev (1961) showed that the foraging ratio was not a reliable estimate, with nonlinear dependence on the ratio of prey in diets and the field. To correct for this problem, Ivlev developed his famous electivity index,

$$E = (r_i - p_i)/(r_i + p_i)$$

which changes linearly with the ratio of prey in the field and in the diets. A value of +1 means that the electivity is high for the prey relative to its abundance, 0 indicates no preference, and −1 means that the forager elects to avoid the prey. A simpler index is the *Strauss linear index* (Strauss 1979),

$$E'' = r_i - p_i$$

This index behaves very similarly to the Ivlev index but is more biologically intuitive, because it simply finds the proportional difference between prey contribution in the diets and the field (Box 7.10). The problem with both of these indices is their statistical properties. Variance depends on the mean value of electivity with values being constrained near values of positive or negative electivity. A forager that consistently elects to select a prey item in the field will generate values near +1, but the variance cannot exceed the value. Conversely, a forager that has neutral electivity for an item will have an index value that varies around 0 and is unconstrained. Because parametric statistics rely on predictable, normal distributions around the means, making comparisons among mean index values is impossible.

The search for a parametrically valid electivity index was concluded by the development of the *Manly–Chesson alpha* (Chesson 1978, 1983). The innovation here is that this index provides a distinct, nonzero target for determining whether a forager is selecting or avoiding prey in the environment. In other words, the null hypothesis is that the index value does not differ from the target value. How the index is calculated depends on whether the prey are being depleted by the forager. If prey are replaced or are so abundant that the predator has no impact on their abundance, then the index for each prey item, i, is calculated as

$$\alpha_i = \frac{r_i}{n_i}\left(\frac{1}{\sum_{i=1}^{m}\frac{r_i}{n_i}}\right)$$

where there are m prey items (Box 7.10). The proportion of prey in diets and in the field is r_i and n_i, respectively. The expectation for no selection for a prey item i is $1/m$. A mean alpha value plus confidence intervals can be generated for each prey item and then determined whether they overlap with the $1/m$ value. If the confidence intervals exceed the value, then the forager is electing to consume prey more than expected based on their abundance. The converse is true for values below the $1/m$ threshold for neutral electivity.

Under many circumstances, prey density changes appreciably as individuals are consumed, emigrate, or die. This also occurs when prey are not replaced in a foraging experiment. The index is calculated as

$$\alpha_i = \frac{\log P_i}{\sum_{i=1}^{m} \log P_i}$$

where P_i is the proportion of each prey item remaining at the end of the foraging trial. $P_i = e_i/n_i$, where e_i is the number of prey at the end of the trial, and n_i is the number of prey at the start of the trial. Again, no selection occurs when $\alpha_i = 1/m$. Any base of log can be used. Note that prey cannot be completely eliminated in the environment or the log transformations will not work in the analysis.

7.5.3 OTHER ANALYSES

All of the index-based approaches in the previous sections attempt to take complex data and simplify them in a standardized way. They are useful for assessing field and experimental patterns as long as their assumptions and mechanics are understood. Many alternative approaches to the analysis of these data exist (Chipps and Garvey 2007). In many of these, the *raw* data are analyzed to search for patterns, for example using ordination techniques like principal components analysis (Chipps and Garvey 2007). Frequency data of prey in diets can be compared using contingency tables (Cortes 1997; Box 7.11), which are used to test the null hypothesis that there is no significant variation in the frequency of prey categories in the diets. Forager diet components are typically in the columns of the table and forager types are in the rows. If a significant pattern is found, then further analysis can reveal which items are more frequent than others in foragers and thus likely contributing more to energetics (Box 7.11).

We may be interested in more than the potential energy contribution of prey to predator diets. The size and shape of the prey consumed also might be important (Chapter 6). However, the diet items may be in bits and partially digested. Forensic work is necessary to reconstruct the diets. Fortunately, the sizes of anatomical parts of prey are typically related (Hansel et al. 1988). Thus, if a relatively indigestible part is found such as the mandible of a beetle, its dimensions can be used to estimate the entire size of the item (e.g., see Scharf et al. 1997, 1998b). The relationships between prey body size and predator size can be compared in many ways. Typically, the relationship is not strictly linear. Rather, the sizes of prey consumed increase with forager body size; however, the variation increases as well (Figure 7.10), making conventional linear regression inappropriate. An ecologist might be interested in determining the maximum and minimum sizes of prey that the consumer eats. To do this, least absolute values or quantile regression can be used to identify the upper and lower portions of the distribution (Box 7.12; Scharf et al. 1998a). From this, models that incorporate the variation in sizes of prey to foragers can be developed.

BOX 7.11

A two-way contingency analysis was conducted to determine the difference in diets between foraging snakes in two patches. The first null hypothesis to be tested was that there was no difference in the frequency of diet items consumed. The second was that there was no difference between patches. And the third was that there was no interaction between the patches and diet items chosen.

Patch	Grasshopper	Beetle	Ant	Cricket
1	12	3	300	1
	15	5	700	0
	20	7	400	4
	0	2	330	3
	16	4	810	0
	11	3	675	0
2	88	1	80	1
	22	2	90	0
	45	4	77	0
	22	3	77	0
	33	1	88	0
	45	3	100	0

All three null hypotheses were rejected by the contingency analysis. The frequency of items consumed differed between patches ($df = 1$, $X^2 = 8.2$, $p = 0.004$), with three times more items being consumed in patch 1. The analysis also showed a difference in the items consumed ($df = 3$, $X^2 = 1686$, $p = 0.0001$) and an interaction between patch and food item ($df = 3$, $X^2 = 488$, $p = 0.0001$). The interaction means that the difference in the frequency of food items depended on patch. In patch 1, ants were the dominant item in diets. In patch, 2, both ants and grasshoppers were frequently consumed.

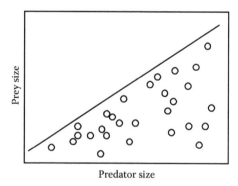

FIGURE 7.10   Relationship between a gape-limited predator and the sizes of prey it consumes.

BOX 7.12

Quantile or least absolute deviation regression is useful for delineating the upper and lower bounds of prey sizes consumed by foragers. This technique differs from traditional least squares regression in that the quantile of the line can be chosen. In least squares regression, only the 50% quantile is determined. Confidence intervals found around the intercept and slope parameters for the 95% and 5% quantiles, which should capture the largest and smallest prey appearing in diets.

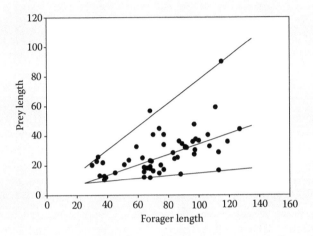

We plotted the relationship between the lengths of several salamanders in a population and the sizes of the tadpole prey they consumed. The average least squares regression line was calculated, which is the same as the 50% quantile line in the quantile regression (middle line). This estimated the length of the average tadpoles consumed but provided no information about the maximum and minimum vulnerable sizes. The quantile regression estimated the lines at the 5% (bottom) and 95% (top) intervals. Although these provided lower and upper boundaries, the accuracy of the estimates for these lines depended on the number of values. The 95% confidence intervals for the parameters were very high for the 5% and 95% quantiles, suggesting that more data were needed at those prey sizes.

| Quantile | Parameter | Estimate | 95% Confidence Limits | |
			Lower	Upper
50%	Intercept	−0.57	−3.3	14.1
	Slope	0.35	0.17	0.4
5%	Intercept	5.9	−146	9.2
	Slope	0.09	−0.007	0.49
95%	Intercept	−1.1	−4.2	162
	Slope	0.79	−0.01	0.82

7.6 CONCLUSIONS

Development of quantitative tools for assessing the diet choice of organisms began in earnest in the 1960s and continues to the present. The goals may be to test evolutionary theory or to predict energy flow, but the techniques are largely the same. Many indices have been developed to simplify the complex data that arise in organisms' diets. Because it is typically simpler to reliably quantify a forager's diet than the array of available prey, indices of prey in diets that do not incorporate the prey in the field may be more accurate for assessing foraging choices. When prey availability data are known and accurate, then indices of electivity may be used. In our view, a goal for trophic ecology is to better understand how variation in the forager's perception of prey availability and our ability to sample prey influence index results. The statistical properties and biological relevance of the indices need to be further assessed.

Ultimately, the data that are generated from analyses are used to test hypotheses of evolutionary biologists, population and community ecologists, and ecosystem ecologists with very different approaches. Using these techniques allows us to determine how decisions made by individual foragers scale to population dynamics and ultimately to trophic transfer of materials and energy. An evolutionary biologist may be interested in the optimal choice of mussel that an octopus consumes by using an electivity index. A community ecologist interested in the mussels determines that the octopus population has a disproportionate impact on mussel species composition. The ecosystem ecologist uses the index data to hone in on the differential effect of mussel species on energy flow into the octopus population and ultimately to the octopuses' predators. The index is used in different ways here from hypothesis testing to providing a road map to species interactions. We have only touched upon the vast number of tools available for quantifying diets. This chapter should guide you to more specific information.

QUESTIONS AND ASSIGNMENTS

1. What were the contributions of Ivlev and Schoener to analysis of diets in trophic ecology?
2. Is foraging behavior easy or difficult to observe in the wild? Explain why.
3. Describe some techniques that may be used to directly observe foraging.
4. Use the warbler example in a tree to explain how determining whether a forager is a specialist or a generalist is a matter of perspective.
5. How does stomach capacity change with increasing body size?
6. Compare how O, N, and W differ in the information they produce.
7. Explain how Strauss linear index, Manly–Chesson alpha, and Ivlev's electivity index differ in their ability to show choice of prey by consumers.

8 Bioenergetics, Ecosystem Metabolism, and Metabolic Theory

8.1 APPROACH

We show that basic physical laws of energy conservation are important and scale from the cellular to the ecosystem level. Respiration or metabolism is consistent among organisms, and this allows ecologists to develop general accounting models that can be applied to predict growth and production of organisms and trophic levels. The efficiency of the transfer of energy also can be estimated. A related approach, the metabolic theory of ecology (MTE), suggests that energy transfer among trophic levels should scale as a function of fundamental relationships among body size, temperature, and metabolic rate. We will compare and contrast these approaches, assessing how they may be applied to understanding trophic relationships in ecosystems.

8.2 INTRODUCTION

The laws of thermodynamics apply to living plants and animals as they do to all physical processes. Energy consumed by organisms can only be transformed, not destroyed as per the first law of thermodynamics. Thus, energy can be mapped as it makes its way from being an eaten prey item to its final form, whether it ends up as heat, body tissue, or waste. These ideas are not new. Scientists such as Antoine Lavoisier and Joseph Priestley deduced that heat was the by-product of the *fire of life*, or *metabolic respiration*, in the eighteenth century. All living organisms generate heat because heat is the end result of using chemical energy to manufacture and run cellular machinery, which is predicted by the second law of thermodynamics. Thus, dividing organisms into categories of *cold* and *warm* blooded is a bit of a misnomer. A single microbe does not produce much measurable heat. But a large mass of *ectothermic* bacteria in a compost pile can generate enough heat to make steam. Endotherms are multicellular organisms that use a combination of insulation, heated internal organs, denser mitochondria, and larger mitochondria to elevate their body temperature relative to their ectothermic counterparts, although some large ectotherms are quite adept at maintaining a constant core body temperature under a variety of environmental conditions (Ruben 1995).

The study of respiration was given the name *bioenergetics* by Lehninger (1960), with the term focusing primarily on processes regulating energy transformations in cells. From the perspective of trophic ecology, these cellular energetic pathways are scaled up to the entire organism, the population, and ultimately the ecosystem.

By the 1970s, the term bioenergetics had been adopted by ecologists to describe macroscale transformations of energy (i.e., ecosystem processes). Whereas at the cellular level bioenergetics focuses on the highly conserved enzymatic machinery used to break down the biological fuels of carbohydrates, protein, and fat into energy, typically with the assistance of oxygen (Figure 8.1), bioenergetics at the ecological level focuses on the budgeting of energy as it passes through organisms (Winberg 1960), ultimately driving trophic interactions through ecosystems (e.g., ecosystem metabolism).

Trophic ecology depends on the foundations of bioenergetics. Herein, we first briefly review the cellular processes underlying energy transformations at the scale of the individual. Energy changes may be measured as a balance between energy consumed and its eventual fate as storage in body components, loss as waste, or physical work. We review these potential pathways by exploring the basics of energy budgets at the scale of an organism. Developing methods for quantifying the transformations

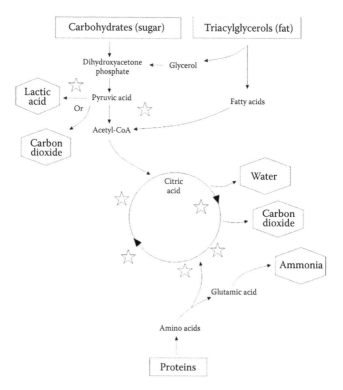

FIGURE 8.1 Simplified conceptual model of glycolysis and tricarboxylic (citric acid) cycle. Starred locations are where energy is harvested in the cycle. In each turn of the cycle, one ATP, three NADH (nicotinamide adenine dinucleotide; an electron carrier), one ubiquinone (another electron carrier), and two carbon dioxide molecules are generated. Although complex, the carbon dioxide produced, oxygen used, and heat produced from this cycle are nearly ubiquitous among all living things, making it possible to scale these cellular molecular energetic transformations to the organism, population, and ecosystem scales.

is critical for developing sound budgets. Because energy budgets must balance, models of *bioenergetics* have been developed to estimate growth, reproductive output, and consumptive demand of organisms. These models are useful to trophic ecology for predicting how energy moves from one population to another in ecosystems. The consumptive impact of predators on prey populations also can be estimated. Because energetics is conserved across taxa, processes at the individual scale may drive ecosystem-scale patterns, culminating in what is called the MTE (Brown et al. 2004). We explore the benefits and promises of these approaches as well as their shortcomings.

8.3   ENERGY TRANSFORMATIONS

When fuel is consumed by fire in an oxygen-rich environment, the exothermic reaction is typically rapid and violent. Organisms manage the chemical conversion of fuel in their cells in a controlled way, allowing them to efficiently squeeze as much energy from the reaction as possible. Electrons are transported (*falling*, think of a ball falling down a flight of stairs) through a chain of reactions, with each transfer between molecules yielding energy. At the end of the chain, an *electron acceptor*, usually the highly electronegative element, oxygen, needs to be present to let the electrons *fall* through the respiration process from higher to lower potentials. The potential energy yielded from the electrons trapped in carbohydrates (i.e., sugars), fats, and proteins is related to the composition of the molecules and their entry into the respiratory process. Carbohydrates, fats, and proteins yield about 17, 39, and 24 kJ/g. Each fuel type enters the respiratory pathway at different points (Figure 8.1). The conversion (i.e., aerobic respiration) of a simple carbohydrate molecule, glucose, to its by-products of water and carbon dioxide is often depicted as a simple exothermic reaction where $C_6H_{12}O_6 + 6O_2 \rightarrow 6H_2O + 6CO_2 + ATP$, where ATP is adenosine triphosphate, the primary energy-storage molecule in cells. Carbohydrates are initially processed in the cytoplasm of cells via glycolysis, where they are converted to dihydroxyacetone phosphate and then pyruvic acid by a series of enzymatic reactions that cost energy and yield carbon dioxide (Figure 8.1). In the absence of oxygen, anaerobic respiration yields energy by converting pyruvic acid to lactic acid (Figure 8.1). If oxygen is available to accept electrons, then the energy yield is much greater. Pyruvic acid is further transformed to acetyl co-A, which is synthesized into citrate and enters the citric acid cycle in the mitochondria (Figure 8.1). The by-products of this process are carbon dioxide and water. Fat molecules enter the glycolytic pathway by being split into glycerol and fatty acids (Figure 8.1). Protein enters the respiration process much later than carbohydrates or sugars, requiring energy to produce ammonia as a toxic waste product through deamination (Figure 8.1). Aquatic organisms simply excrete ammonia into the surrounding medium at minimal energy cost. Terrestrial animals must use additional energy (ATP) to convert the ammonia in the serum to nontoxic forms such as urea or uric acid, which are excreted via organs such as the kidney.

The biochemical transformations occurring in the cells of organisms are highly conserved across taxa, meaning that these energetic relationships are generalizable at the whole individual scale. Oxygen and food consumed are typically the inputs into the organism. This is not always the norm—some microbes use other molecules than sugar, fat, and proteins as energy sources, such as nitrogen and sulfur compounds (e.g., ammonia and hydrogen sulfide), and, of course, cyanobacteria and plants use photons for energy. Depending on the level of aerobic versus anaerobic respiration occurring and the fuel consumed, the outputs are lactic acid, carbon dioxide, nitrogenous waste, solid waste, and heat. The same is applicable to autotrophs when light is unavailable for energy. Each of these inputs and outputs can be quantified with standard techniques. Carbon dioxide production is relatively easy to quantify for terrestrial organisms because this gas makes up such a small proportion of the atmosphere. In water, however, this gas is highly soluble and converts to other forms (e.g., carbonic acid, carbonate), making it difficult to measure the carbon dioxide produced by respiration. The converse occurs for oxygen production. Oxygen is so abundant in the atmosphere that detecting small changes due to respiration is difficult for terrestrial animals. In contrast, oxygen is poorly soluble in water and relatively easy to quantify in respiring aquatic organisms. Heat produced by organisms also may be measured as a by-product of metabolism. However, quantifying this, especially for ectotherms, is difficult because it is hard to tease apart heat changes due to respiration versus thermal variation in the environment. The other components of energy transformations are relatively smaller than the production of oxygen and the production of carbon dioxide and heat. Lactic acid from anaerobic activity can be quantified by measuring the amount of lactate dehydrogenase, the enzyme responsible for breaking down pyruvate into lactate and yielding energy (Selch and Chipps 2007). The energy content of waste products such as uric acid and feces can be easily quantified by burning them completely in a *bomb calorimeter* (Box 8.1).

Measuring respiration of organisms at the individual level is relatively simple to accomplish in the laboratory (Box 8.2). The *resting metabolic rate* (RMR) is the energy consumption of an organism at rest and fasted. The *standard metabolic rate* (SMR) is a similar value, but applies to organisms that cannot remain completely at rest, such as a fish maintaining its position in the water column. The SMR is calculated by estimating the metabolic rate at various levels of activity and then extrapolating the relationship back to the intercept at zero activity. The metabolic rate increases with activity and can be quantified by making the organism exercise. For ecologists, the problem lies in the ability to quantify the *true* daily, integrated respiration cost of living in the environment, which includes the RMR or SMR and being active. Terrestrial ecologists estimate this as the *field metabolic rate* (FMR) using doubly labeled water. Both the hydrogen ^2H and the oxygen ^{18}O in the doubly labeled water are heavy isotopes of the lighter, common forms of these elements (^1H and ^{16}O; Figure 8.2). Both heavy isotopes are nonradioactive and occur in the organism but at very low concentrations (Figure 8.2). A researcher injects an organism with a known quantity of water, greatly increasing its body concentration of these heavy isotopes, and releases it back into the field. As the organism goes about its business, both heavy oxygen and hydrogen are lost as water is removed via excretion (Figure 8.3). Additional heavy oxygen is lost as it is combined with carbon and exhaled as carbon dioxide. After some time, the organism is recaptured, and the amounts of heavy hydrogen and oxygen remaining in the blood serum are quantified. By modeling how hydrogen and oxygen are removed from these two pools, FMR can

BOX 8.1

Calorimetry is the process by which the energy content of biological material is quantified by measuring the heat (calories or joules; 1 cal = 4.184 joules) it produces when combusted. The combustion process should be complete; in other words, no uncombusted material should remain. The standard method that has been used for decades is oxygen bomb calorimetry. A sample of food or tissue is dried and the wet mass quantified. The dried sample is compressed and then burned completely in a casing (i.e., a bomb) with pure oxygen. The heat expressed is quantified and expressed as calories or joules per gram dry mass. The energy produced by the bomb is equivalent to that available for catabolic processes occurring in the tissues of organisms.

Bombing is a labor-intensive process. A minimum amount of sample is necessary to quantify a temperature change in the instrument. If insufficient sample is available, multiple samples may be pooled. Because the average energy densities of lipids, carbohydrates, and proteins are known (Table 8.1), an alternative to bombing is to determine the proximate composition of the item and then convert each component to its predicted energy equivalent. Lipids are extracted using a solvent (e.g., two parts chloroform to one part methanol, by volume) followed by elutriation of the solvent–lipid layer (Folch et al. 1957; Bligh and Dyer 1959). The solvent is then evaporated under a fume hood using neutral gas such as nitrogen, and the remaining total lipid residue is weighed. The known energetic value of lipid is then used to estimate its contribution to the item.

The century-old Kjeldahl method is used to quantify protein. Items are chemically digested, protein nitrogen is released as ammonia, the ammonia is distilled, and the total nitrogen content is determined by titration (AOAC 2016). Because protein contains about 16% total nitrogen, 100% protein divided by 16% nitrogen yields a conversion factor of 6.25 grams protein per gram nitrogen (FAO 2002). The carbohydrate component of a diet can be estimated as the unaccounted portion of the sample minus its ash free dry mass (AFDM). The AFDM is determined by combusting a sample of the item completely in a muffle furnace, with the ash representing the mineral content of the sample.

For some organisms, a simpler way to estimate energy density is by determining the percent of water in the tissues. Water content (and its converse, dry mass) often is related to the relative percent composition of fat, protein, and ash content, making it relatively easy to estimate proximate composition of prey after simple keys are developed (Hartman and Margraf 2008).

BOX 8.2

Respirometry is the process by which the consumption of oxygen is quantified. Most respirometry systems are *open*, meaning that air or water with oxygen is pumped into a holding chamber with the organism. The organism uses a portion of the oxygen for metabolism, and the decline in concentration is quantified by an oxygen sensor. These systems are quite complex involving a pump or impeller that moves air or water through the chamber. The oxygen concentration must be precisely and accurately quantified both at the incurrent (i) and excurrent (e) locations. The consumption (VO_2) of oxygen is measured as

$$VO_2 = F(O_i - O_e)$$

where O_i and O_e represent the oxygen concentrations in mL/L entering and exiting the chamber, respectively. The rates (F in L/s) of the air or water and thereby the oxygen replenished in the system need to be included.

A tadpole is in an open respirometry chamber. The oxygen concentration entering the chamber (O_i) is 8 mL/L, while that exiting (O_e) is 7.95 mL/L. The flow rate is 0.01 L/s. Thus, the oxygen consumption is 0.0005 mL/s. Knowing that 21 J of heat is produced for each milliliter of oxygen consumed, then the tadpole is generating about 0.11 J/s or about 38 J/h. Quantifying this small amount of heat produced in the water would be very difficult relative to determining the amount of oxygen consumed.

The business of building and maintaining an open chamber respirometer is complicated. The flow of the media needs to be quantified using a variety of tools from simple bubble counters in a terrestrial setup to laser flow velocimeters in aquatic systems. The chambers for the organisms need to be tightly sealed. For aquatic organisms, the flow passing through the chambers needs to be nonturbulent; otherwise, it will likely increase activity and metabolic rate. For terrestrial systems, the water in the air entering the chamber and produced by the organism needs to be removed because oxygen will dissolve in it. The concentration of oxygen both entering and exiting the chamber can be quantified with many kinds of probes. A common *Clark-type* probe uses a gas-permeable membrane, a potassium chloride solution, a gold or platinum anode, and a silver cathode. Chloride is attracted to the anode and gives up electrons in the process. The oxygen travels past the membrane into the probe and is consumed at the cathode, producing OH^- and allowing the free electrons to flow as the oxygen is reduced (i.e., a net gain in electrons). There is a direct positive relationship between the oxygen concentration and the current generated between the electrodes. Because oxygen concentration at the membrane is constantly declining, stirring is needed. Using a more recent technology, fiber optics measure the interaction of oxygen with an oxygen-sensitive probe embedded in a sol-gel (i.e., a colloid-like substance that is hydrophobic).

Oxygen makes the probe fluoresce, and the light emitted is measured through the fiber optic cable. With these probes, oxygen is not consumed and thus stirring is not needed.

An alternative to an open respirometer is one with a closed chamber (i.e., closed respirometry). An organism is placed in a chamber with a known amount of oxygen. The chamber is sealed, and the organism remains in it as oxygen declines and carbon dioxide accumulates. The problem with this method is obvious. The impacts of changing gas concentrations may alter physiology and bias results.

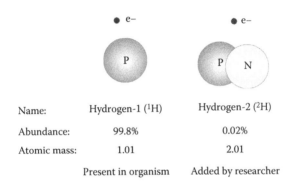

Name:	Hydrogen-1 (^{1}H)	Hydrogen-2 (^{2}H)
Abundance:	99.8%	0.02%
Atomic mass:	1.01	2.01
	Present in organism	Added by researcher

FIGURE 8.2 Two stable (i.e., nonradioactive) isotopes of hydrogen. Hydrogen-1 is the most common isotope, with one proton and no neutron. Most water (H_2O) in the bodies of organisms is made up of this light isotope. The heavy stable isotope, hydrogen-2 (^{2}H), and the heavy stable isotope of oxygen, ^{18}O, both with more neutrons than the light, common isotopes, are added to an organism, combined as heavy water. The heavy hydrogen is washed out as water in excretion. Conversely, some of the water is split for metabolism, and heavy oxygen ends up being washed out as both excreted water and carbon dioxide. The difference between heavy hydrogen and heavy oxygen rates yields the net carbon dioxide produced.

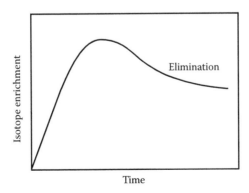

FIGURE 8.3 Pattern of enrichment of heavy hydrogen and oxygen isotopes in the body and their elimination. The portion of the curve that is elimination is used to calculate the rate of excretion.

be estimated (Schoeller 1999). The basic equations for calculating the rate of carbon dioxide produced are a function of the rates of ^2H and ^{18}O removal:

$$r_{H2O} = Nk_{2H} \text{ and}$$

$$r_{H2O} + 2r_{CO2} = Nk_{18O}, \text{ leading to}$$

$$r_{CO2} = (N/2)(k_{18O} - k_{2H})$$

where N is the total water pool (in mols) and k is the rate (r) of elimination of water and carbon dioxide (Schoeller 1999). Estimating the rates used to calculate FMR is still a matter of debate among physiological ecologists, and estimates for endotherms vary as much as 9–15 times RMR daily (Speakman 2000). Unfortunately, FMR cannot be quantified in aquatic organisms using doubly labeled water because the internal water pool of these animals is always in flux. Estimating the contribution of activity metabolism to aquatic organisms continues to be a source of inquiry in trophic ecology.

8.4   ENERGY BUDGETS

Central to *macro* bioenergetics in trophic ecology is the development of an individual's energy budget, similar to the budget an accountant creates for a business. Rather than dollars, the currency of the energy consumed by a consumer is usually broken into its heat content in calories or joules. Assuming that respiration is aerobic, the heat produced is directly proportional to the oxygen consumed, depending on the fuel being catabolized. For carbohydrates (i.e., sugars), about 21 kJ of energy is released for every liter of oxygen consumed (Table 8.1). Fat and protein produce less heat per liter of oxygen (Table 8.1). Because of this relationship, oxygen can be exchanged for heat in the energy budget. Carbon dioxide also is a candidate for use in the budget. However, it is important to note that different fuels generate different *respiratory quotients* (RQs), the ratio of mols of carbon dioxide respired to mols of oxygen consumed. Recall that mols is the unit used to standardize the number of atoms or molecules in chemical reactions, equal to Avogadro's number. The RQ for

TABLE 8.1

Energy Equivalents and Oxygen Consumed per Metabolic Fuel Type

Fuel	kJ/g	L O_2/g	kJ/L O_2	RQ
Carbohydrate	17.6	0.84	20.9	1.0
Fat	39.3	2.0	19.6	0.71
Protein	17.9	0.96	18.8	0.80

Source: Schmidt-Nielsen, K. 1997. *Animal Physiology: Adaptation and Environment.* Cambridge, MA: Cambridge University Press.

Note: Respiratory quotient (RQ) is the ratio of carbon dioxide formed to oxygen consumed.

carbohydrate is equal to 1 because 6 mols of oxygen are used to produce 6 mols of carbon dioxide (Table 8.1). Both fat and protein produce less carbon dioxide per unit oxygen, with RQs < 1 (Table 8.1). Measuring RQ indicates what metabolic fuel is consumed by the organism but makes the use of carbon dioxide as currency in an energy budget difficult to interpret.

The organizational scale of the boxes in an energy budget depends on the level of the ecological question being asked. Ultimately, an energy budget can be divided into the individual cells, tissue types, or organs within an organism, each having an energy demand, respiration rate, and outputs of heat and waste. Of course, developing an energy budget at such a fine scale is infeasible, so the budget is typically divided into physiologically and ecologically relevant categories (Figure 8.4). After expending energy to capture and consume a prey item, the food enters the consumer as a known quantity of energy (Figure 8.4). The total potential energy content of

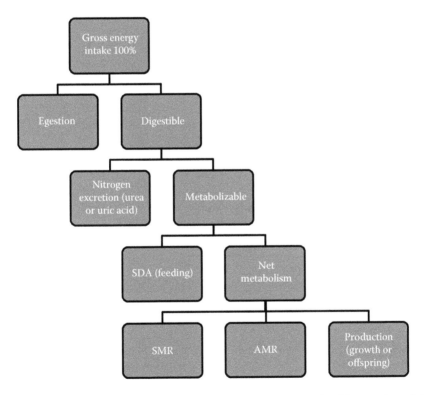

FIGURE 8.4 Energy budget of an organism. A meal is consumed, with a portion of the energy being egested and the remainder being absorbed into the tissues. Absorbed energy is either lost through removal of nitrogen or assimilated as digestible energy. This remaining energy is either used to digest food as SDA or to fuel metabolic needs. SMR is used to run the cellular machinery of an organism at rest. The activity metabolic rate (AMR) is the metabolic cost of moving, finding food, etc. The remainder of energy is available for production of growth or offspring. Each energy transformation in the budget produces heat as a measurable waste product.

the item is determined using calorimetry (Box 8.3). Energy is broken down into two components once it enters the alimentary canal of the organism. The energy is either assimilated into the organism or passed as waste through egestion. The composition of the prey affects the amount of energy lost through egestion. A nonruminant herbivore may have egestion losses as great as 80% of consumed energy (see Chapter 7). In contrast, a carnivore has much lower egestion loss and higher absorption or assimilation efficiency. Digestible energy enters the organism's tissues or circulatory system, with some energy lost due to nitrogen transformation and excretion and the rest being metabolically available (Figure 8.4; Box 8.4). The metabolically available energy comes at the additional cost of *specific dynamic action* (SDA), otherwise known as the heat of digestion (Beamish 1974). Processing food is not cheap,

---

BOX 8.3

The proportion of consumed energy that enters the bloodstream or tissues is called the absorption efficiency (AE), which ranges from 0 to 1. We might assume that simply estimating the energy content of food consumed and feces egested equates to the energy absorbed. However, feces often contain much more than the undigested portion of the meal. Cells, bacteria, mucus, and digestive enzymes also are produced. Thus,

$$AE = [R - (F - O)]/R$$

where R is the energy consumed, F is the energy content of the feces, and O is the energy content of the nonfood component.

The AE is not the same as the assimilation efficiency of the organism (AsE), which is the energy that is available for metabolism after losses due to nitrogen excretion. The proportion value is calculated as

$$AsE = [R - (F + U)]/R$$

where U is the energy content of nitrogen excreted as ammonia, urea, or uric acid.

AE is usually determined experimentally by incorporating an inert marker into food. Chromic oxide is a common marker, giving the feces a characteristic green color. Acid-insoluble ash, chitin, cellulose, and other indigestible materials may be used. The markers have to be sufficiently abundant to be detected but rare enough not to affect digestion. Assimilation efficiency is typically determined by collecting nitrogen excreted and quantifying its energy content. There is an additional energy cost to nitrogen excretion that is not incorporated into the traditional assimilation equation. The energy costs of converting protein into waste products vary but may be about 25–30 kJ/g of the nitrogen component of protein consumed. The true net energy available for metabolic activity is AsE minus this cost.

BOX 8.4

Nearly 40 years ago, the availability of computer modeling and much organismal-level physiological data reached a threshold, allowing Kitchell et al. (1977) to develop a mass-balance bioenergetics model for the fish, yellow perch. The *energy in must equal energy out* concept of the model was simple (see Figure 8.8). However, even then, it was apparent that the internal workings of the simulations were complex, with 17 interrelated parameters used.

The model revealed how sensitive growth predictions should be relative to the various parameters. The proportion of maximum consumption, typically the parameter being estimated by these models, was the most sensitive factor affecting growth. The activity multiplier in the model also was an important parameter affecting patterns of growth. As we note, the impact of these factors on model predictions are still being explored today.

Kitchell et al. (1977) used their model to estimate growth of yellow perch in two basins of Lake Erie. The western basin remains warmer than the eastern basin during most of the year. The model predictions for growth fit observed patterns in the lake. Yellow perch from the western basin grew faster because of the greater temperatures and scope for growth. For fish and other poikilotherms, the importance of temperature to trophic interactions was becoming clearer.

This figure shows the growth of yellow perch in the warmer western basin and the cooler eastern basin of Lake Erie. Growth increased during each summer and then stopped during winter, when feeding ceased. (From Kitchell, J. F. et al. 1977. *Journal of the Fisheries Research Board of Canada* 34:1922–1935.)

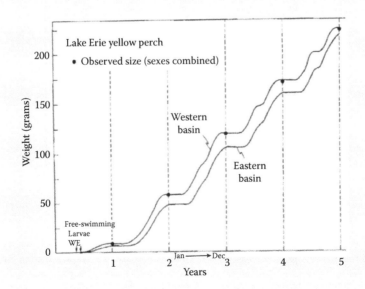

requiring the production of digestive enzymes and accessory compounds (e.g., acid, mucus, salts) as well as the energy of mobilizing the digestive organs. The net energy remaining is used by the organism for its daily needs, starting with its RMR or SMR. Activity is an additional metabolic cost that varies among organisms and is typically treated as a multiplier of the RMR or SMR. The energy remaining in the budget is available for the production of body tissue (Figure 8.4). Body tissue is divided into body parts (e.g., organs, tissue), storage (e.g., fat or glycogen), or reproduction (e.g., gametes), depending on the life history needs of the organism. If all the energy is lost to metabolism and waste, the organism will not grow or reproduce. Starving organisms will catabolize tissue to meet metabolic needs, typically using glycogen and fat while sparing protein.

Because of the laws of thermodynamics, the energy budget must balance, with the heat value of the energy consumed equaling the total heat of all the possible end points. The challenge for researchers is to accurately account for all of these inputs and outputs in organisms that are constantly responding to their environment and trying to meet their life history requirements. Some basic patterns occur among organisms making predictions possible. The relationship between body mass in organisms and metabolic rate is well known (Figure 8.5). Mass increases exponentially as organisms grow, because body volume increases as the cube of the body length. Given that metabolism scales with mass (i.e., bigger organisms need more energy), the relationship between these two is exponential as well. When converted to a logarithmic scale to make this exponential relationship linear, the slope between mass and metabolic rate across taxa scaling from microbes to vertebrate endotherms is about 0.75, leading to what is called Kleiber's 3/4 mass relationship (Kleiber and Rogers 1961). The slope

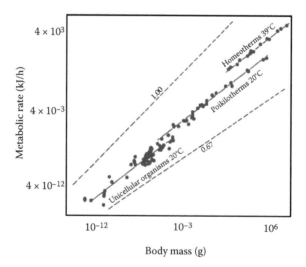

FIGURE 8.5 Relationship between body mass and metabolic rate for organisms. The slope is less than 1 for all taxa, showing that larger organisms have lower mass-specific metabolic requirements. Still, larger organisms have much higher total energy needs than smaller counterparts. (From Schmidt-Nielsen, K. 1997. *Animal Physiology: Adaptation and Environment.* Cambridge, MA: Cambridge University Press.)

is less than 1, which shows that, on average, bigger animals have less energy requirements per unit body mass than smaller counterparts. This useful relationship allows trophic ecologists to estimate roughly what the basic costs of respiration will be as a function of body size and refine these predictions within taxa.

Although the slope is consistent, there is variation around the regression lines and the intercepts differ among taxa (Schmidt-Neilsen 1997). The largest difference occurs between poikilotherms and endotherms (Figure 8.5), with endotherms needing more energy per unit mass given the high metabolic demand of maintaining a constantly elevated body temperature. Even at equivalent body temperatures, endotherms use greater than four times energy than poikilothermic ecotherms (Ruben 1995). Poikilotherms regulate their body temperatures and metabolic rates behaviorally by choosing their environment. Although their energy costs are lower on average, individuals can select temperatures that elevate their metabolisms. The impact of acclimation, more properly termed acclimatization for organisms in the field, is important to consider. An organism experiencing an abrupt change in body temperature will experience a significant change in metabolism. A 10°C rise in the body temperature of a poikilotherm leads to an increase in the metabolic rate called the Q_{10} *response*:

$$Q_{10} = (R_2/R_1)^{10/(T_2-T_1)}$$

where R_2 and R_1 are the two metabolic rates, and T_2 and T_1 are the temperatures. In this case, $T_2 - T_1 = 10$ because the change is 10°C. Thus, Q_{10} is simply the ratio of change of metabolism. If metabolism doubles, $Q_{10} = 2$; if it triples, $Q_{10} = 3$ (Figure 8.6). Endotherms also respond to temperature changes by expending energy to maintain body temperature by cooling or heating the body. These energetic responses may

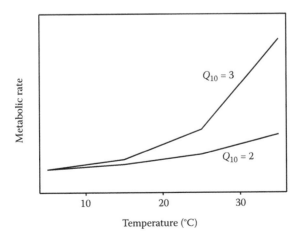

FIGURE 8.6 Immediate temperature-dependent metabolic response of poikilothermic organisms. The rate of increase is called the Q_{10}, usually ranging between a doubling and tripling of metabolic response with each 10°C increase.

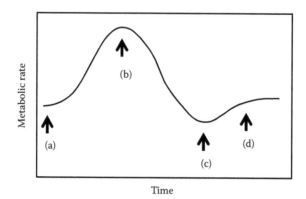

FIGURE 8.7 Process of acclimatization in an organism. (a) A perturbation (e.g., temperature increase, stress) occurs causing an increase in metabolism. (b) Metabolic needs peak. (c) Through time, the organism acclimatizes to the perturbation, with metabolic rate declining and perhaps overcompensating. (d) Metabolic rate returns to a level similar to pre-perturbation. The temporary metabolic response may influence patterns of trophic interactions.

be short term, because animals acclimatize to their surroundings, with metabolic rates returning to rates similar to those before the temperature change (Figure 8.7). This is accomplished in many ways through changes in behavior, physiology, and morphology. Both the short-term energetic costs of abrupt changes in the environment as well as the response of acclimatization need to be incorporated into models of individual energy budgets.

8.5   ORGANISMAL MODELS

Because of the apparent generalities in energetic requirements of organisms, ecologists in the mid-twentieth century began to explore whether growth and consumption at the individual and population scales could be estimated based on simple energy budgets (Winberg 1960; Ivlev 1961). These modeling ideas were developed primarily in the fisheries biology discipline and continue to predominate in this field today. The premise is deceptively simple. Fish biologists do not have a good means of measuring consumption of fish in the field. This is important for determining the impact of fish predators on their prey population, because fish populations that outstrip their prey will grow poorly and perhaps collapse. In aquaculture, knowing the food requirements of fish ensures that underfeeding or overfeeding does not occur. To estimate consumption with the model, fisheries ecologists quantify growth of fish, which is relatively simple. Growth is incorporated into a balanced energy budget for fish, using estimates of respiration and waste. The remainder should be the consumption by the fish (Figure 8.6).

Early models developed for fish populations were oversimplistic and were not particularly effective. With the advent of desktop computer modeling in the 1970s came the promise of sophisticated models, incorporating species-specific parameters for all of the boxes within the energy budget. Environmental factors, primarily temperature, were incorporated into the models to predict consumption as a function of growth under a host of conditions. The ultimate test of these models is their ability to predict growth based on observed consumption, which is difficult to accomplish (Chapter 9). Because this general model was first formalized by Kitchell et al. (1977), a group of researchers at the University of Wisconsin Madison, it is now called the *Wisconsin model* (Ney 1990; Figure 8.8; Box 8.4). This modeling framework has been applied to many organisms other than fish, including invasive mussels (Schneider et al. 1998), *Mysis* (Chipps and Bennett 2002), and marine mammals (Noren et al. 2012).

Critical to the Wisconsin model are the parameters used in its many mathematical functions. Perhaps one of the most important functions is the relationship between temperature and maximum consumption. For most poikilotherms, maximum consumption increases with temperature until it reaches a critical maximum (Figure 8.9). Beyond this temperature, consumption declines to zero because the organism's digestive system no longer works. Thus, the consumption side of the energy budget, which is typically the unknown being solved in the equation (Figure 8.8), relies on this temperature-dependent function of maximum consumption. Maximum consumption also scales positively with body size. In endotherms, maximum consumption is a simple function of body size, with environmental temperature having little impact. In Chapter 9, we explore how these relationships are quantified.

The quantities on the other side of the energy budget equation in the model are usually known (Figure 8.8). Growth is the most robust parameter in the model, with mass converted to energy equivalents. Respiration as RMR or SMR is another response that relies on sound mathematical relationships, depending on body mass and temperature (Figure 8.8). Activity metabolism is often included as a multiplier of SMR (Figure 8.8). For example, in many Wisconsin models of fish, the activity

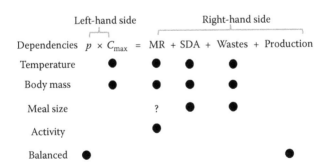

FIGURE 8.8 Wisconsin mass-balance bioenergetics model, with parameters and their dependencies. Consumption on the left-hand side of the equation is predicted using known rates of maximum consumption (C_{max}) and a tuning parameter, p, a proportion that ranges from 0 to 1. Consumed energy is allocated to metabolic rate (MR), digestion (SDA), excretion and egestion (Wastes), and growth and reproduction (Production), all on the right-hand side of the equation. Black dots represent factors that influence parameters in the model. A question mark suggests factors that may affect parameters. Ultimately, the parameters that must be modified to balance the two sides of the model are the p-value and the observed growth and reproduction.

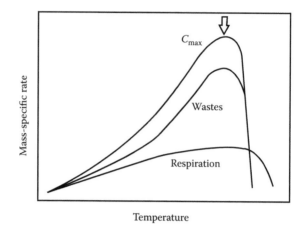

Temperature

FIGURE 8.9 Relationships between temperature and major parameters of the mass-balance bioenergetics model for poikilotherms. All rates increase with temperature to a threshold point. When this temperature is exceeded, both maximum consumption (C_{max}) and waste products drop. A further rise in temperature causes metabolism to cease and organisms to die. The point of greatest difference between C_{max} (income) and wastes and respiration (costs) is the temperature of optimal growth, depicted by an arrow.

multiplier is often assumed to be about twice the SMR (Kitchell et al. 1977), based on the finding by Winberg that most fish use about double their SMR to meet their daily needs. The cost of the SDA is usually modeled as a constant proportion of the size of the meal (Figure 8.8). Both loss of waste as feces and as excretion are modeled as a function of temperature and meal size (Figure 8.8).

Using the model to estimate consumption requires the ability to balance the two sides of the equation. Because consumption is a parameter that is both being estimated on the left-hand side and also influences several parameters on the right-hand side (Figure 8.8), the equation cannot be solved by simply plugging in the parameter values and yielding a result. Rather, iteration must be used where estimates of consumption are inserted into the two sides of the equation and subsequently tweaked until the two sides of the equation balance. The consumption estimate is constrained by the value of C_{max} for the body size and temperature of the organism. Thus, the value of consumption in the Wisconsin model is actually a proportion of C_{max}, often called a p-value (Figure 8.8).

As with all models, the Wisconsin model is used for two purposes. The first and probably most robust use is for hypothesis testing and generating new ideas (i.e., deduction engines; Hilborn and Mangel 1997). The p-value is a robust biologically relevant and interpretable variable for comparing the consumptive response of two or more populations. p-Values provide an alternative to growth as a response in experiments. Individuals may grow differently in two experiments—the p-values from the model allow the investigator to compare how much the individuals had to consume to reach their growth end points. Environmental effects such as temperature can be incorporated into models to predict how consumption and growth may vary among populations. These predictions can generate testable hypotheses for experiments.

The second use of the models is to predict actual patterns of growth and consumption in the lab or environment. Few studies have successfully tested the model in this fashion, even given its widespread use (Chipps and Wahl 2008). The reason this approach is so tricky is because consumption is very difficult to reliably estimate (Chapter 9), leading some investigators to suggest that model estimates of consumption from the field may be more robust than their empirical counterparts. According to Chipps and Wahl (2008), only 17 field tests of the Wisconsin model have been conducted for fishes in its nearly 35 year history, with errors (i.e., model predictions minus field data) ranging from −84% to 770%. Laboratory tests were similarly scarce and fared little better.

Even given the poor fits of the Wisconsin model, its use in trophic ecology has increased exponentially over the last three decades (Chipps and Wahl 2008; Hartman and Kitchell 2008), leading to clear needs for refinement for fish and other organisms, including endotherms. These models are complex, with many interdependent parameters (Figure 8.8). The relative impact of the parameters on model behavior may be assessed using sensitivity analyses. This approach involves varying model parameters by a known amount and determining how this affects model output. Small changes to parameters that cause large changes to model output indicate that these parameters are the most *important* and require robust data. Problems with sensitivity analysis are that the selected changes to the model parameters are arbitrary and that the parameter responses are not independent. A better approach is to use parameter selection approaches from information theory such as Akaike information criteria (AIC) (Burnham and Anderson 2002) or developing statistical techniques such as hierarchical Bayesian analyses (Fordyce et al. 2011).

Arguably, the parameter that has caused the most controversy in the use of the Wisconsin bioenergetics model is the activity multiplier in respiration. For terrestrial organisms, this parameter is available by quantifying FMR. In aquatic organisms, the value of this parameter is usually assumed because FMR is difficult to ascertain. For some fishes, this assumption may lead to model error (Boisclair and Leggett 1989; Boisclair and Sirois 1993). Since this issue was raised, techniques for quantifying activity metabolism of fish have been developed using biotelemetry (Briggs and Post 1997; Lowe 2002). Body movement and muscle contractions are measured by an implanted detector and transmitted to a receiver. Swimming activity, which correlates with activity metabolism, also can be quantified continuously in the field with tagged fish. Field data collected to date confirm that activity metabolism in fish is much less than that of terrestrial endotherms, averaging between 2 and 3 as Winberg (1960) suggested.

Several additional factors contribute to the accuracy and precision of bioenergetics models. Consumption-dependent error is likely prevalent (Bajer et al. 2004). Recall that egestion, excretion, and SDA are modeled as a constant proportion of consumption in most forms of the model. However, in many organisms, absorption efficiency declines with large meals and increases for small meals (Jobling 1994; Chipps and Wahl 2008). This effect is particularly prevalent when ration is low or fluctuating, leading to a phenomenon known as compensatory growth

(Figure 8.10; Whitledge et al. 2006). The higher absorption efficiency causes models to overestimate the consumption necessary for the observed growth to occur. Given that ration of most organisms in the field likely trends toward low daily intake, compensatory mechanisms may be common in the field. In confirmation of this, consumption is often overestimated in the field by the Wisconsin model (Hartman and Kitchell 2008).

The Wisconsin model requires considerable refinement, even after decades of use. Quantifying the necessary parameters is time consuming, expensive, and, frankly, not going to generate funding or accolades for researchers. Generalities in physiological parameters do exist. When applicable, the practice of *species borrowing* (Ney 1990, 1993), where known parameters from related species are used in models, occurs. When parameters and their interactions are not well understood, then researchers must roll up their sleeves and collect more robust data. In the case of rare species with poor data and scarce opportunities to collect more information, the best course of action is to attempt to use parameter fitting exercises to find the *best* parameter values that allow the predicted and observed growth and consumption curves to match (Paukert and Petersen 2007). This is accomplished by running thousands of simulations where each parameter value is varied. The best model fit reveals the appropriate adjusted parameter values. The problem with this approach is that these values are not at all based on biological data and thus do not capture the true underlying mechanisms. Modeling at the individual scale with bioenergetics will continue to increase in trophic ecology, with more taxa being included as well as previously modeled species being used. The goal in the future is to test these models and improve their accuracy and precision. If not, then the models will have

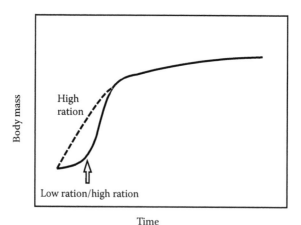

FIGURE 8.10 Compensatory growth in some organisms. At the outset, one individual is fed a high ration (dashed line) and the other a low ration (solid line). At a later time (arrow), the organism receiving the low ration is switched to the high ration and grows more rapidly, *catching up* with its counterpart and perhaps temporarily outgrowing it. Eventually, the two organisms reach the same size. This phenomenon suggests that the relationship among consumption, assimilation, and metabolism is complex. Organisms acclimatized to low ration may be able to process meals more efficiently and grow faster.

to be relegated to their use as deduction engines and not actual tools for quantifying interactions leading to trophic flux between organisms or among trophic levels in ecosystems.

8.6 OPPORTUNITIES FOR POPULATION BIOENERGETICS

Even given the current limitations of the models, they lead to many interesting, compelling approaches for use in trophic ecology. Some models do accurately and precisely predict trophic interactions (Rice and Cochran 1984) and are used to generate conservation decisions. In Lake Powell at the Utah–Arizona border of the United States, cycles of consumption of striped bass predators and the abundance of their threadfin shad prey were modeled accurately with the Wisconsin model (Vatland et al. 2008), showing that striped bass controlled their prey during most years. These data allow managers to understand why this sportfish population exhibits strong boom and bust cycles as it follows the variation in abundance of threadfin shad. Overconsumption by striped bass is partially responsible for these fluctuations. Predator–prey cycles are phenomena of great importance to trophic ecology (Chapter 5). As this example shows, these models lend considerable insight into underlying mechanisms.

An ideal use of bioenergetics models is in individual-based, agent-style models (IBMs). Each individual within the energetics simulation is modeled separately allowing for individual variation to influence behavior of the entire population. In the absence of variation among individuals, the aggregate response will always be the same (Figure 8.11). Variation in growth and reproduction of individuals can lead to dramatic differences in population-scale growth and consumer effects (Crowder

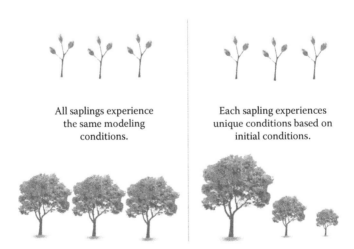

All saplings experience the same modeling conditions.

Each sapling experiences unique conditions based on initial conditions.

FIGURE 8.11 Individual-based modeling of organisms with indeterminate growth (e.g., trees, fish, invertebrates). (Left) A bioenergetics model of growth for saplings treating all as an aggregate will fail to capture individual differences, leading to homogenous size distributions. (Right) A model following the response of each sapling to unique environmental conditions is more complicated but will capture the variation in growth that occurs in populations.

et al. 1992; DeAngelis and Mooij 2005). This should be particularly prevalent in communities consisting of species with indeterminate growth and phenotypic differences in energetics, because this variation leads to different growth trajectories with fitness consequences (Figure 8.11). The biomass accumulated will affect the energetic needs of the consumers and their impact on prey populations.

How energy is partitioned in the energy budget and the associated trade-offs is an area of important inquiry in trophic ecology. Selection for allocation decisions that lead to differences in growth and reproductive output will have important effects on trophic interactions. A common pattern in species is that organisms living at high latitudes have large bodies relative to counterparts at low latitudes, otherwise known as Bergmann's rule (Lindsey 1966). Populations at high latitudes also may have faster growth rates, likely due to higher consumption rates (Conover and Schultz 1995). These differences in energetic needs may alter trophic interactions at geographic scales (Garvey et al. 2003; Figure 8.12). Populations also respond to perturbations,

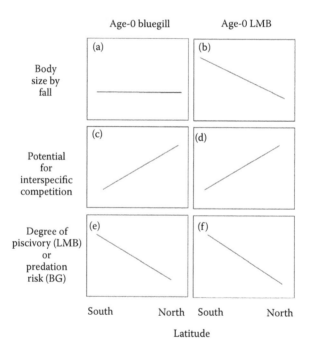

FIGURE 8.12 Predictions for latitudinal interactions between bluegill, a prey fish, and largemouth bass (LMB), its major fish predator in North America. These predictions are for these fish during their first year of life, when interactions likely set the stage for future trophic interactions as adults. Empirical data show that bluegill growth rates do not vary with latitude (a) while those of largemouth bass decline (b). Both species compete for the same prey during the beginning of life; shorter growing seasons at higher latitudes should intensify competition (c, d). At lower latitudes, largemouth bass should become sufficiently large to consume bluegill because of a longer growing season, whereas in the north, largemouth bass should be less likely to reach piscivory (e, f). These relationships show that the interaction among temperature, growth, and body size should affect trophic interactions. (From Garvey, J. E. et al. 2003. *Bioscience* 53 (2):141–150.)

with individuals adjusting their energetics. Toxicants reduce growth and reproductive output because organisms must use energy for detoxification. Factors such as harvest that select against large body size and rapid growth can lead to small individuals that have reduced consumptive impact on lower trophic levels (Conover et al. 2005). The intersection between evolutionary responses and their implications for trophic interactions is particularly important given the large scale of environmental change and its impact on individual traits of survivors. Climate warming in the oceans has been implicated as a cause of smaller fish, with implications for ecosystems (Perry et al. 2005).

Given the complexity of individual-based bioenergetics models and their spotty track record, scaling them to entire ecosystems is impossible. Science will never produce a complete set of bioenergetics models for each individual within each species that can then be aggregated to the ecosystem. However, natural selection operates at the scale of the individual within species. Thus, ultimately, the emergent properties of ecosystems are the results of the literally billions of energetic responses of the component organisms and their evolutionary responses to the environment. How do we rectify this seemingly huge expanse in processes occurring at very different scales?

8.7 ECOSYSTEM METABOLISM

Quantifying the respiration and growth of ecosystems is similar to that of single organisms, although the energy accounting is for the aggregate of all the organisms in the system. Growth is based on the production of primary producers (i.e., autotrophs). The metabolic costs quantified as calories, carbon, or oxygen consumed of the ecosystem include those of the primary producers, the consumers, and the detrital food web (see Chapter 3). Thus, net primary production (NPP) = gross primary production (GPP) – metabolism – waste. Waste is organic material that is not metabolized and lost, for example, organic matter falling into anoxic sediment of a lake and being buried. Since metabolism includes the production of the autotrophs, the consumers, and the detritivores, further investigation is needed to quantify secondary production and metabolism in the microbial loop (see Chapters 3 and 11 and Section 8.8). These data can then be used to quantify ecological efficiencies between trophic levels (Chapter 11).

One of the simplest, yet most informative uses of the ecosystem metabolism approach is to quantify the ratio of GPP to metabolic costs, often called the production-to-respiration ratio or P/R (Odum 1969). Ecosystems with $P/R > 1$ accumulate organic matter; those with $P/R < 1$ are heterotrophic, consuming more organic matter than what is being fixed. Ecosystems with $P/R = 1$ are metabolizing carbon as rapidly as it is fixed by GPP. Odum (1969) predicted that *mature* ecosystems would have a P/R near 1, because the successional stages present would balance energy fixed versus used. This is obviously more complicated because ecosystems are constantly experiencing disturbances, temporal variation in climate, and other factors that may cause P/R to vary. Many freshwater lakes have $P/R < 1$ (Hanson et al. 2003), meaning that these ecosystems are respiring more carbon than they are accumulating. However, inputs of nutrients may increase $P/R > 1$, causing carbon to accumulate as the lakes become eutrophic. Recall from Chapter 2 that Vannote et al. (1980) predicted in the river continuum concept that the importance of autotrophy versus heterotrophy would differ depending on the location within a stream. In headwaters where not much light can penetrate through the trees, they predicted that $P/R < 1$, as leaf matter and other material washed into the stream. At middle reaches, sufficient light and nutrients would cause $P/R > 1$, as algae and plants could grow. At large, lower reaches, the accumulation of organic matter would be so great and water so stained that heterotrophic activity would again overtake autotrophic production leading to $P/R < 1$.

Quantifying ecosystem metabolism has unique challenges relative to estimating the respiration of individual organisms. Using either radioactive ^{14}C or stable ^{13}C isotopes of carbon (Chapter 13) is a way to estimate primary production and metabolism by quantifying their uptake into plants and consumers relative to the common isotope ^{12}C. We will discuss this in detail in Chapter 13. Terrestrial and aquatic ecosystems differ in metabolism techniques. Because carbon dioxide is rare in the atmosphere (0.04% of gas), uptake during the day and production at night by terrestrial ecosystems is fairly easy to measure by enclosing an area and measuring carbon dioxide concentrations. Carbon dioxide taken up during the day is carbon fixed as GPP. Carbon dioxide respired at night is due entirely to respiration. Thus, NPP is calculated by subtracting the respiration from the GPP. These estimates assume that metabolism by the organisms in the plants and soil have negligible respiration costs. The contribution of these heterotrophs to total ecosystem respiration can be quantified over areas with no plants.

As mentioned previously, carbon dioxide is highly soluble in water and difficult to quantify. Thus, the production and consumption of oxygen is quantified instead. The same approach applies as with terrestrial ecosystems, where water samples containing algae and/or plants are first placed in a clear bottle and the oxygen produced is quantified (GPP). Replicate samples are placed in dark bottles and the oxygen consumed is then quantified (respiration). Oxygen consumed is converted to carbon using

$$mgC = mgO_2 * \frac{1}{1.2} * \frac{12}{32}$$

The number 1.2 is called the photosynthetic quotient, which is the moles of O_2 released by photosynthesis for each mole of CO_2 consumed, while 12 is the atomic weight of C and 32 is the molecular weight of O_2. From this NPP can be estimated. This approach again assumes that all oxygen consumption in dark bottles is due to autotrophs, although heterotrophic organisms, especially bacteria, can consume considerable oxygen. The contribution of algae and plants to oxygen consumption can be determined by filtering them from the water and quantifying their contribution to oxygen consumption. Another method was introduced by Odum (1957) to quantify metabolism in large, open systems like streams or rivers by comparing diel oxygen curves. Light/dark bottles are often not appropriate for these systems because water is continuously being exchanged. At the time when Odum suggested this method, there was no access to accurate and precise equipment for quantifying oxygen. With the advent of sophisticated oxygen probes and computer loggers, diel curves are more commonly being used to estimate the GPP during the peak of the day and respiration at night of these ecosystems. The central problem with these estimates is that the movement of the water causes reaeration, inflating oxygen concentrations. A tracer gas like propane or SF_6 can be released at the collection site, and the concentration dissolved in the water can be quantified with a gas chromatograph, although estimates can vary widely (Riley and Dodds 2013). Other methods of accounting for reaeration include modeling empirical rates of oxygen uptake as a function of the physical conditions including velocity, stream slope, and water depth. A recent technique using Bayesian model fitting including light levels and temperature as variables has been shown to tease apart the contribution of reaeration to diel oxygen curves relative to GPP and respiration (Grace et al. 2015).

8.8 ECOSYSTEM MASS BALANCE

As we saw in the previous section, the production and respiration of entire populations within ecosystems can be quantified in a manner similar to that of a single individual (Odum 1969). Thus, bioenergetics-based mass-balance approaches can be used at the scale of populations or trophic levels. Similar to individual-based bioenergetics, the productivity of the prey trophic level must balance the production of the consumer trophic level (*Ecopath modeling*; Pauly et al. 2000). Each trophic level has a production (*P*) to biomass (*B*) ratio, which is equivalent to the amount of *surplus* biomass added to the unit through reproduction and tissue accumulation:

$$(B_i)(P/B)_i([P-L]/P)_i = \sum_j (B_j)(S/B)_j(D_{ij})$$

The biomass (*B*) of prey trophic level *i* on the left-hand side of the equation depends on the ratio of production of prey biomass $(P/B)_i$, which will be greater than 1 if the prey trophic level is growing. The *P/B* ratio accounts for the respiratory costs within the trophic level. Loss of production to environmental loss, emigration, or removal by humans is *L*. If *L = P*, then no excess production is available for the next trophic level *j*. On the right-hand side of the equation is the production of biomass within the consumer trophic level. *S* is the consumption of food per unit biomass of *j*. *D* is the proportion contribution of B_i to the diet of the consumer *j*. If the prey trophic level contributes completely to the consumers, then *D* = 1.

Similar to the individual-based bioenergetics models, solving the two sides of the equation are relatively straightforward, using the known *P/B* ratio of the prey trophic level and the standing biomass of the consumers. This allows trophic ecologists to estimate *S*, which is a measure of the trophic flux of biomass or energy. Because all the trophic levels within the ecosystem can be modeled similarly, the primary production needed to sustain the ecosystem can be determined. Changes in primary production (e.g., through enrichment, climate change, biochemical processes) can predict changes in food web structure (Pauly and Christensen 1995).

As with the individual-based bioenergetics model, this modeling approach has been used primarily in fisheries where these sorts of data are readily available and trophic levels are fairly easy to categorize. Biomass lost to fishing mortality is incorporated into the model as a loss of biomass available to the consumer trophic level. Pauly et al. (1998) showed that these models explain the *top-down* impact of fishing in marine ecosystems. As the apex consumers in the ecosystem are removed by harvest, the biomass of the lower trophic levels will increase as the energy is no longer constrained by the inefficiencies of transfer to the highest trophic levels. For fisheries, this means smaller fish and more invertebrates (Figure 8.13). This approach is applicable to a variety of ecosystems and helps to form a quantitative framework for food web theory, such as the trophic cascade hypothesis (Chapter 10).

The ecosystem-level mass-balance approach has been incorporated into a computer simulation system called Ecopath (http://www.ecopath.org), which has largely been used in fisheries research. There are a couple of reasons this model focuses on questions in fisheries, especially marine systems. Abundant information on the demographics, vital rates, and abundance of fish exist in many fisheries, making it fairly simple to populate the models. Probably more important is that these models were developed by fisheries ecologists, particularly those working in marine ecosystems. The applicability to other ecosystems is high. An example of the use of Ecopath is in the applied evaluation of predictable anthropogenic food subsidies (PAFS) on ecosystems (Fondo et al. 2015). Humans discard waste as PAFS, and these materials can serve to subsidize ecosystems. In fisheries, bycatch (i.e., fish not targeted) are often returned to the ocean, most likely as dead organic matter. This

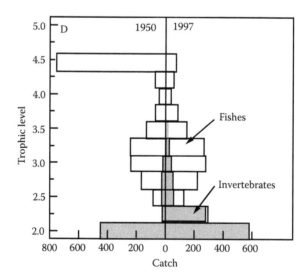

FIGURE 8.13 Change in the relative distribution of biomass in a marine food web before (1950) and after (1997) intensive fishing of the apex predators. Invertebrate biomass increases as less energy is transferred to higher trophic levels. This can be modeled using mass-balance bioenergetics techniques. (From Pauly, D. et al. 2000. *ICES Journal of Marine Science* 57 (3):697–706.)

PAFS can alter food web interactions and has been banned in some fisheries. The authors determined that PAFS were important to scavenger pathways, and banning them would negatively affect those trophic levels. This is but one of hundreds of recent examples of using Ecopath and related models to assess ecosystem responses to management and conservation.

8.9 METABOLIC THEORY

Up to now, we have focused on empirically based predictions that cater to specific individuals or ecosystems, especially fisheries. However, trophic ecology seeks generality. In our view, bridging the expanse between evolutionarily relevant processes and their ecosystem consequences across taxa is a marked challenge for trophic ecology in the future. As organisms adjust energetically to environmental change, their roles in ecosystems will change, affecting the transport of energy and materials and ecosystem function and structure. General patterns do exist, underlying ecosystems and culminating in the MTE that was formalized by Brown et al. (2004). In the following section, we review the basic tenants of metabolic theory and assess its utility for framing research in trophic ecology. This approach is markedly different from the mass-balance approaches described previously, which focus on pairwise energetic transactions between individuals or trophic levels. In short, metabolic theory is based on the entire ecosystem, whereby energy is transferred among organisms continuously, dispensing with mass-balance accounting.

Scaling metabolic patterns from individual bioenergetics to the population, community, and ecosystem forms the basis for the MTE (Brown et al. 2004), which has generated much debate (Clarke 2004). As we noted earlier, energy transformations within cells are nearly universal among life on Earth. Similarly, photosynthetic fixation of energy from photons follows the same energetic rules, only in reverse (Farquhar et al. 1980). Recall that patterns such as the predictable scaling of metabolism with body mass hold across taxa (Figure 8.5). The MTE takes these statistical relationships and attempts to form generalizations that unite trophic ecology.

Incorporating temperature into predictions for metabolism requires including a constant that accounts for the effect of temperature on kinetics in metabolism. The Boltzmann's factor is common in biochemistry where the rate of chemical reactions is equivalent to

$$e^{-E/kT^*}$$

where T^* is temperature in units of Kelvin. The coefficient k is Boltzmann's constant of 8.617×10^{-5} electron volts/Kelvin. Thus, multiplying the two values converts the denominator to the unit of electron volts (eV). The activation energy E, in eV, is the energy necessary to fuel cellular metabolism and is about 0.62 eV (Gillooly et al. 2001; Brown et al. 2004). Dividing the fraction in the exponent yields a unitless number that scales metabolic rate to temperature.

To develop the MTE, we must translate bioenergetics at the individual scale to that of populations. Populations grow in a logistic fashion, with the behavior of growth driven by two primary variables. The intrinsic rate of increase, r, is driven

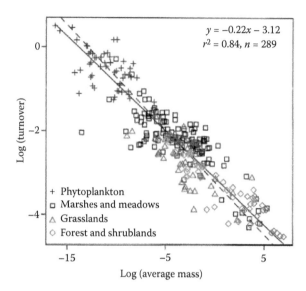

FIGURE 8.14 Relationship between the individual size of organisms in a community and their intrinsic rate of increase (*r*). The larger an organism, the slower the turnover of biomass, which means that biomass will be *stored* longer in the ecosystem. (From Brown, J. H. et al. 2004. *Ecology* 85:1771–1789.)

by the rates of birth and death in the population. The carrying capacity, *K*, is the maximum density of individuals that an environment can support. As the mass of individuals within a population increases, *r* declines (Figure 8.14). This relationship can be extended to biomass at the population scale:

$$K \propto (R)M^{-0.75}e^{E/kT^*}$$

The limiting resource (e.g., food), *R*, limits *K*. Because the density that a population can support declines with individual mass, *M*, the exponent is −0.75. The Boltzmann's function scales the relationship with temperature. This relationship predicts that *K* declines with increasing temperature, which generally holds true in the field (Allen et al. 2002).

The standing biomass (*W*) of a population should be related to the metabolic capacity of the individuals. Again, the mass of individuals, body temperature, and resource availability affect the prediction:

$$W \propto (R)M^{0.25}e^{E/kT^*}$$

This relationship predicts that increasing the temperature should reduce *W*, because the turnover of tissue will be high. Increasing individual body mass will increase standing biomass. Large individuals require longer times to turn over tissue, mature, and reproduce, allowing biomass to be *stored* and accumulate in the population. This relationship allows us to determine total production (P) of biomass as it is related to standing biomass,

$$P/W \propto M^{-0.25}e^{-E/kT^*}$$

Larger individuals within populations yield a lower rate of production relative to population size. Increased temperature increases production.

The metabolic predictions for populations translate to predictions for population-level consumption and trophic transfer. Ample evidence supports the supposition that energy transfer between prey and their consumers is only about 10% (Chapter 2; Lindeman 1942). Becoming large increases metabolic efficiency within a population, but the size advantage of being a large predator consuming a small prey item is offset by the large energy loss during trophic transfer. Higher trophic levels cannot have higher metabolic needs than their prey. Similarly, metabolic theory that we reviewed earlier predicts that the density within populations should decline disproportionately with a slope of −0.75 with increasing body size (Figure 8.15). However, when scaled to trophic levels, the expected decline in density should be closer to a slope of −1 because of the inefficiencies in energy transfer (Figure 8.15). Metabolic constraints of trophic transfer place a severe limit on the number of trophic levels that can occur (Chapter 2).

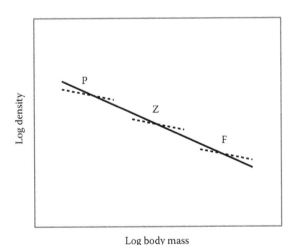

Log body mass

FIGURE 8.15 Prediction for abundance of organisms in a pelagic food web composed of phytoplankton (P), zooplankton (Z), and planktivorous fish (F). Within populations of each organism, the density of individuals is predicted to decline with individual body size with a slope of −0.75, because biomass is *stored* for longer periods in larger organisms (see Figure 8.14). Trophic transfer energy from one trophic level is inefficient, resulting in lower predicted density as a function of body size at the whole ecosystem scale, with a slope = −1. (From Brown, J. H., and J. F. Gillooly. 2003. *Proceedings of the National Academy of Sciences of the United States of America* 100 (4):1467–1468.)

The metabolic theory is therefore a way to incorporate metabolic processes at the cellular level to the patterns of energy transfer predicted by trophic pyramids. However, the accuracy and precision of this technique depend on several factors, which are currently open to debate. Algar et al. (2007) attempted to fit predictions of biodiversity of ectotherm populations along a latitudinal gradient based on metabolic theory predictions (Allen et al. 2002). The theory predicted that temperature, which is warmer at the equator, should promote biodiversity by reducing the carrying capacity and standing biomass of populations, allowing more species to *pack* within communities. However, the empirical data from Algar et al. (2007) deviated from theoretic predictions. There are many reasons why global patterns might not fit the theory. The body temperatures of ectotherms likely do not track the temperature of the environment—many species can regulate body temperature well. Allen et al. (2002) generated the activation energy of metabolism (E) statistically from data across many taxa. There is variation around these data, and this variation may result in deviations among populations, perhaps due to factors such as acclimatization and deviation from the 3/4 scaling law of metabolism with mass. Clearly, MTE is a strong starting point for looking for general patterns. And we will learn more about metabolic scaling by picking apart the details.

8.10 PROMISES AND PITFALLS

The individual- and ecosystem-based bioenergetics mass-balance approaches and MTE are based on sound physiologically based principles. They hold much promise for framing questions about ecology and evolution. Mass-balance bioenergetics models provide more information about trophic ecology than consumption and growth. They can be used to assess fluxes of contaminants and other materials (e.g., nutrients) between populations (He et al. 1993). Modeling energy allocation into various physiological components (i.e., fat, tissue, reproduction) within individuals as a function of environmental conditions, physical state, and biotic interactions is a useful way to ask relevant evolutionary questions about energetic trade-offs that can be scaled to the ecosystem (Munch and Conover 2002). Similarly, MTE provides predictions for the transfer of energy between trophic levels. Unlike the bioenergetics approach, it is not based on mass-balance expectations. Rather, the expectation is that energy transformations are limited by trophic inefficiencies that can be incorporated into statistical models. Energy flows from the base of the food chain up through the biomass accumulated within each trophic level and is then transferred to the next trophic level via consumption or into the decomposer web through mortality. These relationships can be incorporated into simulation models to assess how size-dependent shifts in biomass within populations affect patterns of energy transfer in ecosystems.

The parameters used in bioenergetics models and MTE are statistically derived, relying on relationships among parameters that reflect processes occurring in cells. Body mass, the central driving factor in both approaches, integrates a variety of physiological responses. Temperature is the other primary factor explaining a significant portion of variation in metabolism from the individual to the ecosystem. Although both of these fundamental models hold promise, they fit an average line through relationships among metabolism, mass, temperature, activation energy of metabolism, and other parameters, not accounting for the variation around these lines.

We have to remember that the majority of relationships used in both the mass-balance and MTE approaches are plotted using data that have been transformed to their logarithmic equivalents, greatly reducing the apparent variation. It is in this unexplained variation that the interesting ecological and evolutionary flexibility lies. Tackling bioenergetics at the individual level is challenging because individual metabolism and other components of the energy budget change as a function of factors beyond temperature and mass. Meal size, behavior, interspecific interactions, photoperiod, oxygen concentration, exercise, and many other environmental factors plus individual genetic traits contribute to variation within species. Similarly, in the MTE, variation among individuals contributes greatly to uncertainty in the theory's underlying statistical models. Brown et al. (2004) suggested that more than 50% of this variation can be explained by the stoichiometric needs of organisms, which we cover in Chapter 12 (Allgeier et al. 2015). In short, stoichiometry is the balance of elements within the organism that must be met to ensure proper physiological performance. If stoichiometric needs such as the balance of phosphorus and nitrogen are not met, an organism will fail to grow at rates predicted purely by the availability of consumable energy. Simulation models can be developed to explore how variation in environmental factors including limiting nutrients (i.e., stoichiometric constraints; Chapter 12), size-based interactions, and evolutionary trade-offs of individuals within key populations influence patterns of trophic transfer.

The MTE as it is currently set forth does not test energetic questions in the same way as the mass-balance bioenergetics approach. The mass-balance approach constrains predictive models by demanding that energy entering the organism equals the energy out. This constraint allows the models to be refined with more data as the two sides of the equation are balanced via iteration. In contrast, the equations underlying the MTE are open, meaning that the energy moving from one *box* to another, whether it is between body sizes within a population or between prey items and consumers across trophic levels, is only constrained by its expected loss or gain as per the MTE equations. With individual- and ecosystem-based bioenergetics, the measure of *success* is a properly balanced energy budget model, depending on no *a priori* metabolic expectations. These models will produce output even if the solution is completely wrong. For the MTE, success is measured if the data explain residual variation around the predictive lines generated from the accepted empirical relationships among mass, temperature, and metabolism as formalized by Brown et al. (2004).

The future of trophic ecology will proceed forward with ecologists continuing to adopt the mass-balance and MTE approaches, in apparent parallel. It is important to understand the benefits and limitations of both. Individual- and ecosystem-based mass-balance models are data-intensive. Because they rely on specific information about individuals or communities, they are not particularly portable. They can be used to test questions about trophic ecology, but the results are only generalizable to the population or ecosystem in question. When the models are inaccurate and imprecise, the sheer number of parameters makes it difficult to assess which underlying empirical relationships need refinement. The benefit of the mass-balance technique is that it generates real values (e.g., consumption or productivity) that make biological sense and can be directly applied to management or conservation questions. Because the equations must balance due to the laws of thermodynamics, the error has to be explained by internal error in the model. The MTE is based on a general underlying principle of ecology that meets the primary litmus test of science—that being falsifiability. If the pattern of trophic transfer of energy does not meet the expectation of the MTE, then research is needed to better understand the ecological and evolutionary reasons for why the data do not fit the model. Presumably, searching for incongruities between the data and the model predictions through experimentation will lead to advancements in trophic ecology. Of course, as with any theory in science, if the empirical data and theoretical predictions cannot be reconciled, the MTE will fail.

QUESTIONS AND ASSIGNMENTS

1. How do the first and second laws of thermodynamics apply to metabolic respiration?
2. Are any organisms truly ectothermic?
3. What does the term *bioenergetics* mean? Who defined the term for the first time?
4. Which energy source yields the most energy per unit mass? Which yields the least?
5. How are sugar, fat, and protein metabolized differently?
6. Why is carbon dioxide simpler to quantify in terrestrial than aquatic organisms?
7. How would you design a study to quantify field metabolism in a terrestrial organism?
8. What is a respiratory quotient and how would it potentially affect respirometry estimates used in bioenergetics models?
9. What is specific dynamic action and how does this affect energy budgets?
10. If energetic models are based on the responses of non-acclimatized organisms, will this affect model results?
11. Define the Wisconsin bioenergetics model. Explain how the p-value matters.
12. What is an activity multiplier in the Wisconsin bioenergetics model? Does it matter to have a robust estimate of this value?
13. Explain how whole-ecosystem metabolism is quantified. Does this differ between terrestrial and aquatic ecosystems?
14. What is P/R? Should this differ among ecosystems?
15. How does the Ecopath mass-balance accounting model for ecosystems work? Provide an example of how the two sides of the equation need to balance.
16. What does the metabolic theory of ecology predict for metabolic needs of organisms as they become larger?
17. How does temperature affect carrying capacity, standing biomass, and production of populations within ecosystems relative to temperature?
18. What are the trade-offs between mass-balance approaches such as Ecopath versus the open-ended statistically based approach of metabolic theory?

9 Consumption and Nutrition

9.1 APPROACH

Both macroscale (gross consumption) and microscale (nutrient intake) patterns of food intake are important to understand in trophic ecology. The approaches described herein complement those described in other chapters, including mass-balance bioenergetics (Chapter 8) and secondary production (Chapter 11). We will discover that not only total consumption of food but also the nutritional content are important.

9.2 INTRODUCTION

Trophic ecology seeks to quantify the amount and quality of materials fluxing from prey to consumer. The previous chapter focused on the flux of energy between trophic levels. Depending on the questions being asked, the gross intake of food and its nutritional content may be important to know. Intake of food by consumers is often limited by morphological constraints and prey composition. At the heart of the bioenergetics approach (Chapter 8) lie accurate and precise models of maximum consumption. Collecting robust data and developing functioning consumption models are critical. Energy transfer is not the only factor limiting trophic interactions. A candy bar contains a great deal of energy, but it is certainly not sufficient in a human diet to sustain life. Similarly, raw energy within prey is not enough for a predator, leading to the familiar topic of limiting nutrients in ecosystems. This has led to stoichiometric ecological concepts that will be covered in Chapter 12.

We begin this chapter by exploring some of the ways that organisms consume food and process it, because patterns of consumption, evacuation, and egestion affect trophic interactions. We then provide a template for generating models of consumption for use in predicting consumer impact in ecosystems. There are surprising commonalities among organisms in their nutritional needs. We review some of the elements and molecules necessary to sustain life, placing their importance and availability in the context of trophic interactions. A useful approach for assessing whether key nutrients are limited will be explored.

9.3 CONSUMPTION MECHANISMS

Guts vary and have been present in some form since consumption began as a strategy. Many single-celled organisms have organelles such as specialized lysosomes for engulfing and absorbing food (e.g., phagocytosis). The first multicellular digestive systems probably arose nearly the same time that defined tissue layers appeared. In fact, the first endoderm-derived guts predate the rise of the mesoderm (i.e., the third tissue layer) by about 40 million years (Stainier 2005). A gut can be a sac with one opening, as in corals and anemones, or a tube with a mouth and anus that is quite common among organisms including many simple worms, arthropods, echinoderms, and vertebrates. Other models do exist. Planarians ingest food into a simple gut that branches throughout the body. Egestion occurs via specialized ciliated flame cells. Other organisms such as some helminth parasites absorb nutrients directly from the environment.

Regardless of the organism, the process of ingesting a prey item and determining the time until the consumer will eat again is limited by the capacity of the digestive system (Figure 9.1). Food typically enters some opening called a *mouth*. How food arrives at this juncture is addressed in Chapter 5. For filter feeders and grazers, the arrival rate may be nearly continuous. In ruminants, the food is entering not only from the outside but also back from the gut for further processing. Carnivores may use their mouths more rarely. Beyond the mouth, there may be a first processing point called a *buccal cavity* or *vestibule*. The dimensions and structure of this anatomical location are the first bottleneck to food intake. Food items are held in the buccal cavity whole and undergo some level of initial breakdown. In some consumers, little modification to the food occurs as it waits to be transferred further downstream. More often, the food is crushed and shredded. The ability to chew evolved in many mammals, further aiding initial processing. Assessment of the quality of the food also occurs in the buccal cavity, whereby the organism may taste the food and evaluate its texture. If the food is accepted, the bolus is washed with enzymes and lubricants and sent down the pharynx and esophagus to the gut. Swallowing is

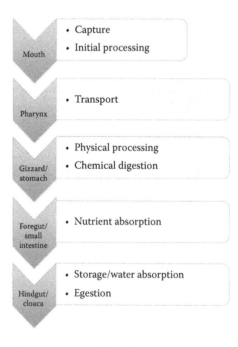

FIGURE 9.1   Process of consuming and digesting a meal.

TABLE 9.1

Common Enzymes Involved in the Denaturation of Nutrients in the Guts of Organisms

Group	Name	Substrate	End Product
Sugars	Amylase	Starch	Dextrins (sugars)
	Lactase	Lactose	Glucose
	Sucrase	Sucrose	Glucose
Fats	Lipase	Triacylglycerides	Fatty acids
Proteins	Pepsin	Protein	Polypeptides
	Trypsin	Protein	Peptides and amino acids
	Aminopeptidases	Peptides	Amino acids
	Nucleases	Nucleic acid	Nucleotides

typically aided by peristaltic contractions of tissue and perhaps accessory organs such as tentacles in cnidarians and tongues in vertebrates. Some organisms like birds use gravity to aid the movement of food into the gut.

When food enters the gut, most organisms no longer have the choice of rejecting it. Further physical processing may occur in a muscular gizzard-like organ (Figure 9.1). The food may enter a stomach-like organ, which has some mechanical action but is mostly a chemical digestor, which is typically acidic. The material in the stomach is converted to a liquid chyme, which is then released into the lower gut where the contents are neutralized and several enzymes are mobilized (Table 9.1). Most nutrient absorption occurs at this point. In vertebrates with stomachs, the process of transferring the chyme to the hindgut (e.g., an intestine) is called *gastric evacuation* (Jobling 1994). This is a different process than transit time, which is the time from ingestion to egestion (Worthington 1989).

For a consumption model to work effectively, evacuation or transit time needs to be coupled with feeding activity. Some grazers and filter feeders feed semicontinuously, with the amount of food entering the gut being limited only by available space. Most organisms, however, need to experience some physiological cue that we perceive as hunger to feed. Thus, assuming that feeding is continuous and limited only by space in the gut as it opens up may not apply to many organisms. Hunger is driven by many complex factors including increased production of the hormone ghrelin in the gut (Chen and Tsai 2012) and a drop in leptin, which is produced by adipose (i.e., fat) cells. A drop in the concentrations of sugars, amino acids, and fat in the gut may trigger hunger (Sepple and Read 1989). Because sugars are typically absorbed more rapidly than amino acids and fats, diets composed primarily of carbohydrates may evacuate faster, rapidly triggering hunger.

9.4 MODELING CONSUMPTION

As we pointed out in Chapter 8, quantifying consumption through time is difficult because of the complexity of the consumption, digestion, and egestion process.

Developing a model of consumption requires an understanding of the motivations for foraging. Models are typically statistically based, where experiments are conducted and relationships are used to develop parameters. Lab experiments provide the most precise and accurate information about the process. However, as with all experiments, the realism depends on the variables being controlled such as the food used (e.g., sugar, fat, and protein content), temperatures, light levels, and consumer condition. A typical experiment involves feeding an organism an *ad libitum* diet, meaning that food availability exceeds that which can be consumed at one time. After feeding occurs, the remaining food is removed. Uneaten food may deteriorate quickly or be consumed by other organisms. Thus, it needs to be removed immediately. If *evacuation* from a digestive organ (i.e., stomach) is the goal, then diets of replicate organisms are collected through time using standard methods (Chapter 7). Evacuation is the average rate that guts empty. If transit time is the goal, then the feces produced needs to be collected. Methods vary for collecting egesta. The critical issue to remember is that feces will decompose quickly, particularly in aquatic environments. Collection and processing should be immediate.

Quantifying consumption in the field is complicated and fraught with conflicting variables. The best method is to observe a point of feeding and collect diets at known times after this occurs. Many organisms consume food during a specific time of day (e.g., crepuscular, nocturnal). Diet contents of multiple consumers are sampled through time after the feeding period. The evacuation or transit time then is assumed to be the period spanning from when guts are full to when only empty guts are collected. To quantify evacuation of a fish called gizzard shad in lakes, Dettmers and Stein (1992) collected fish through time after dark, when the fish stopped feeding. From this, they were able to determine how quickly guts emptied.

The resulting patterns of evacuation or transit vary in shape. The shapes of these curves affect how models of consumption need to be developed. The pattern of gut evacuation or transit may be a negative exponential (light, solid line; Figure 9.2a). This may occur because certain diet items digest more rapidly than others. If the rate of evacuation or transit is the same for all items, then the pattern will be linear (dark, solid line; Figure 9.2a). If a meal is held in specialized organs for processing (i.e., a gizzard or stomach), then there may be a delay in evacuation, followed by a linear decline in chyme (dashed line; Figure 9.2a). Resulting patterns of consumption are assumed by most models to be continuous during the feeding period (dark line; Figure 9.2b). In fact, this line is likely the average of a more complex pattern as consumption (gastric intake) begins at a level where the gut becomes partially empty, feeding occurs, and the gut is filled (light line; Figure 9.2b). Each day, this cycle of feeding and partial evacuation occurs until the feeding period ceases (Figure 9.2b).

The simplest model of daily consumption assumes that evacuation is linear through time:

$$C = S(24/n)$$

Consumption (C in g/d) depends on the average mass of food in guts (S). Feeding is assumed to occur all day (24 h) and evacuation or transit time is n (in h). If

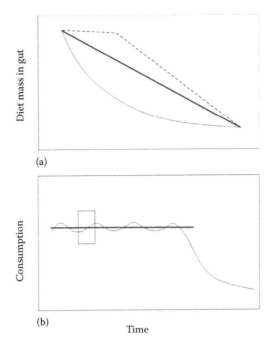

(a)

(b)

Time

FIGURE 9.2 Process of gut evacuation or transit of a consumed meal (a) and the resulting pattern of total consumption by an organism through a day (b). (a) Reduction of a meal in the gut may be exponential (light line), linear (solid, dark line), or delayed (dashed line). The resulting pattern in consumption will vary through time (light line, b). Models of consumption *smooth* this variation by assuming that guts fill up with more food as soon as they empty (dark line, b). When an organism stops feeding during the day, then consumption stops (b).

consumption only occurs for a limited time during the day, then the hours in the model can be changed to reflect the daily feeding period. An earthworm consumes organic matter (i.e., dirt) continuously. The transit time for the meal is $n = 2$ h. The average mass of the soil in the gut is 0.5 g. Thus, daily consumption per earthworm is estimated as 6 g. If there are 20 earthworms/m², then the earthworms are processing 120 g of soil per m² each day.

Unlike the model described above, the most versatile and widely used model of total consumption (C) over a period of time (t) directly incorporates an explicit non-linear function for evacuation or transit time (Elliott and Persson 1978):

$$C_t = \frac{(S_t - S_0 e^{-kt})kt}{1 - e^{-kt}}$$

Consumption is a function of the average mass, S, in diets at time t, S_0 is the mass of the diets at the beginning of consumption, and k is the instantaneous rate of gastric emptying. The rate k is derived experimentally by measuring evacuation or transit;

k is the slope of the relationship. If an organism is fed with a single meal and gut contents quantified through time, then

$$k = (\log_e M_{end} - \log_e M_0)/t$$

where M is the average meal size at the start and end of the experiment. t is the time elapsed during the experiment. Because k is the slope of a declining exponential curve, it will be a negative number. The absolute value of k is used in the consumption model. The major assumption of this model is that feeding is constant.

Depending on the rate of gastric emptying, k, the size of the initial meal and the average mass in diets during the interval t may vary in unique ways. During a period of $t = 1$ h, a snail initially consumes $S_0 = 35$ mg of algae. As it continues foraging for the hour, the average gut mass grows to $S_t = 125$ mg, meaning that mass has accumulated in the gut. Instantaneous evacuation rate estimated from experiments is $k = 0.1$. Using the equation above, $C_{1h} = 98$ mg of algae. If grazing is continuous during the day, then the snail is consuming $C_{24} = 98 \times 24 = 2{,}353$ mg of algae daily.

Many organisms do not feed continuously. Rather, a meal is consumed, approaching some maximum (C_{max}), which is determined experimentally. As C_{max} is approached, feeding rate declines. Consumption at the end of time t (C_t) is thus a function of C_{max}, the initial size of the meal S_0, and some constant b that slows feeding as C_{max} is approached (Elliott and Persson 1978):

$$C_t = C_{max} - ae^{-bt}$$

where $a = C_{max} - S_0$, and the exponent b is determined iteratively from the equation used to estimate the food present (S_t) in the gut after time t:

$$S_t = S_0 e^{-kt} + \frac{abe^{-kt}}{b-k}(1 - e^{(k-b)t})$$

Empirical values of S_t, S_0, and k need to be collected. As with all iterative approaches, different values of b are plugged into the right-hand side of the equation until the two sides balance and an estimate of b is reached.

An individual skink can consume a maximum of 500 mg of food daily. At the start of feeding, it consumes 20 mg. At the end of the feeding period $t = 3$ h later, its stomach contains $S_3 = 285$ mg of food. If the instantaneous evacuation rate $k = 0.2$, then b can be estimated using iteration to equal 5. From this, the consumption during the 3 h is $C_3 = 500$ mg. Thus, the skink is consuming its maximum daily ration of 500 mg, according to the model. Conversely, if less food is found in the stomach after 3 h, $S_t = 103$ mg, then $b = 0.1$ and the total consumption is less than the daily maximum at $C_3 = 144$ mg. Placing a daily cap on consumption greatly reduces the potential consumptive impact of predators on prey resources but is likely more realistic for many organisms given anatomical and behavioral constraints.

Any model of consumption assumes that all consumers are eating prey. However, empty guts are a common occurrence in populations, because organisms simply do not eat all the time. Internal state such as satiation, energy reserves, reproductive readiness, and fear may trump feeding. Consumption models that sum total consumption across all individuals in the population will overestimate consumer impact on prey. Population level models need to incorporate the probability of nonfeeding individuals. Dealing with *zero-inflated data* statistically involves developing appropriate statistical models (Martin et al. 2005), which are becoming commonly available in statistical packages like R. Zhang et al. (2011) developed a multivariate tool to account for absence data in human nutritional studies. Begona Santos et al. (2013) applied zero-inflated additive models to assess the diets of dolphins off the coast of Spain. These animals should be expected to consume high-energy prey frequently. The statistical approach was able to determine whether the abundance of these prey was due to selection or simply because many dolphins had empty stomachs.

9.5 NUTRITION

9.5.1 WATER AND MINERALS

We have focused on the intake of bulk material or energy as the primary factor influencing consumption in organisms. However, there are a host of specific materials that must be consumed either in large or minute quantities to maintain life and reproduction (Table 9.2). Nutrients that are required in large concentrations are classified as *macronutrients*. A functioning organism (i.e., one that is not in a dormant state) needs the simple molecule, water. Water hydrates cells, provides structure (i.e., turgor), transports materials internally and externally, provides cooling, and plays a key role in metabolism. Most animals and plants are composed of greater than 60% water. In terrestrial systems, water is continually lost and must be replenished, usually through consumption, although some organisms (e.g., kangaroo rats) can meet most of their water needs by generating metabolic water. Aquatic organisms that regulate their internal environment are either continually gaining water in freshwater or losing water in marine systems. Water is undeniably the dominant macronutrient for life, and intake/regulation of this substance may be as important to model as the consumption of food for understanding population dynamics and ecosystem interactions. Both aquatic and terrestrial organisms capture many of their essential (i.e., those that cannot be synthesized) nutrients such as calcium, magnesium, and chloride from dissolved minerals in water. A few organisms can survive for long periods in the complete absence of water. Tardigrades, microscopic aquatic organisms, can survive complete desiccation by replacing the water in cells with the sugar trehalose (Figure 9.3). These organisms made history by surviving for short periods exposed to the vacuum of space (Jonsson et al. 2008).

Thanks to autotrophs, which contain high concentrations of carbohydrates, the macronutrient carbon is readily available to most consumers and is often not limited. In animals, carbon may comprise greater than 20% of biomass by wet mass, but is largely found in fat molecules rather than sugars. Obtaining carbon as a food resource may be a problem in areas where carbon dioxide and other carbon sources are limited, such as in soil in grasslands or the sediments of aquatic ecosystems. Heterotrophic bacteria and fungi benefit from carbon inputs from sources such as plant roots and decaying

TABLE 9.2
Common Nutrients Found in Plants and Animals

Class	Nutrient	Plants	Animals	Use
Liquid/gel	Water	X	X	Structure, solute, metabolism
	Trehalose		UNK	Dormancy
Energy	Carbohydrate	X	UNK	Energy
	Linoleic acid		X	Fat precursor
	Linolenic acid		X	Fat precursor
Protein	Nitrogen	X		Amino acid precursor
	Arginine		X	Protein precursor
	Histidine		X	Protein precursor
	Isoleucine		X	Protein precursor
	Leucine		X	Protein precursor
	Lysine		X	Protein precursor
	Methionine		X	Protein precursor
	Phenylalanine		X	Protein precursor
	Proline		X	Protein precursor
	Threonine		X	Protein precursor
	Tryptophan		X	Protein precursor
	Valine		X	Protein precursor
Elements	Boron	X	X	Micronutrient
	Calcium	X	X	Macronutrient
	Cobalt	X	X	Micronutrient
	Copper	X	X	Micronutrient
	Chromium	X	X	Micronutrient
	Chlorine	X	X	Micronutrient
	Fluorine	X	X	Micronutrient
	Iron	X	X	Micronutrient
	Iodine	X	X	Micronutrient
	Magnesium	X	X	Macronutrient
	Molybdenum	X	X	Micronutrient
	Phosphorus	X	X	Macronutrient
	Potassium	X	X	Macronutrient
	Selenium	X	X	Micronutrient
	Silicon	X	X	Micronutrient
	Sodium	X	X	Macronutrient
	Zinc	X	X	Micronutrient
	Aluminum	X		Micronutrient
	Bromine	X		Micronutrient
	Cesium	X		Micronutrient
	Strontium	X		Micronutrient
Vitamins	Vitamin A	X	X	Pigments
	Vitamin C	X	X	Metabolism
	Vitamin D	X	X	Calcium regulation

(Continued)

TABLE 9.2 (CONTINUED)
Common Nutrients Found in Plants and Animals

Class	Nutrient	Plants	Animals	Use
	Vitamin E	X	X	Free radical removal
	Vitamin K	X	X	Blood clotting
	Vitamin B12	X	X	Enzyme synthesis
	Biotin	X	X	Enzyme synthesis
	Choline	X	X	Neurotransmitters
	Folacin	X	X	Metabolism
	Niacin	X	X	Enzyme synthesis
	Pantothenic acid	X	X	Enzyme synthesis
	Pyridoxine	X	X	Enzyme synthesis
	Riboflavin	X	X	Enzyme synthesis

Note: UNK = need is unknown for most organisms; X = essential.

FIGURE 9.3 Tardigrade, which can survive complete loss of water in its cells. (Courtesy of Shutterstock.)

organisms (Hooker and Stark 2012). The carbon allows these organisms to grow and capture other nutrients such as nitrogen that may be in forms that are not readily available to plants and multicellular consumers such as worms and insect larvae.

Many other macronutrients exist (Table 9.2). Nitrogen is a key component of organisms, usually found in amino acids that comprise the proteins that comprise everything from connective tissue (e.g., collagens) to the enzymes that regulate metabolism. Phosphorus is another important component of all living organisms, forming the molecular backbone of DNA, RNA, and ATP. Bony organisms store phosphorus in their bones, making this resource less likely to be limiting than for soft-bodied counterparts. In Chapter 12, we will show how the stoichiometric ratios of carbon, nitrogen, and phosphorus can be used to predict trophic interactions in ecosystems. Potassium, sulfur, sodium, chlorine, bicarbonate, and other ions are typically maintained at precise concentrations in the intracellular and extracellular fluid of organisms (Figure 9.4) and often depend on dietary intake.

Micronutrients also are essential for organisms (Table 9.2). However, excessive quantities of these elements may be toxic. The need for each of these nutrients depends on the physiology of the organism. Copper is an important component of the blood of mollusks and many arthropods, providing oxygen transport. Similarly, iron carries oxygen in hemoglobin of vertebrates and many other organisms. Both of these elements are toxic in high concentrations, interfering with functions such as metabolism and protein synthesis. Molybdenum (Mo) is essential for the function of specific enzymes but, again, can be toxic in high concentrations. In western North America, Mo is concentrated in soils and can lead to toxicity of some organisms (Figure 9.5a). Iodine is an element essential to the synthesis of thyroxin. When

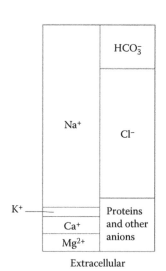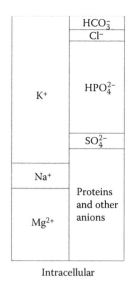

FIGURE 9.4 Cations and anions present outside the cells and within the cells of organisms that regulate their internal environment. Sizes of bars depict their relative concentrations. Nutrition is critical for maintaining these concentrations.

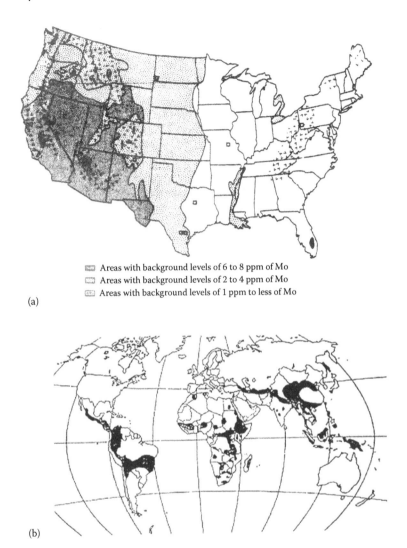

(a)

☐ Areas with background levels of 6 to 8 ppm of Mo
☐ Areas with background levels of 2 to 4 ppm of Mo
☐ Areas with background levels of 1 ppm to less of Mo

(b)

FIGURE 9.5 Distributions of the micronutrients (a) molybdenum (From Pond, W. G. et al. 1996, *Basic Animal Nutrition*, Wiley) and (b) iodine (Adapted from WHO, Iodine status worldwide: WHO global database on iodine deficiency. B. de Benoist, M. Andersson, I. Egli, B. Takkouche, and H. Allen (eds.). Department of Nutrition for Health and Development, World Health Organization, Geneva, 2004). (a) Dark areas are where molybdenum is present. (b) Dark areas are where iodine is absent.

limited, it causes the common condition goiter in vertebrates, where the thyroid organ becomes enlarged and cellular metabolism is impaired. The distribution of iodine in soils varies globally (Figure 9.5b) and can alter trophic interactions, including having negative impacts on human consumers. Interestingly, the influence of microelemental concentration in land and water on the organisms that inhabit them has not been a major focus of the zoogeography of trophic ecology.

9.5.2 MOLECULES

Even if all the required macro- and microelemental nutrients are available in the environment, a heterotroph will starve to death if the elements are not available in the proper molecular configuration. The most obvious group of essential molecules that cannot be synthesized by consumers is carbohydrates. The ability to generate other molecules varies among organisms. We will explore a few of the molecules critical to consumption, growth, and reproduction, including lipids, amino acids, and vitamins.

Lipids are common in plants and animals, serving as both structural and energy-storage molecules. A fat is a type of lipid composed of *fatty acids* attached to a molecular backbone. Fatty acids are responsible for most of the structural characteristics of a fat molecule. The basic structure of a fatty acid is a carboxyl group (COOH) on one end with a chain of carbon and hydrogen atoms attached, called an aliphatic tail (Figure 9.6). The chains of fatty acids vary in length and in the number of double bonds that occur between carbon atoms. If double bonds are present, then the fatty acid is unsaturated. If no double carbon–carbon bonds are present (i.e., the maximum number of hydrogen atoms is attached to the carbon atoms), the fatty acid is saturated (Figure 9.6a). The *C:D n-x* nomenclature for fatty acids is the most common. The number of carbon atoms in the molecule is denoted by *C* and the number of carbon–carbon double bonds by *D*. The location of the carbon atom of the first double carbon–carbon bond (if present) from the terminal methyl end (*n*) is depicted by *x*. Many fatty acids also have common names that are unrelated to their structure.

Many kinds of fatty acids are possible. Stearic acid is a long-chained, saturated fatty acid with 18 carbon atoms and no double bonds (Figure 9.6b). The nomenclature

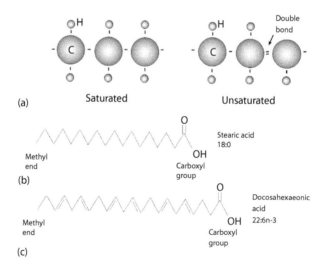

FIGURE 9.6 Chemical configuration of fatty acids in lipids. (a) The basic configuration of carbon and hydrogen atoms in a saturated and unsaturated fatty acid. (b) Stearic acid is a common saturated fatty acid. (c) Docosahexaeonic acid is a nutritionally critical unsaturated fatty acid.

for this fatty acid is 18:0, with no *n-x* included because of the absence of a double bond. Docosahexaenoic acid (DHA) is unsaturated with 22 carbon atoms and 6 double bonds, beginning at the third carbon atom from the methyl end (Figure 9.6c). The official nomenclature is 22:6n-3. Saturated fatty acids have higher melting points than unsaturated fatty acids, and thus tend to *pack* in a solid form at normal temperatures. The position of the double bonds in unsaturated fatty acids affects properties such as their shape and their melting points.

The fatty acid composition of organisms varies quite widely as a function of physiological needs (e.g., energy storage, vitamin uptake, and cell membrane fluidity) and diet. In Chapter 14, we will explore how the composition of fatty acids in organisms reflects those of their prey, making these molecules useful trophic markers. Most fatty acids can be synthesized by organisms from sugars and other fatty acids. Two essential fatty acids that are found in plants and some fish that cannot be synthesized by most other organisms are linoleic acid (18:2n-6) and linolenic acid (18:3n-6 or 18:3n-3). These fatty acids are important in cell membrane function and are precursors for other important fatty acids such as eicosapentaenoic acid (EPA; 20:5n-3) and DHA (Figure 9.6) that can be synthesized at energetic cost to the organism. Their limitation in the environment can hinder growth and reproductive success of organisms (Arts et al. 2009).

Lipids are broadly classified in three categories. *Triacylglycerols* (TAG) are typically used for energy storage. The fatty acids are attached to a glycerol base (Figure 9.7a), like three fingers on a hand. These fatty acids are nonpolar, meaning that there is no difference in electrical charge from one end to the other. Because of their lack of polarity, TAGs are hydrophobic, not dissolving in water. All parts of an organism may contain TAGs. Adipose tissue is specialized in synthesizing and storing TAGs. *Phospholipids* vary in structure, having a phosphorus-based functional

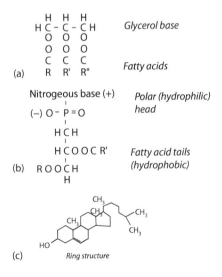

FIGURE 9.7 Molecular structure of (a) triacylglyceride, (b) an amino acid, and (c) cholesterol. R depicts the location of functional groups (e.g., fatty acids).

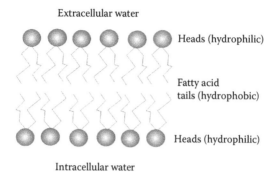

Extracellular water

Heads (hydrophilic)

Fatty acid
tails (hydrophobic)

Heads (hydrophilic)

Intracellular water

FIGURE 9.8 Configuration of a cell membrane composed of phospholipids.

end with a positive and negative pole (Figure 9.7b). Because of this electrical charge, this functional part of the molecule will be attracted to water (i.e., hydrophilic). However, the fatty acid tails of the phospholipids are nonpolar and hydrophobic. These properties make these molecules ideal components of cell membranes, allowing the polar ends to orient toward the exterior and interior of cells and the fatty acids to be sandwiched in between (Figure 9.8). Sterols are the third trophically important lipid molecular group, of which cholesterol is the most common (Figure 9.7c). Rather than forming a chain, these molecules are in the form of simple, flat rings of carbon and hydrogen. They are typically synthesized from acetyl coA or through the dietary uptake of low-density lipoproteins. Cholesterols are important in cell-membrane function, and these and other sterols form the basis for hormone production.

9.5.3 AMINO ACIDS AND RNA

Proteins are composed of amino acids. More than 200 naturally occurring amino acids exist, although only about 20 are common in proteins (Pond et al. 1996). The common structure of an amino acid is a carboxyl group (COOH) with an amino group (NH_2) on the adjacent carbon atom (Figure 9.9). The remainder of each molecule provides its unique structural characteristics. Tissue synthesis is costly or impossible for about 10 amino acids in many multicellular organisms, meaning that they must be obtained from the diet (Figure 9.9). These essential amino acids are

Essential	Non-essential
Arginine	Alanine
Histidine	Aspartic acid
Isoleucine	Citrulline
Leucine	Cystine
Lysine	Glutamic acid
Methionine	Glycine
Phenylalanine	Hydroxyproline
Threonine	Proline
Tryptophan	Serine
Valine	Tyrosine

FIGURE 9.9 Basic molecular structure of an amino acid with a R functional group attached. The ten essential and non-essential amino acids are listed.

produced by some plants, algae, and bacteria. The production of these amino acids is important in all ecosystems, including remote ones like hydrothermal vents (Amend and Shock 1998). Amino acids such as arginine can be limited in some ecosystems, affecting growth and reproduction (Koch et al. 2012). Amino acids may be used as specific tracers in food webs, similar to fatty acids.

Synthesis of amino acids and production of proteins is a primary function of life. Ribonucleic acid (RNA) is coded from deoxyribonucleic acid (DNA) in the ribosomes of cells. The three types of RNA used in protein synthesis are ribosomal RNA, transfer RNA, and messenger RNA. Ribosomal RNA combines with proteins to make ribosomes. Messenger DNA codes (via codons) for specific amino acids; transfer RNA brings the amino acids building blocks to the ribosome where they are combined to form polypeptide chains. The chains fold into the appropriate proteins that have specific binding or structural properties.

Because ribosomes and RNA are intimately linked to protein synthesis and thereby growth and maintenance, quantifying the RNA concentration in consumers is a way to assess real-time growth rates. The content of DNA in cells remains constant, whereas RNA synthesis increases when growth and reproduction are occurring. Thus, the ratio of RNA:DNA correlates positively with growth (Rhee 1978). In the Pacific Ocean intertidal of the United States, RNA:DNA ratios in mussels were strongly positively correlated with phytoplankton concentrations, showing the trophic importance of this food resource on suspension feeders. The technique has been used widely in the marine and microbial disciplines and has applicability to all organisms, particularly those where it is difficult to quantify growth of individuals by collecting multiple samples through time.

9.5.4 VITAMINS

Vitamins are a unique group of molecules that are essential for biological function and that cannot be readily synthesized. They are typically categorized as either fat or water-soluble. Fat-soluble vitamins are capable of accumulating in tissues, and some may become toxic if overconsumed. Vitamin A is produced by carotenoid pigments found in plants. Carotenoids are converted to vitamin A by the consumer and serve many purposes such as vision, respiration, and anti-mutagenic function. In vertebrates, greater than 90% of vitamin A is stored in the liver. Vitamin D is a group of sterols that are converted to their biologically active form from exposure to sunlight. The sterols can be found in plants and in skin cells of some organisms, including reptiles and humans. This vitamin is critical for regulating blood calcium concentrations, and a deficiency leads to problems such as bone deformations. Vitamin E is a tocopherol (i.e., containing methylated phenol[C_6H_5OH] groups) that is critical as an antioxidant in cells of organisms. Free radicals are molecules with atoms missing electrons that react with other molecules by oxidizing them. Vitamin E scavenges these damaging molecules, rendering them harmless. This vitamin likely plays other roles in metabolic processes. The availability of all of these fat-soluble molecules in the environment can affect populations and trophic interactions. Because they can be stored within individuals, the concentration within the organism as well as the environment needs to be quantified to assess patterns.

Water-soluble vitamins are *washed out* of organisms and need to be continuously replenished through diet. Ascorbic acid, known popularly as vitamin C, is not synthesized in all organisms including humans and some fish. It is important in many metabolic processes, and deficiencies (i.e., scurvy) will eventually lead to death. Choline is an important essential vitamin that is found widely in many plants and animal tissue. Structurally, this molecule is important in nerve function (i.e., precursor to acetylcholine) and in forming the polar functional end of phospholipids. Other water-soluble vitamins include biotin, vitamins B6 and B12, pantothenic acid, niacin, and riboflavin. All play roles in the production of enzymes for cellular metabolism. Thiamine is an important vitamin that has been identified as a major limiting factor in some ecosystems. The invasive fish species alewife in the North American Great Lakes produces thiaminase, which breaks down thiamine in tissues. Predatory fishes consuming this nonnative prey become deficient in thiamine, particularly during early development (Mills et al. 2003).

9.5.5 SYMBIONTS

It may seem curious that so many multicellular organisms, including many plants, are unable to synthesize all the molecules needed for housekeeping and growth using the basic building blocks in their environment. No organism is completely alone, and most appear to have developed relationships with commensal or mutualistic symbionts that provide essential nutrients. Ruminants have taken the process of symbiont-aided digestion to an extreme level. Non-protein-derived ammonia from low-quality plant material is synthesized by the gut flora (e.g., bacteria and protozoa) into essential amino acids. Similarly, essential fatty acids are synthesized from carbohydrates and other carbon-based molecules.

The loss of symbionts can lead to malnutrition or even death for the host. Bleaching of corals is a classic example, as zooxanthellae are lost during times of stress. Many other organisms can experience similar problematic losses. Stinkbugs exposed to high temperatures may lose important gut symbionts reducing their growth and reproduction (Prado et al. 2012). Similar to claims about warming events causing bleaching in corals, warming may negatively impact these and other species indirectly reducing nutrient availability from important commensals or mutualists.

The landscape of symbionts is largely hidden, tucked away in the organisms that depend on them. However, how they are obtained and maintained is an important factor influencing trophic ecology. Most organisms do not obtain their symbionts from the environment. Rather, vertical transmission from parent to offspring is the typical pathway (Cary 1994), although examples of repeated *recruitment* of symbionts from the environment do exist (Scheuring and Yu 2012). Factors that disrupt this relationship may have negative implications for populations.

9.5.6 PARENTAL TRANSFER

The survival of offspring is key to trophic ecology. Offspring allow consumers and prey to respond numerically to changes in the environment and ultimately drive the flux of energy and materials between trophic levels. Parental care via feeding is certainly one way that offspring may benefit. Mammals have evolved the unique mechanism of lactation. Many insects, honeybees in particular, feed larvae as they develop. Another contribution by parents does not rely on direct parental interaction. Rather, vertical transmission of nutrients to offspring in embryos can be important in organisms that reproduce both externally and internally. For organisms that reproduce clonally, the nutritional condition of the parent will likely mimic that of its clones. The nutritional composition of eggs may influence the success of offspring via direct effects such as survival as well as influencing sex ratio of young (Nager et al. 1999). In walleye, the nutritional condition of fish prey for mature females affects the production of walleye offspring in the future (Madenjian et al. 1996). For animals that brood or rear their young internally, adult nutritional condition dramatically influences the growth and survival of offspring, with effects that can last through life.

9.6 ANTINUTRITIONAL COMPOUNDS AND FOODBORNE DISEASE

Not all food is safe to eat. Plants and animals have evolved specific chemical defenses against predation (see Chapter 6). Conversely, some plants and animals contain compounds that have characteristics that reduce the nutritional quality of foods by inhibiting uptake. Other threats that come from eating are diseases and parasites that hitchhike with food items. These problematic *mines* in the dietary landscape cannot always be avoided and affect population dynamics and ultimately trophic interactions.

An antinutrient is a compound that blocks the uptake or metabolism of certain essential nutrients when consumed. Thiaminase is one such compound that we introduced earlier. More broadly, antinutritional compounds are categorized by the nutritionally important molecule or element they block. Protease inhibitors such as those found in some legumes can prevent the digestion of specific proteins. Lipase inhibitors prevent the catabolism of fats and occur naturally in many plants. These compounds are being isolated by pharmaceutical companies from some plants to reduce fat uptake in humans to control obesity. Phytic acid, found in many nuts and seeds, inhibits the uptake of minerals including calcium, magnesium, and zinc. Oxalic aid is common in plants such as spinach. It forms salts with calcium, blocking its uptake. The importance of these materials is well known in the nutritional literature but not well understood in trophic ecology.

Parasites and diseases often use food as a vector to infect their hosts. Parasites enter tissues that are consumed. Tapeworms have larvae that encyst in muscle tissue of their herbivore hosts. Carnivores (including humans) consume the muscle and serve as a secondary host, where the adult tapeworms insert in the intestine and reproduce. Disease pathogens such as bacteria, viruses, and fungi also spread via food. An ecologically significant disease that is prevalent in some populations of deer and other herbivores is caused by prions, which are misfolded proteins that are quite stable in the environment. The source of these agents is still a matter of debate (Aguzzi and Polymenidou 2004). However, it is accepted that prions are transferred among members of a species and more likely to accumulate in dense populations.

An animal infected with prions dies in the environment, and the prions accumulate in the soil and vegetation. Consumers are infected by directly consuming infected soil on food (e.g., a blade of grass) or eating infected tissue (e.g., cannibalism). The prions are replicated and cause irreversible damage to the nervous system resulting in diseases such as scrapie or chronic wasting disease.

9.7 CONCLUSIONS

Identifying whether essential nutrients or energy intake is limiting populations and trophic interactions is a critical question that will dominate much of the remainder of this book. Liebig's law of the minimum predicts that one essential nutrient will limit populations and ecosystem processes. This is similar to the theory of constraints in business management, in which a small number of factors limit business growth (Naor et al. 2012). Nutrients can be colimiting, meaning that more than one material or their interaction is negatively affecting growth, reproduction, and perhaps survival (Koerselman and Meuleman 1996). Gross energy intake also constrains population development and can be a major issue. Teasing apart the relative contribution of limiting nutrients and energy intake requires experiments where the limiting nutrient and food availability/energy density are varied. How organisms respond to these variables will determine the relative importance of these trophic components.

The availability and limitations of nutrients and energy availability may depend on ontogeny (Showalter et al. 2016). During early life when tissue accumulation is rapid and organs are developing, the need for essential nutrients may be particularly acute. Some of these materials may be provided from parents either directly or via vertical transmission (Figure 9.10). Young organisms typically have limited access to food resources and little room for energy stores. For plants, seedlings have small root systems and may have difficulty reaching critical nutrients in the soil. Small animals often have small mouths and limited ability to capture or find prey. Macronutrients and energy sources may be limited as well. For adults, large body size and greater capacity to search the environment may ensure that essential micronutrients and

Early life

Source	Parents/environment
Micronutrients	Limited
Macronutrients	Limited
Fate of nutrients	Growth

Adulthood

Source	Environment/reserves
Micronutrients	Usually adequate
Macronutrients	Limited
Fate of nutrients	Reproduction/maintenance

FIGURE 9.10 Ontogenetic differences in nutrient sources and requirements.

macronutrients are obtained and stored. Adults need nutrients and energy for tissue maintenance and reproduction. Major limitations are likely the availability of energy and raw materials for protein synthesis (Figure 9.10).

Given the broad nutritional and energetic needs of plants and animals, it is surprising that organisms meet their dietary needs, especially during early development. Natural selection provides some assurances that plants and animals are successful in a trophic context. Parental care and vertical transmission of nutrients help offspring get the nutrients that they need. Adult parasitoids ensure that their larvae are successful by placing them in nutrient- and energy-rich hosts. Senses such as taste, smell, and vision play a critical role in allowing animals the ability to discriminate foods containing key nutritional characteristics. Energy reserves have evolved in organisms to provide insurance against times of scarcity. Fat also provides the ability to store fat-soluble vitamins; adipose stores likely evolved to reduce the need to find a consistent nutrient source. If critical nutrients become unavailable, some organisms have evolved the extreme tactic of entering a dormant state to the extreme of completely becoming dehydrated such as tardigrades. This chapter is a stepping point for understanding how larger-scale patterns at the level of the community or ecosystem are driven by the interaction between the life histories of component individuals and the distribution of food in both time and space in the environment.

QUESTIONS AND ASSIGNMENTS

1. Are guts relatively new or old in evolutionary history?
2. Describe a generalized gut. Why is it important to understand how guts are structured for trophic ecology?
3. What is the definition of gut evacuation?
4. Under what circumstances might a linear versus nonlinear model of consumption be used?
5. Why are empty guts a problem for trophic ecology?
6. Explain why water is the *master nutrient* for life on Earth.
7. What is the difference between a macronutrient and micronutrient? When are nutrients essential?
8. How are fatty acids classified?
9. What does the RNA:DNA ratio in organisms potentially tell us about their status in trophic ecology?
10. What is a vitamin?
11. Describe antinutritional compounds and why they are important.

Section IV

Community and Ecosystem Concepts

10 Food Webs

10.1 APPROACH

The food-web concept is an important subset of trophic ecology because food webs serve as the mechanisms by which trophic interactions are transmitted among organisms and the environment. Food webs are networks within ecosystems that have quantifiable characteristics such as topologies, connectedness, linkage density, interaction strength, and size (e.g., food chain length). Many books have been written by biologists and mathematics on characterizing, categorizing, and predicting food webs (Pimm 1982; Belgrano et al. 2005; Pascual and Dunne 2006; Moore and de Ruiter 2012). Online databases of food webs exist, most notably http://www .foodwebs.org, to construct and study them. This chapter is intended to summarize the basics for navigating food-web ecology and explore some basic generalizations where they exist. Ecological network analysis is a tool that has arisen. We will find that in the past several decades, there has been a strong push, with mixed success, to determine whether properties emerge from food webs based on their inherent topologies.

10.2 INTRODUCTION

Hungry organisms are everywhere. In Chapter 2, we described trophic pyramids as simplifications of real ecosystems, where the number of species and trophic connections in even the simplest ecosystems is quite high. Lumping species into similar roles within ecosystems makes it possible to describe the flow of materials and energy but obviously fails to capture underlying mechanisms. *Food webs* are an alternative approach to pyramids where the connections between species are mapped as networks and likely the core framework upon which all the chapters of this book are hung. The idea that interactions among species create a tangled web originated from ecologists like Charles Darwin (1859), Stephen Forbes (1887), and Charles Elton (1927). And many authors argued that complex, speciose food webs should be more stable than simple ones (Hutchinson 1959). Food webs are terribly complicated, leading to much enduring debate in ecology. Consider an ecosystem containing only ten species. In the unlikely case that all ten eat each other, then there are 45 links among them and potentially 90 bidirectional interactions. Increasing species numbers leads to an explosion of potential interactions. All species do not eat each other and interactions are often weak, so simplifications of food-web networks may be possible. The goal of food-web ecology is to use all the concepts of trophic ecology to find general patterns of behavior in these networks and determine how they respond to changes in biotic and abiotic conditions (Pascual and Dunne 2006).

The discipline of community ecology attempts to disentangle the complexity of food webs and ascertain important processes underlying emergent characteristics such as species diversity, ecosystem stability, and population productivity. This field has seen its share of scientific turmoil in the past 50 years, leading to important discoveries as well as new directions for research. With the rise of the Internet, the concept of networks has become commonplace in our culture. Community ecologists have grappled with the concept of networks for a much longer time. All networks are characterized by nodes. For ecology, these *nodes* are typically populations of species within communities, although this is not necessary (Cohen and Briand 1984). Nodes can be life stages, functional groups, and yes, even trophic levels. Connections between the nodes are often consumptive, where energy and materials are transferred from one organism to another, similar to electricity moving through circuits between electrodes or information traveling through optical cables between computers. Effects of one node on another such as behavioral modification or chemical communication also serve as connections. The configuration of a food-web network is called its topology. The number, strength, and order of the connections among nodes are all examples of the food web's topology (Figure 10.1).

In this chapter, we briefly explore the different ways that food webs may be constructed to describe trophic interactions in communities. The differences among them depend on the nature of the connections among nodes. We will discover that they are not mutually exclusive approaches. An important take-home point of this section is that not all organisms are created equally in a food-web context. The distinctive nature of some species or particular individuals within species may make them more important in promoting, maintaining, or dampening food-web interactions than others. Theory predicts that food webs should have various properties that make them distinct. We will determine whether these predictions bear weight with the empirical evidence collected to date.

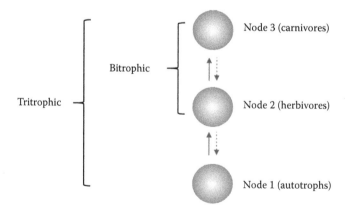

FIGURE 10.1 Nodes (balls) and connections (arrows) in a simplified food web. Solid arrows depict the flow of energy and materials to consumers. Dashed arrows are consumptive effects.

10.3 DEFINING A FOOD WEB

As any community ecologist knows, generating a complete census of all the species in an ecosystem is impossible (Martinez 1991). Resolution is important for developing a food web and tracing the connections between nodes within it. A technique used in community ecology to closely approximate the species present is *rarefaction* (Whittaker et al. 2001). The community is sampled multiple times and the accumulation of species is quantified (Figure 10.2). The rate of new species appearing in the samples will eventually decline as most species are encountered. The number of species (i.e., the *gamma diversity*) is estimated at the asymptote of the curve (Figure 10.2). Under most circumstances, the asymptote continues to climb, but the accumulation of new species is so infrequent that it is considered negligible.

There are many obvious problems with quantifying species to define food webs. Species may vary temporally in a community or ecosystem. If they become dormant, immigrate, or emigrate, they might be discounted from the food web, although they play an important role when present. A classic example is the food web subsidies provided by Pacific salmon immigrating to their spawning streams in North America and Asia. Once they spawn, the thousands of kilograms of adults die, releasing their nutrients into the streams (Helfield and Naiman 2001). These individuals also provide food for a host of other species including birds, mammals, and insects, with food-web effects propagating into the surrounding forest (Helfield and Naiman 2001). This example leads to another food-web problem—where do one ecosystem end and the other one begin? Although rarefaction provides a relatively comprehensive picture of the species composition of an ecosystem, it is not complete. Important species are missed by sampling, perhaps because the gear used is not selecting for them. Because most ecologists do not have the resources to do a complete inventory or it is impossible to reliably conduct one, food-web interactions are often inferred

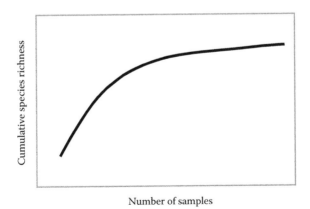

FIGURE 10.2 Rarefaction curve in an ecosystem. As more samples are collected, the number of species encountered decelerates.

by researchers from their personal observations. As we will see later, this approach
may lead to false interpretations of food-web mechanics.

Even if all the species within an ecosystem are known, their feeding relationships, or more generally, the patterns of energy flow among nodes are unknown until
research is conducted. Assessing feeding relationships has many assumptions and
requires effective methods (Chapter 4). Linking consumptive relationships among
organisms depends on observations of diets and of quantifying feeding behavior
in the field and laboratory. Relationships also may be inferred using other methods
including stable isotopes and fatty acids (Chapter 13) as biomarkers. Regardless of the
approach, linkages will be missed; this needs to be recognized and addressed in any
food-web study. Martinez (1991) conducted a food-web inventory of a lake and determined how aggregating species affected inferences about interactions. Characters
such as links per species within food webs decline as aggregation increases, reducing resolution. Aggregation of food-web inventories causes them to behave more
similarly (Martinez 1991).

10.4   FOOD-WEB CATEGORIES

Connectedness food webs require the least information about relationships among
species. If two species or their surrogate nodes are connected by feeding, then they
are connected in the food web, regardless of the strength or nature of their interaction. Historically, theoretical ecologists explored these relationships to ask questions
about emergent topological features of food webs. A generalization that emerged
from much of the early analysis of food webs was that a quantitative measure of
connectedness was possible (Cohen et al. 1990). The number of connections (C) in
a food web is

$$C = L/[S(S-1)/2] \tag{10.1}$$

where L is the number of links among species and S is the number of species. Said
in another way, C is the ratio of the actual number of pairwise links (L) relative to
the theoretical maximum number of links possible (the denominator). For example,
a simple food web with three species and all species interacting would have $C = 3/[3(2)/2] = 1$, because all possible links occur. C varies among ecosystems and likely
depends on the number of species (i.e., being scale dependent). Related to this mea-
sure is linkage density, L/S, which has been estimated theoretically as being close
to 2 (Pimm 1982), or an average of two links between species. Later research summarized in Dunne (2006) and Mittelbach (2012) showed that this was an oversimplification, with linkage density varying among ecosystems and potentially being
underestimated as the number of species increases in food webs.

The number of connections among species within food webs varies dramatically
and depends on many contextual features. One of the most important factors influencing connectedness is the relative sizes of species within food webs (Williams and
Martinez 2000). If consumers are large relative to food items, then a high number of
prey species may be connected to the consumer trophic level, because many species
are vulnerable (Figure 10.3). In contrast, a community dominated by large predators

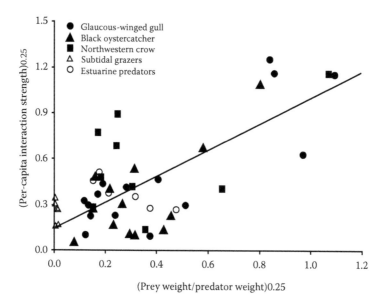

FIGURE 10.3 Relationship between the ratio of prey to predator mass and the interaction strength between the two. As prey become large relative to their predators, interaction strength intensifies. (From Wootton, J. T., and M. Emmerson. 2005. *Annual Review of Ecology, Evolution, and Systematics* 36:419–444.)

and large prey or a food web containing prey with effective defenses against consumers may have considerably fewer trophic connections (Figure 10.3). These size-dependent connections affect the magnitude of interaction strength, which we will explore later. Habitat complexity affects whether species in communities encounter each other. In a complex ecosystem like a tropical rainforest, the community of organisms in the canopy rarely or never encounters the animals and plants in the understory. This does not mean that they do not interact. Some animals dying in the canopy fall to the forest floor serving as food for decomposers and scavengers. All of these patterns can be directly observed, with time and money. Many organisms behave in unpredictable ways, with connections between species depending on internal state, learning, and cognition. The connectedness within a food web may depend on the behaviors of the species involved.

Defining food webs by the number of connections between species or other nodes provides limited information. Quantifying the strength of the interactions provides more insight. The nature of the interactions varies depending on the question being asked. Tracing energy transfer from one organism to another through consumption is a goal of trophic ecology (see Chapter 1). The consumptive impact of the predator on the prey is an additional interaction transmitted by connections between nodes (Chapter 4). Information also can be transferred along food-web connections through pathways such as chemosensation, vision, sound, electricity, and olfaction. Brown et al. (1999) reviewed patterns of vigilance to predators in many animals and showed that this information about predators was transmitted both within and across species. The food-web impacts of the response of multiple species to risk lead to reduced transmission of energy among trophic levels (Majdi et al. 2014). Paine (1980) was among the first to recognize that interactions within food webs have different magnitudes. Determining these interaction strengths is key for understanding how trophic processes influence biodiversity and ultimately energy flow in ecosystems.

The search for reliably quantifying interaction strength is ongoing. Power et al. (1996) created an index based on past efforts by Paine (1980). A trophic interaction should lead to changes in the physical traits of the consumer and prey nodes within a food web. A community trait can be any characteristic of importance that influences trophic structure such as biomass, density, species richness, energy content, body size, and age distribution. The instantaneous change in the *interaction strength* (I) is

$$I = \frac{dt}{dp} \times \frac{1}{t} \tag{10.2}$$

where t is the community trait and p is the proportion of the consumer in the community. If the proportion of the consumer is not changing rapidly but the change in the community trait (e.g., total biomass of prey) is large relative to the total quantity of the community trait, then I will be large. This index allows ecologists to standardize interaction strengths across food webs and compare them on an identical scale, ranging from negative to positive values. The problem with the instantaneous index (I) is that it relies on immediate (e.g., per capita) changes in the community, which are difficult to quantify. Tracking small changes within a single population is difficult because of errors in sampling and variability in the environment. Imagine trying to quantify changes in a whole community composed of complex populations simultaneously. Even if possible, the computational power necessary to do this correctly with today's computers is insufficient.

Experimentation arose as a viable technique in ecology about a half century ago and with it came the ability to determine how communities change with the presence and absence of species (Paine 1966). At this time, the idea that community control emanated from the bottom up was pervasive, with concepts about top-down consumer control of species diversity and productivity being limited to a few researchers such as Brooks and Dodson (1965) and Hairston et al. (1960). Paine (1966) dispensed with the suspense and decided to completely eradicate a top consumer in a community and watch what happened. This approach is difficult to accomplish in many ecosystems because removing all the top predators is nearly impossible. Luckily, the rocky intertidal zone of the ocean is not such a complex place. Paine removed a slow-moving, large starfish predator from enclosures and found that species composition and diversity of prey declined. This sledge-hammer approach to the ecosystem transmitted beyond the prey to change algal productivity, something Paine (1980) later coined, for better or worse, a trophic cascade. We will bridge the trophic cascade concept in more detail later.

Paine started a revolution of removal experiments that led to a simpler way to quantify interaction strength (I^*) via Power et al. (1996),

$$I^* = \frac{N - D}{N} \times \frac{1}{p} \tag{10.3}$$

where N is the community trait (see above) with the consumer present. When the consumer is removed and the community or ecosystem has had enough time to presumably approach equilibrium, then D, the community trait following consumer removal, is quantified. As in the previous equation, p is the proportion of the consumer in the community. If the change in the community trait is large and p is small, the interaction strength for the consumer species will be large. If p is large, then the interaction strength will be reduced because it is the high abundance of the consumer, not its interaction strength, that influences community structure.

The indices of interaction strength by Paine (1980) and Power et al. (1996) are great methods for providing standard comparisons but are not biologically meaningful. An index that explicitly incorporates the energy contribution of the prey to the consumer was developed by Bascompte et al. (2005). This method does not require the removal of the top consumers. In this version, interaction strength (I^{**}) is

$$I^{**} = \frac{C \times E_{pred}}{E_{comm}} \tag{10.4}$$

where C is the total energy consumed by the consumer, E_{pred} is the energy consumed of the prey by the consumer, and E_{comm} is the total energy of the prey. If a consumer is consuming a large amount of energy and it comprises a significant proportion of the prey in the environment, the interaction strength will be large.

A common *topological* characteristic arises from these and other analyses of food webs. Power et al. (1996) compiled data from removal experiments and found that weak interactions were common, while a few were quite strong. Bascompte et al. (2005) compiled data from marine food webs ($N = 249$ species or trophic groups) and found a similar pattern (Figure 10.4). Weak interactions are common in food webs probably because those ecosystems that have many strong interactions are unstable (Wootton and Emmerson 2005). Simply put, a food web with many highly effective predators will likely quickly go extinct and thus food webs assemble to contain (perhaps through coevolution) species that are weakly interacting, through processes such as prey switching, density-dependent feedbacks, and population compensation (Figure 10.4). Occasionally, a food web will contain a strongly interacting species, coined a keystone by Paine (1966). It could be that these rare occurrences that emerge are as important at stabilizing a food web as all the weak interactions are in others. The trick, as we will find in the following sections, is identifying these strong interactors before they are removed by perturbations such as human eradication.

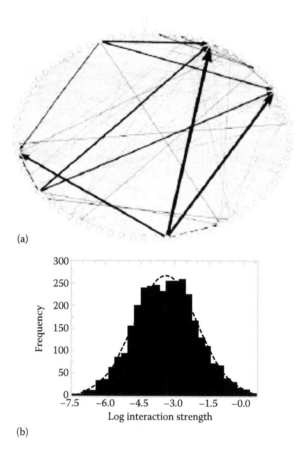

(a)

(b)

FIGURE 10.4   (a) Trophic interactions and their relative strength (weight of lines) in marine food webs. (b) Distribution of interaction strengths within the marine food webs. (From Bascompte, J. et al. 2005. *Proceedings of the National Academy of Sciences of the United States of America* 102:5443–5447.)

10.5   EFFECTS IN FOOD WEBS

Identifying important species or other nodes in food webs is difficult. Calculating interaction strength reveals two kinds of consumptive interactions. The *keystones* are those that have a low density or biomass in an ecosystem but exert a strong per capita impact on their prey and perhaps other, seemingly unrelated species (Figure 10.5). This is akin to the keystone of an arch. When it is removed, the entire arch crumbles. Power et al. (1996) clearly showed that keystones are not the same as numerically dominant species in a food web. A dominant species is abundant in the food web and may exert strong consumptive effects. However, the effect per individual within the population is low (Figure 10.5). Dominant consumers should be a stable force in food

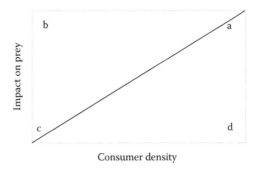

FIGURE 10.5 Effect of consumer density on food-web interactions in ecosystems. (a) The line depicts a directly proportional effect on consumer density. High consumer density has a strong impact on prey and thus this is a numerically dominant effect. (b) Low densities of consumers have a strong impact on prey and trophic interactions. This consumer would be considered a keystone species. (c) Both consumer density and its impact of prey are small. (d) A weakly interacting consumer would have little impact on trophic transfer of energy and materials from prey even at high densities. (From Power, M. E. et al. 1996. *Bioscience* 46:609–620.)

webs because they are dense and thus not as prone to collapse, although degradation through perturbations like overharvest is possible.

What determines whether a species is a keystone was hotly debated in the 1990s. In some ways, the distinction between a keystone and a numerically dominant species is subjective, based on how the magnitude of the consumptive interaction strength is interpreted (Figure 10.6). Some ecologists argued that strong

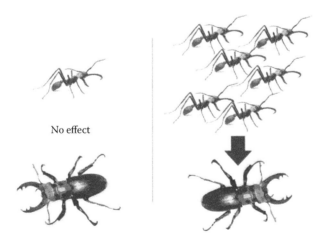

No effect

FIGURE 10.6 Army ants as a numerically dominant and/or keystone species affecting antlions. A single army ant cannot affect prey. However, this species does not hunt as individuals, requiring coordination among hundreds of individuals to impact food webs. Whether this species acts as a keystone or numerical dominant is difficult to assess.

interactors were more likely in relatively simple aquatic food webs but not complex terrestrial ones (Strong 1992). Other ecologists argued that strong interactors were a figment of the scale of inquiry, because subsidies from adjacent food webs were really the factors driving interactions (Polis et al. 1996; Figure 10.7). Whatever the argument, some species do seem to have characteristics that make them particularly important in ecosystems, even when they are relatively rare. Wootton and Emmerson (2005) showed that the body size of the consumer relative to the size of the individuals in the prey assemblage affects interaction strength. If prey items are at similar sizes to the predator, then the likelihood of a strong interaction is high (Figure 10.3).

Size-structured interactions are among the most important consumptive mechanisms affecting ecosystems (Sih et al. 1985; Werner and Peacor 2003). It is not surprising that the prey have to be large for a consumer to have a disproportionate impact. Starfish feed by extending their stomach and digesting their prey externally, allowing them to digest prey that are large relative to their body (Figure 10.8).

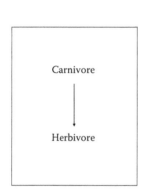
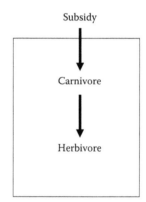

FIGURE 10.7   How top-down effects may really work. In the left ecosystem (box), a carnivore in an ecosystem exerts a weak consumptive impact on herbivores. In the right ecosystem, food subsidies from outside the ecosystem box allow the carnivore population to become dense or efficient, allowing it to have a strong consumptive impact on herbivores.

(a)   (b)

FIGURE 10.8   Starfish with stomach (a) retracted and (b) extended. The ability for this organism to engulf its prey with its stomach allows it to consume large prey that are inaccessible to many other predators.

By consuming mussels and snails, they can open up space in the intertidal zones of the ocean, allowing better protected, less palatable organisms to proliferate. Large mammalian predators like carnivorous cats, wolves, and bears are capable of knocking down prey organisms that are larger than them. Large-gaped fish can consume a wide range of large prey, controlling their density. The loss of all of these predators can cause the large prey they control to become dense and negatively affect other species, usually through their consumptive impact. The commonality among all these systems is the negative relationship between prey size and population turnover rate. Populations of large prey have slow turnover rates and thus are unable to quickly compensate for predation. Predators can control their biomass and thus the effect of these prey on other species through exploitative or interference competition.

Not all food-web connections are consumptive. Potential predators and conspecifics can exert effects on other species that affect energy flow among nodes but do not require direct mortality. Predators can create a *landscape of fear* (Matassa and Trussell 2011; Chapter 5) that reduces the activity and energy intake of their prey. Research with larval insects in streams demonstrates that entry into the drift is timed to avoid predation (Kohler and McPeek 1989). Filter-feeding zooplankton in lakes will only move vertically up the water column to feed on phytoplankton during times when predators are inactive. Many birds and mammals display behaviors or emit chemical signals that reduce or alter activity of their conspecifics; some of this communication also is transmitted between species (Kats and Dill 1998; Lonnstedt et al. 2012; Box 10.1). The presence of predators reduces energy intake, growth, and reproductive output of their prey, affecting community and ecosystem structure and function. These effects are not restricted to predators. Superior competitors may alter food-web interactions. Other population-level processes such as reproduction and rearing of young will affect patterns of energy intake and transfer in food webs.

10.6 FOOD-WEB PATTERNS

Many possible topologies exist for food webs. The typical pattern that we describe in ecology includes two connections between two nodes, which is positive for the consumer node and negative for the prey (Figure 10.9). As noted earlier, the connections typically depict energy flow, but other relevant quantities can be transmitted. Topologies can vary widely, with only the simplest receiving descriptions (Figure 10.9; Menge 1995). In all of these, both direct and indirect effects are involved. A direct effect occurs through direct consumption. An indirect effect is created by a direct effect through the network, but is not directly linked to the interaction. A numerically dominant predator or keystone species transmits an indirect positive effect on one prey species by removing another prey species (Figure 10.9). For this to occur, the two prey species have to compete, and the selective removal of one species enhances the other (Figure 10.9). Symmetric exploitative competition between consumers benefits prey species by reducing their combined consumptive impact (Figure 10.9). Apparent competition is a condition where two prey species appear to have a negative impact on each other (Figure 10.9). In reality, one or both

BOX 10.1

Trophic interactions in food webs are not limited to consumption. Some species communicate socially, to announce either the availability of food or the presence of predators. Within species, some evidence exists that inclusive fitness occurs, where individuals communicate information to closely related counterparts. Both within and across species, there is an individual benefit to decoding the signals produced by conspecifics or other species.

Within ecosystems, social interactions may affect the rate of foraging and trophic transfer. Many organisms forage in mixed species assemblages, where communication may occur. Much literature exists on mixed flocks of birds foraging together, supporting an ongoing debate about whether multiple species forage together to avoid predators or to increase feeding efficiency (Bednekoff and Lima 1998). Regardless of the mechanism, identifying the existence, frequency, and effectiveness of these interactions lends insight into food-web relationships.

Farine et al. (2012) demonstrated that social network analysis is a useful way to determine whether patterns of communication occur in mixed flocks of tits. Imagine three species of birds in a flock (see figure below). If most respond to the behavior of other individuals regardless of species, then the analysis will find that the distribution of pairwise interactions will not deviate from those expected randomly (see figure below). Although multiple species may occur together in space and apparently interact, their social behavior may be isolated to conspecifics (see figure below). The group will respond very differently to external cues that affect food-web interactions under this scenario.

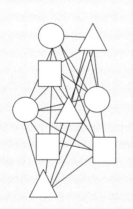
Heterospecific interactions common

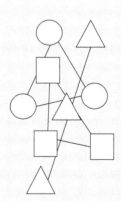
No heterospecific interactions

Two assemblages with individuals of three species (circles, triangles, squares) in space (e.g., a mixed flock of birds by a feeder). Lines depict communication among individuals.

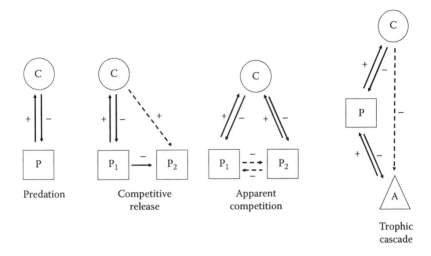

FIGURE 10.9 Bitrophic and tritrophic food-web topologies. Consumers are C, prey are P, and autotrophs are A. Direct interactions and their direction (either consumption or energy flux) are depicted by solid arrows. An indirect effect facilitated by a direct interaction is shown by a dashed line. Effects can be positive (+) or negative (–) on the populations in each trophic level.

of the species enhance the density of the consumer population, resulting in increased consumption of both. Without quantifying the response of the predators, it seems as if the two prey are competing when they are not (Figure 10.9; Box 10.2). These and other possible food-web interactions are bitrophic, occurring between two putative nodes or trophic levels.

Multitrophic interactions including direct and indirect effects are common in food webs. Determining how these translate to energy flow, diversity, and ultimately ecosystem stability is challenging. The tritrophic interaction that has led to much debate is Paine (1980)'s trophic cascade. The consumer reduces the prey trophic level, which in the classic trophic cascade is an herbivore (Figure 10.10). By reducing the biomass of the herbivore prey, the density of the primary producers increases (Figure 10.10). Thus, a trophic cascade is an indirect effect by a species that skips a trophic level (Figure 10.10). Trophic cascades are not limited to three trophic levels,

BOX 10.2

Apparent competition is a confusing term in trophic ecology because it has nothing to do with competition. Rather, as Figure 10.9 shows, two prey species in an assemblage promote the density and consumptive capacity of their predator. Both species have a symmetrical, negative effect on each other. The effect is not always symmetrical. If one of the prey species has high production and is able to sustain the predator, then the other prey species may be consumed to extinction (Wittmer et al. 2013). This kind of apparent competition is common in many ecosystems where a novel species has invaded or a domestic species has been introduced and is cultivated at high densities.

A good example is the apparent competition between domestic sheep and native south Andean deer in Patagonia (see figure below). These two species share common predators, including foxes and occasionally cougars. The sheep outnumber the deer by about 30 times, providing regular prey for the predators and allowing them to dramatically negatively impact the deer. To create a conservation park for deer, planners intended to remove sheep. In the long term, this may be beneficial for the deer. However, in the short term, the inflated predator population caused high mortality of deer (Wittmer et al. 2013).

The point of apparent competition is that competition should not be assumed to be the causal factor for any ecosystem where negative interactions between two species occur. Rather, the impact of a shared predator may be lurking. In fact, they can interact with each other (Garvey et al. 1994). A nonnative species can outcompete a native species and increase predation risk simultaneously. Wittmer et al. (2013) noted that ecologists and managers need to consider both competition and predation when making conservation decisions. They concluded that the best management option for recovery of a target species was simultaneous control of both the apparently competing prey species and the predators.

A South Andean deer, also known as a huemul.

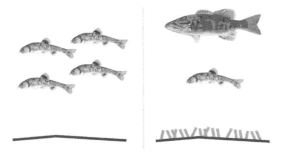

FIGURE 10.10 Effects of herbivorous stonerollers on algae in streams without (left) and with (right) smallmouth bass predators. Smallmouth bass control stoneroller density and suppress their feeding, allowing standing biomass of algae to increase.

but can propagate across multiple species with trophic specializations. Whether trophic cascades are merely a theoretical construct or a reality has been a source of debate, which we will discuss later.

10.7   EMERGENT PROPERTIES

Given the number of topologies that are possible, generalizations for food webs in ecosystems seem difficult and will keep theorists busy for decades to come. We will first explore a very old question about whether the number of connections in a food web affects the stability of an ecosystem. This is more important than ever because biodiversity is declining globally. If species loss causes food-web collapse, then the ability for the Earth to provide ecosystem services to humans will decline. What affects the distance from the bottom to the top of the food web (i.e., food-chain length) continues to be a source of conjecture. Shorter food webs may contain less charismatic top predators, which to many conservation biologists are indicators of healthy ecosystems. Additionally, the distribution of biomass within food webs and how it is affected by species interactions continue to be a focus of research. Bottom-up, top-down, and perhaps middle-out forces are responsible for these patterns of food availability in food webs. This is important for many reasons, including determining the distribution of biomass of trees available for harvest by foresters, the fish available in the ocean for consumption by humans, and, more generally, the prey available for a predator assemblage.

10.7.1   COMPLEXITY

Consider a ball of yarn that is tangled with knots. Trying to unravel it into a single strand is difficult. Early ecologists believed that the more tangled the connections among nodes in a food web, the more stable (i.e., resistant) the ecosystem would be to outside perturbations that removed individuals, populations, or entire species. Stability in this sense is the ability for the food web's component nodes to keep from unraveling. Populations within food webs can fluctuate due to stochastic processes that propagate throughout the network. These effects either promote or reduce the density of populations, which affect other food-web components. Theoretical ecologists model these effects by introducing perturbations that are either rapid and intense called *press disturbances* or gradual called *push disturbances*. As we alluded to earlier, early modelers (e.g., May 1972) found the opposite of the tangled yarn expectation. Adding more species to food webs in models made the communities more prone to collapse during these perturbations. These models were simplistic, assuming that connections among species were random and a process called *runaway consumption* would occur. A runaway means that one species consumes another to extinction, releasing another species, which overcomes another species, and so on. Eventually, the food web loses most or all of its component nodes and subsequently its biodiversity.

Later modeling and empirical information provided more resolution about food webs. As we pointed out earlier, weak interactions are common and thus the possibility for strong, runaway interactions is low. Connections among species within food webs are not random, but are compartmentalized into categories, subsidizing each other (Pimm and Lawton 1980; Polis and Hurd 1996; Box 10.3). These compartments

BOX 10.3

As we describe in the text, early theoretical models in ecology predicted that more complex food webs would be less stable (May 1972). It was already clear at the time that May's assumptions about random connections were unrealistic and that more species in an ecosystem should facilitate stability. Pimm and Lawton (1977) sought to determine how adding internal subsystems or compartments within food webs would influence community stability. In a grassland, a subsystem might be a relatively tight association among owls, mice, and seeds within a much larger ecosystem containing many species of predators and prey. Using models, these ecologists found that adding specific compartments did not include much additional stability relative to the inclusion of nonrandom connections not contained within specific compartments. They did point out, however, that linking the ecosystem with external compartments increased stability.

With this finding, Pimm and Lawton (1977) showed how compartmentalized linkages among ecosystems may subsidize species within them and alter interactions. Polis and Hurd (1996) provided a classic example of how one ecosystem may support and perhaps stabilize food webs in another. The ocean provides an almost constant supply of organic matter to some islands. These investigators found that marine detritus washing up on shore and being deposited as feces from marine-feeding birds was consumed by invertebrates at the beach. These invertebrates were consumed by spiders, allowing these predators to both support spider-eating lizards and control herbivorous insects. The ocean subsidies stabilized island ecosystems by facilitating spider production. In contrast, spider densities were much lower in mainland locations remote from the ocean. No ocean subsidies were available to support spiders, resulting in fewer lizards and allowing herbivorous insects to prosper. Presumably, herbivorous insect populations fluctuated markedly, increasing variability in autotrophic production available to other consumers (i.e., destabilizing nonmarine food webs).

tend to stabilize food webs as well. The generalization that food webs with more species are likely less prone to collapse with outside disturbances holds under most circumstances (but see empirical data from Mittelbach et al. 2001). Important to this point is that these are food webs that have had time to develop naturally. Ecosystems that are *fabricated* by humans have not necessarily developed these emergent properties because the component species are novel and not the products of past interactions (Lindenmayer et al. 2002). An example of a fabricated ecosystem is a garden containing horticultured species with no interactions. If not intensively managed by the gardener, the garden will become overgrown by native and invasive weeds plus a few of the aggressive gardened species.

10.7.2 FOOD-CHAIN LENGTH

Although food webs are intended to dispense with the subjective classification of trophic levels, ecologists continue to debate about how many trophic levels can be supported by a food web (see Chapter 3). In later chapters (Chapters 12 and 13), we will determine ways to quantify trophic levels using tracer substances like fatty acids and elements. The number of trophic levels (i.e., food-chain length; Figure 10.11) within food webs averages at about three or four (Ulanowicz et al. 2014), depending on how you classify an ecosystem. Many food webs have more trophic levels, although six seems to be near an upper limit. The classic Eltonian argument for why food-chain length in a food web is limited is because of the inefficiency of energy transfer from one trophic level to another. Recall that energy is lost as heat into space as it moves from prey to predator. This loss limits the total number of species in food webs, especially carnivores. The logic here is that adding energy and increasing the primary productivity of the basal species in a food web (i.e., enrichment with nutrients) should counter the loss of efficiency and increase species richness plus the length of the food chain. If true, humans could make food chains longer by adding more plant fertilizer on a global scale. We have learned that this is certainly not true.

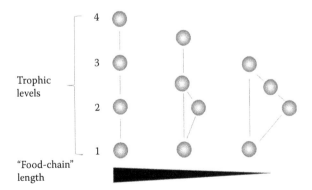

FIGURE 10.11 Relationship between connections within food webs and food-chain length. As top consumers feed at lower trophic levels, food-chain length declines because more materials (e.g., carbon and nitrogen) fixed at the base are present in the upper trophic levels.

Humans have long known that fertilizing ecosystems increases the biomass of plants. So, if we can make nutrients unlimited, where are the super food webs, with *apex carnivores* occupying trophic levels to the *n*th degree? In the 1970s, ecologists discovered that some nutrients added to ecosystems in high concentrations promoted species within food webs to an apparent threshold concentration. Past this concentration, species richness and food-chain length decline (Figure 10.12). One reason for this decline is explained by models that show that longer food chains are less stable in the face of perturbations (Pimm and Lawton 1977). Species that occupy higher trophic levels are less able to rapidly respond because they tend to have life stages of *k*-strategists, with larger body size and low reproductive rate. These life history types are unable to respond effectively to disturbances because they are rare and unable to reproduce rapidly enough to avoid extinction. Top consumers within food webs will eventually become limited because of the very characteristics that make them good predators—long lives, parental care, large bodies, and time to learn to hunt efficiently and effectively.

The hump-shaped curve between nutrient concentration and species richness also occurs because it is impossible to increase all limiting resources simultaneously (Figure 10.12). Enriching one or more common resources will make others limited (see Chapter 12). Thus, apparent enrichment tends to favor competitive dominants rather than releasing species from competition (Huston 1979). These species reach high densities and cause other species to decline. A terrestrial example is a manicured lawn that is fed with a steady diet of nutrients from the lawn shop. The grass cultivar forms a tight grip on the landscape and excludes most native species because

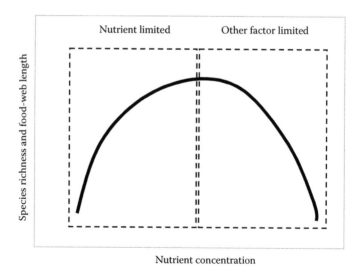

FIGURE 10.12   Relationship between nutrient enrichment in an ecosystem and the number of food-web connections. A single nutrient or multiple nutrients limit species richness and trophic connections until a peak is reached (left side). After this limit is exceeded, some other factor (e.g., space, another nutrient, a dominant competitor, poor environmental quality) limits species richness and food-web structure.

of its tight network of stolons and rhizomes, making space a limited resource. Lakes and streams enriched with the nutrient phosphorus become choked with cyanobacteria that are able to outcompete other species for nitrogen that becomes limited (Chapter 12). In aquatic ecology, this effect of high nutrient loads is called *eutrophication*. It is not limited to aquatic ecosystems. Nitrogen deposition due to burning of fossil fuels has become a global problem likely altering terrestrial ecosystems in complex ways (Fenn et al. 1998).

Food-chain length may be related to ecosystem size. The theory of island biogeography (MacArthur and Wilson 1967) posits that larger islands or habitats have more niches for species to occupy. Thus, more species accumulate in these areas through immigration and fewer species decline due to emigration or extinction. This concept can easily be extended to food webs in ecosystems (Post et al. 2000). More species or nodes should lead to more connections and an increase in the probability that species at higher trophic levels will emerge and reach equilibrium with prey at lower trophic levels. A larger ecosystem should provide more heterogeneity for prey, allowing them to avoid being consumed to extinction by the consumers they support. As ecosystems shrink due to fragmentation, food-chain length should decline due to instability as species are consumed to extinction, causing their predators to starve or emigrate to better foraging patches.

10.7.3 TROPHIC TRANSMISSION

Whether the distribution of biomass in food webs is controlled by primary production from the bottom-up or by consumers from the top-down continues to be debated (Terborgh and Estes 2010). This matters because it affects how an ecosystem *looks* to humans and other species. It also may affect the stability of ecosystems. As any gardener knows, enrichment by nutrients enhances plant growth and makes the flowerbeds look green and lush. Logic dictates that the biomass of gophers and insects should respond similarly consuming the flowers thereby making the gardener upset. These herbivores need to be controlled, unfortunately all too often with the use of pesticides. The gardener in this case cannot rely on predators to control the pests in the garden because their biomass is assumed to be limited by energetic constraints. This is the classic bottom-up view of food-web structure (Figure 10.13).

There might be hope for the gardener. As we touched upon earlier in this and other chapters, the contrasting top-down view came to light with the publication of Hairston et al. (1960). Recall that these authors posited that the biomass of herbivores should be kept low by carnivores, allowing plants to remain abundant (Figure 10.13). Otherwise, the herbivores would grow out of control and crash the ecosystem through overgrazing. Paine (1966) found that the removal of predators reduced the diversity of entire communities. Bolstering Paine's observations, early theoretical food-web models produced puzzling predictions about bottom-up effects called the *paradox of enrichment* (Rosenzweig 1971). Enhancing the production of autotrophs transmitted up the food chain in the models, causing the top consumers to become overabundant; herbivores were overconsumed and the modeled food web collapsed. These issues created a camp of ecologists who hypothesized that consumer control was important for maintaining a stable distribution of biomass in food webs.

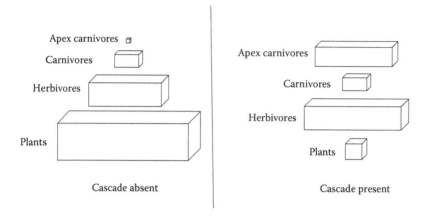

FIGURE 10.13 Distribution of biomass depicted by the size of boxes of trophic levels without (left) and with (right) a trophic cascade present. Low densities of plants and carnivores in the cascade example may render an ecosystem vulnerable to density collapse within those trophic levels, making the stack of blocks collapse and impairing food-web structure.

Not only was primary production important, but the characteristics and quantity of the top consumers (i.e., those that did not cause a collapse as predicted by the Rosenzweig model) also played a critical role.

Paine's 1980 creation of the term *trophic cascade* energized the debate, with the idea that consumer effects transmit across multiple nodes within food webs all the way down to the primary producers. With the exception of a few ecologists, consumer effects were thought to be bi-trophic at best. In this view, consumers cull weak and vulnerable prey, but would be unable to overcome the prey's productivity to cause a change in the prey's standing biomass. The entry of the *trophic cascade hypothesis* into ecology revitalized predictions by Hairston et al. (1960) that the presence of a keystone species or numerically dominant consumers in a food web could transmit across multiple nodes in the food web and change the world from green to brown. This idea is intriguing and potentially disturbing. Food webs, rather than being the sturdy pyramids built of trophic blocks held up by a broad base of nutrients and sunlight, might be vulnerable to changes in densities of rare consumers seemingly far removed in the trophic network (Figure 10.13).

How common are trophic cascades in nature? From a theoretical standpoint, the jury is still out. Food-web models with more realistic density-dependent feedbacks within populations showed that bottom-up effects within food webs do not jeopardize stability (Arditi and Ginzburg 2012). Populations reach equilibrium and respond to perturbations by temporarily declining and then returning to the equilibrium point. Pyramid shapes of biomass should be the norm (Figure 10.13). However, it is possible that different populations can coexist at different equilibrium densities, whereby the consumer level holds an intermediate prey node at a low equilibrium biomass, allowing primary producers to reach high equilibrium biomass (Figure 10.13). This can be extended beyond three trophic levels, creating patterns of biomass that look nothing like pyramids (Figure 10.13). This is a troublesome prediction because populations at low densities are notoriously susceptible to collapse through depensation, even if their productivity is high (see Chapter 6). An analog would be a tower of biomass blocks held up by alternating large and small blocks. With no disturbance to the system (a jiggle), the tower is stable. But any shift in the obviously unstable small blocks causes the entire tower to fall.

For trophic cascades to transmit downward from the top of the food web and remain stable, the apex consumers have to be limited in some fashion independent of the density of their prey. As alluded to earlier, density-dependent processes internal to these top consumer populations may be at play to prevent them from overreaching. Slow reproductive rates, late maturation, and other k-selected characteristics must be more important than the intake of resources. Top consumer populations also may be regulated by external, physical factors that keep them in check. This is similar to the intermediate disturbance hypothesis, which posits that an occasional disturbance is necessary to keep the top consumer density from becoming so high that it reduces the prey species and its competitors to extinction (Connell 1978; Menge and Sutherland 1987; Box 10.4).

In the 1990s, the debate about trophic cascades heated up. Aquatic ecologists appeared to be more receptive to the idea, because some empirical evidence existed. One of the earliest ecosystem-level experimental approaches was summarized in Carpenter et al. (1987), where they manipulated the food-web structure of three, north-temperate lakes, with the help of a winterkill that removed the apex carnivore, largemouth bass, in one of the lakes. Winterkill occurs in lakes when ice and snow cover the surface, causing oxygen to decline. Only fish tolerant of low oxygen like minnows survive. In the winterkill lake, a planktivorous minnow became dense. In the other two lakes, largemouth bass was added in high densities. The largemouth bass presumably kept the minnow population from rising, allowing large filter-feeding zooplankton called *Daphnia* to rise to dominance. The filtering capacity of these planktons was so high that they were able to control the standing biomass of the algae in the lakes, making the water clear. This phenomenon has long been known to occur in the springtime, called the *clear-water phase* where zooplankton overgraze their algae resource and increase clarity. What Carpenter et al. (1987) and later Mittelbach et al. (1995) showed was that this control of algae was facilitated by top predators and that it was stable through time, not just in the spring. The *control* lake with minnows only had low zooplankton density and green, opaque water.

Empirical evidence aside, doubts still exist. The likelihood of transmission of effects across multiple nodes of food webs may be relegated to a subset of very specific circumstances (Figure 10.14). Herbivores must be very efficient at controlling autotroph biomass, which is unlikely for many ecosystems because plants have defenses (see Chapter 2). Primary consumers must be vulnerable to apex predators, but able to maintain production to support predation (Figure 10.14). All trophic levels below the apex predator must have high production (i.e., a high intrinsic rate of increase, r) to support the high predation (Figure 10.14). However, as we noted earlier, the apex predator must self-limit and have a low r. An apparently physically simple ecosystem like a lake should have tighter connections among nodes, inducing strong transmissions (Strong 1992). Terrestrial systems, with greater heterogeneity,

BOX 10.4

A debate continues in ecology about whether food-web structure is driven by stable, equilibrium-based processes (e.g., populations near their carrying capacities) versus abiotic or biotic perturbations that knock them regularly out of equilibrium. Like everything in ecology, the answer is, *it depends*. The intermediate disturbance hypothesis (IDH) was posited to explain how perturbations in ecosystems keep populations from reaching equilibrium density, and in turn, promote species richness. Connell (1978) in his classic paper showed that regular disturbances (e.g., hurricanes) in tropical forests and coral reefs reduced species that may rise to dominance as competitors within food webs, allowing other species to coexist with them. This greatly altered the view that both of these systems were the result of highly stable conditions that promoted diversity and food-web connections.

Both experiments and field data generally support the IDH. If the disturbance is weak or infrequent, then a dominant competitor will exploit the majority of resources, leading to few species and limited food-web connections (see figure below). If the disturbance is strong, persistent, and frequent, then all species including the competitor will be reduced to low densities, greatly curtailing trophic transfer (see figure below). Intermediate strengths and frequencies of abiotic disturbances keep the competitor from reaching its equilibrium density and maximizing trophic connections (see figure below).

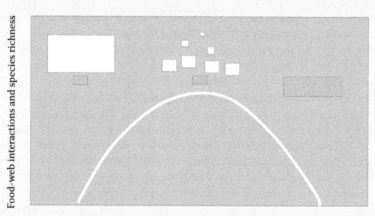

Disturbance intensity plus frequency

Effect of disturbance on the number of species and food-web connections in an ecosystem. With few, weak disturbances (left), most biomass will be contained within one competitive dominant species (large, clear box) that will monopolize algal resources (small, shaded block). With frequent, strong perturbations, no consumer species and all biomass will accumulate as algae (right, shaded block). Moderately frequent, intermediate disturbances promote multiple species and trophic levels by preventing the competitive dominant from reaching high densities (center).

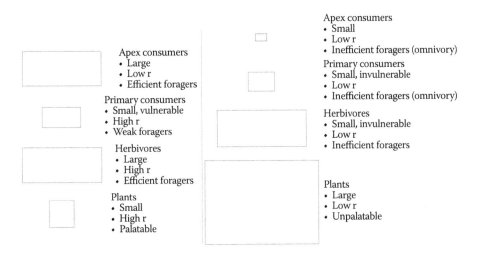

FIGURE 10.14 Hypothetical conditions necessary to maintain a trophic cascade or trophic pyramid in a food web. For a successful trophic cascade to occur (left), all lower trophic levels must have a sufficiently high population productivity rate (i.e., intrinsic rate of increase, *r*) to support high biomass of the top consumer. The top consumer must be self-limiting to prevent it from consuming the prey to extinction and causing the food web to collapse. A trophic pyramid (right) does not require high productivity to support high biomass at alternating trophic levels above the base.

should have less intense connections among consumers and prey, preventing trophic cascades. However, terrestrial ecologists like Terborgh and Estes (2010) hold that cascades do occur on land (also see Hanley and La Pierre 2015). The presence of omnivory should attenuate consumer effects because they act on multiple nodes of the food web rather than on a single one strongly connected to lower trophic levels (Power 1992). Many other arguments against the commonness of trophic cascades have been made (Figure 10.14).

Not all trophic transmissions have to *cascade* from the top to the bottom of the food web. Transmissions may be exerted from the *middle out* by many species (Stein et al. 1995), and this impact may be pervasive in ecosystems. These species or nodes have characteristics that allow them to control prey through strong interactions and regulate their consumers through their population dynamics (Figure 10.15). This may be a common phenomenon in nature that still requires exploration. Before the northward migration of humans, herbivorous megafauna (e.g., wooly mammoths) dominated much of the landscape (Terbogh and Estes 2010). These species likely strongly affected plant biomass and structure while being largely impervious to apex predators. Predator biomass would be regulated by the availability of these prey when they produced offspring and old, dying individuals. In the central United States, a fish species called gizzard shad is dominant in reservoirs. It modifies these ecosystems through consumptive effects on zooplankton and phytoplankton (Vanni et al. 1997). Dynamics of its population that are independent of the predators affect the predator assemblage. Many species with strong defenses such as lobsters, noxious herbivorous insects, and porcupines may affect

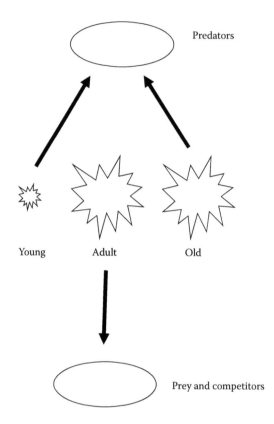

FIGURE 10.15 Ontogenetically driven effects of a middle-out consumer on both higher and lower trophic levels as well as individuals within the same trophic level. Variation in the availability of old and young individuals in the middle trophic level drives population dynamics of top consumers. Mature, reproductively viable adults are large-bodied and largely free from predation. Consumptive impacts of these species negatively affect prey as well as consumers through exploitative competition.

dynamics of both the food they consume and the predators that seek to consume them.

The ability for consumer control to come from the top down as a cascade or emanate from the middle out may rely on an important, often-missed factor called *subsidies* (Polis et al. 1996; recall Box 10.3). Food webs do not exist in isolation. Lakes receive materials and nutrients from the surrounding watershed. Much of stream ecology is predicated on the assumption that energy and nutrients come as subsidies from outside the ecosystem (Vannote et al. 1980). Polis et al. (1996) suggested that consumer control in ecosystems was only possible if subsidies were provided to stabilize important nodes in the food web. High densities of consumers may only be maintained in an ecosystem if subsidies are provided from the outside. To illustrate, the top consumer, largemouth bass, in the trophic cascade experiments conducted by Carpenter et al. (1987) were maintained at high densities by stocking them in lakes to *sledgehammer* the ecosystem. The largemouth bass top predator in Mittelbach et al. (1995) may have been able to maintain its cascade by feeding on other prey resources from outside the lake such as small mammals, reptiles, birds, and amphibians. The middle-out impact of gizzard shad in reservoirs is subsidized by its ability to consume detritus on the bottom. As its zooplankton and phytoplankton prey decline, it switches to feeding on the organic ooze until the prey populations recover (Vanni et al. 2005). Similarly, access to alternative food resources allows it to quickly outgrow its predators. The algae and thus the snails in our fish tank rely on a regular income of fish food to support their production. As the predaceous fish eat the food, they excrete nutrients, which supports the primary producers and herbivores. Similar examples of nutrient recycling in aquatic and terrestrial systems are common (Vanni 2002; Box 10.5).

Detecting the presence of trophic cascades and middle-out transmissions in food webs is very much a matter of scale and resolution within ecosystems. But undoubtedly, they do exist. As Paine (1980) and others (Power et al. 1996) pointed out, there are species that emerge within ecosystems with special characteristics that make them especially important in maintaining structure and function. A topic in ecology that has recently gained attention is that of alternative stable states (Scheffer and Carpenter 2003). This is nothing new for modeling. Depending on where a community begins, there may be multiple equilibrium points to which the relative densities of species are attracted. Most of these models are fairly simplistic and do not account for inherent phenotypic differences of the species involved. Incorporate the characteristics that make a strongly interacting consumer important in the food web and the predictions are dire. The presence of this special species causes the ecosystem and its inherent processes to exist in one state. Like Paine's starfish, if you remove the species, the ecosystem may jump to a different state. Both states are stable. This means that if you reintroduce the special interactor, the ecosystem might not necessarily respond the way it did before the removal. It will resist the change because of inherent changes to the food web while the consumer was absent. This lag is called *hysteresis* and is a problem for the restoration of ecosystems (Box 10.6).

BOX 10.5

Similar to the effect of apex predators on trophic cascades, the contribution of organisms to nutrient dynamics in aquatic and terrestrial ecosystems was discounted for many years. Nutrients were assumed to come from the physical environment and drive autotrophic production from the bottom up. Vanni (2002) wrote a review showing that animals ranging in size from small insects and crustaceans to whales are important to nutrient cycling in aquatic and terrestrial ecosystems. He identified their direct and indirect effects (see following figure). Direct effects include the nutrients lost from the organism via egestion, excretion, and eventually death. Indirect effects are from affecting the nutrient cycling of prey and modifying the physical environment to release (or sequester) nutrients (see figure below).

Many examples of animals affecting nutrient dynamics exist. Crustacean zooplanktons not only consume phytoplankton in lakes; they also modify the nutrient environment in which the phytoplankton live (Schindler et al. 1993). In Chapter 13, we will show how modified concentrations of nutrients affect phytoplankton assemblages and in turn change zooplankton. Bison urine creates significant heterogeneity in prairie plant and invertebrate communities (Frank et al. 2004). Daily movements of fish translocate nutrients from one zone of lakes to another (Schindler and Scheurell 2002). If the behaviors and life histories of the organisms in the ecosystems are not understood, then pathways of nutrients and ultimately trophic interactions cannot be predicted.

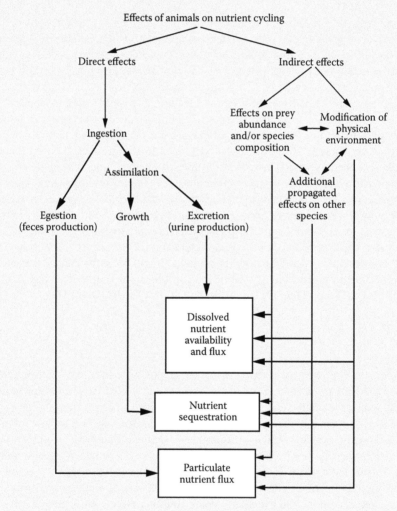

Direct and indirect effects of organisms on nutrient dynamics in food webs. (From Vanni, M. J. 2002. *Annual Review of Ecology and Systematics* 33:341–370.)

BOX 10.6

Recovering species within ecosystems may not restore the complexity of food webs. This is a challenge for emerging disciplines such as restoration ecology, which seek to restore ecosystems to their previous state. To illustrate how ecosystems might not recover, imagine an ecosystem where a keystone species is reduced to very low densities in three years (see figure below). Within the next four years, densities of the former keystone species are restored to levels higher than before their reduction (see figure below). However, the total number of food-web interactions is much less than before the perturbation reduced the keystone. The keystone species was clearly important in maintaining diverse trophic interactions. After its removal, other components of the food web changed. Perhaps a new, dominant competitor invaded or a predator rose to dominance. This lag in the food web's recovery is called *hysteresis*.

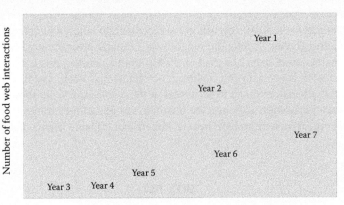

Density of keystone species

Depiction of the response of a food web to the removal of a keystone species starting in Year 2, to the complete recovery of the keystone species by Year 7.

These changes to ecosystems with the loss of apex species have been called *regime shifts* (Steele 1998). A classic example is the loss of Atlantic cod from the eastern coast of North America. This species was once so abundant that managers, fishers, and policy makers could not fathom that the population would collapse. However, this indeed happened and recovery may never occur because the ecosystem that the Atlantic cod once shaped changed dramatically once its density was reduced. A good introduction to this topic is "Cod: A Biography of the Fish that Changed the World" by Mark Kurlansky.

Restorations of some ecosystems are underway, with the reintroduction of lost top predators being a priority. One of the most dramatic examples is the restoration of sea otter populations of the Pacific coast of North America (Estes et al. 1998; Steneck et al. 2002). This species was hunted nearly to extinction by the early 1900s for its waterproof pelts. Kelp, a large aquatic plant that produces extensive beds off the coast, declined dramatically. It was not until the sea otter populations were restored that the trophic cascade linking sea otters to kelp was confirmed. Sea otters consume sea urchins, voracious herbivores that prevent kelp from emerging from the seabed. Urchins are protected from most predators by their sharp spines. Sea otters have developed a unique behavior where they bring the urchins to the surface, crack them open with a rock on their bellies, and consume the rich internal organs. As predicted for most keystone consumers, sea otters have developed unique characteristics that cannot be duplicated by other species such as the hundreds of species of fish in the kelp beds. When sea otters reduce urchin biomass, urchin consumption on kelp is diminished and the beds return.

The introduction of keystones or numerical dominants is not always positive. Many invasive species negatively affect ecosystems through food-web effects. A classic example is the rusty crayfish in north temperate lakes, which is an omnivore able to remove aquatic vegetation, reduce other aquatic invertebrates, and also suppress reproduction of fish (Lodge et al. 1998). In this manner, it is a strong middle-out interactor in lakes, exerting multitrophic effects up, down, and laterally within the food web. Recent removal experiments at the whole-lake scale showed that partial recovery is possible, although the crayfish was resistant to removal (Box 10.7). Many other invasive species have trophic effects that transmit across multiple nodes.

BOX 10.7

The irony of conservation of food webs in trophic ecology is that both loss of biodiversity and invasive species (i.e., an increase in unwanted biodiversity) can reduce trophic interactions. An attempt to restore invaded ecosystems is accomplished by removing the invader, which is an arduous but possibly successful task (Zavaleta et al. 2001). Some invaders such as the invasive salt cedar in the western United States have legacy effects that removal cannot ameliorate. This species increases soil aridity and salinity, making recovery of native species difficult. It is further problematic that this invasive provides habitat for an endangered native bird (Zavaleta et al. 2001). Thus, removal of salt cedar may have detrimental, unintended consequences for the protection of some desirable species.

Removals treated as experiments provide insight into food-web responses. Sparkling Lake in northern Wisconsin was the site of a massive removal of the invasive rusty crayfish with trapping from 2002 through 2005 (Hein et al. 2007). Crayfish were attracted to baited traps and were removed from sites around the lake (see figure below). By the fourth year, trap densities of crayfish were reduced by 95%. As shown by a multilake study (Lodge et al. 1998), the removal of rusty crayfish in Sparkling Lake led to increases in benthic invertebrates and greater trophic interactions (McCarthy et al. 2006).

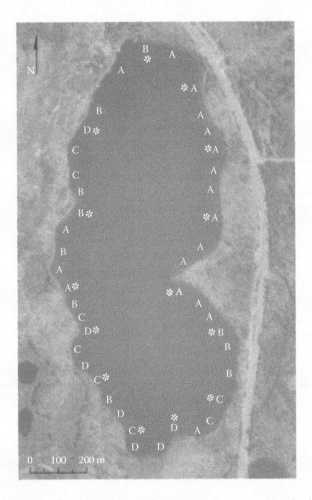

Sparkling Lake, Wisconsin, where rusty crayfish were removed using baited traps during 2002 through 2005 in the shoreline. Each letter and asterisk are sampling sites depicted in Hein et al. (2007).

10.8 TOPOLOGY, ECOLOGICAL NETWORK ANALYSIS, AND FOOD-WEB PROPERTIES

Theoretical developments in food-web ecology are thriving, although the patterns that emerge are often more confusing than enlightening. Central to the notion that the topology of food webs underlies their structure and function is the idea that food webs are responsible for the stability of ecosystems and therefore maintenance of biodiversity. Food webs are self-organized systems that have to be stable to persist. The central question to the theoreticians is, how does the topology maintain this stability and keep food webs from collapsing? Simple assembly rules likely govern the complexity.

Pascual and Dunne (2006) provide an excellent summary of concepts of stability arising in food webs, showing the historical development of techniques in food-web modeling. The cascade model based on Lotka Volterra, predator–prey dynamic predictions (see Section 10.7.2), was the earliest model exploring stability (May 1972). As mentioned earlier, this model incorporates the consumptive effects of predators on prey populations and searches for stability in population cycles between them. This model is a useful construct but has largely been abandoned by theoretical ecologists for other more realistic approaches. The niche model (Williams and Martinez 2000) models feeding relationships between consumers and prey based on each species niche value, which is the range of all possible trophic interactions. The model then determines how data fit to this pattern. This model has been modified further to include nested interactions (Cattin et al. 2004), where niche breadth (i.e., feeding linkages) in the food webs can be adjusted to improve model fit.

These theoretical approaches are largely based on the pattern of linkages within food webs and do not incorporate factors such as energy flow and interaction strength. *Ecological network analysis* is a powerful tool (Ulanowicz et al. 2014; Fath 2007; Fath et al. 2007) that allows food-web ecologists to quantify the path of energy or some other relevant currency through food-web linkages. We touch upon this approach in Chapter 8. This approach involves more data than simple pairwise linkages within food-web inventories. A powerful aspect of this approach is that it is based on mass-balance constraints. The energy entering the food web must balance among all the species or nodes (Chapter 8). Thus, the model should provide robust results within the constraints of the structure of the food web and the precision and accuracy of the data (Box 10.8).

The jury will remain out for quite some time about which model approach or suite of approaches will provide the best fit to real food webs. As important as the appropriate model to use is the relevant metric by which food-web properties are quantified. There are dozens of definitions of food-web stability in the literature. *Resilience* is the concept that a food web can rebound from a perturbation, while resistance is the measure of how much of a perturbation a food web can take before a change in structure and function occurs (Pimm and Lawton 1977). Similar to this is the concept of *reactivity*, where a perturbation *propagates* through the elements (i.e., nodes) of a food web. A small perturbation may have effects that push component populations from equilibrium and cause the ecosystem to change its state (Chen and Cohen 2001).

BOX 10.8

Ecological network analysis (Ulanowicz 2004) is a technique that has simi-
larities to mass-balance bioenergetics at the individual and ecosystem scale
(Chapter 8). These models, unlike the models in Chapter 8, are open, allowing
for inputs from exogenous sources as well as internal dynamics. These models
produce simultaneous solutions for ecosystem flows, whereas those in Chapter 8
are largely discrete simulations.

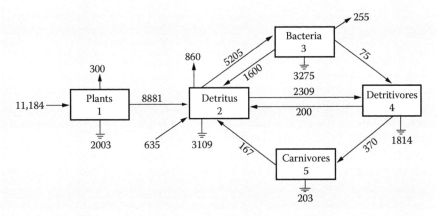

These models account for the movement of energy, elements, or molecules
among their compartments. The transfer among organisms (e.g., trophic levels, guilds, species, predators, prey) is balanced as the biomass within each
box and the flow of materials among them. The general assumption of these
models is that through time, the flow (the arrow) and the materials in each box
reach a steady state.

$$X_i + \sum_{j=1}^{n} T_{ji} = \sum_{k=1}^{n} T_{ik} + E_i + R_i$$

This equation shows that the sum of energy or materials across all the taxonomic boxes (T_i) in a food web is the result of the income X and the loss
through consumption E or internal losses R. Models in ecological network
analysis balance the two sides of the equation by finding under what mathematical conditions this occurs using linear, matrix algebra. This is similar to
stage-structured modeling in population biology to assess how simple population characteristics like growth rate and reproductive ability affect population
behavior. In these models, the rates of flow and the capacity for each node to
accumulate biomass affect food-web behavior.

Theoretical developments will incorporate both the mechanisms described throughout this book and combine new techniques for measuring food-web interactions such as tracers to improve predictions about the stability of food webs. Although we noted that compartments exist in food webs (Box 10.3), theoreticians have had a hard time coaxing these relevant locales of strong interactions within food webs from their models (Pascual and Dunne 2006). As Polis and others have pointed out, the frontiers of food-web theory include incorporating species richness and spatial scale into models. Further, the adaptability of webs to changes in internal and external forcing factors will need to be explored.

10.9   CONCLUSIONS

Food webs are attempts to bridge the gap between population/community aspects of ecosystems with patterns of materials and energy flux. These constructs comprised of nodes and connections in a network framework vary from capturing simple feeding connections to the intensity of energy flow and indirect effects. Regardless of the resolution of the food webs and their complexity, they, like all models, are simplifications of the trophic relationships occurring in ecosystems.

Food webs have been studied both empirically and theoretically for several decades. Given the sheer complexity of interspecific and intraspecific relationships in nature, generalities seem elusive. However, some general patterns do seem to exist. Most interactions between organisms are weak, meaning that changes in the density or behavior of one population do not exert much of an effect to the other. In some ecosystems, consumers emerge—likely through coevolutionary processes—that affect characteristics of the food web such as the distribution of biomass, relative productivity, and perhaps the species present. These consumers either have disproportionate effects relative to their density and are considered keystones or have strong impacts because of their high density and consumptive demand (i.e., numerical dominants).

What drives the structure and ultimately the function of ecosystems has generated much debate among ecologists. Perhaps one of the oldest questions in ecology is whether food-web size and stability are related. The answer is complicated but generally yes. The food-chain length within ecosystems, which is the distance from the basal producer to the most distant consumer, is limited on Earth. The amount of energy available to food webs is finite, and inefficiencies in trophic transfer place an upper limit on feeding steps. Growing complexity of consumer life histories with higher trophic levels and limited space available for trophic interactions also likely limit food-web size.

The most recent, pervasive discussion about food webs revolves around the relative roles of donors (prey), midtrophic species (both prey and consumers), and apex consumers on trophic interactions. Some consensus appears to have been reached that all ecosystems are ultimately limited by the amount of energy available at the base. However, how food webs are structured also relies on the impacts of consumers from both the middle and the top of the trophic pyramid. Consumers exert strong effects that propagate across multiple trophic levels, affecting patterns of autotrophic biomass and production. Predicting these consumer impacts is difficult and depends on tight, potentially evolutionarily driven linkages with food webs. Not all consumers are created equal, underscoring the need to preserve biodiversity to avoid dramatic changes in ecosystem structure and function.

QUESTIONS AND ASSIGNMENTS

1. What is a node in a food web?
2. How are food webs and the World Wide Web similar?
3. What are some common topologies of food webs?
4. Can food webs be used to find trophic levels in ecosystems?
5. How can the connectedness within food webs be calculated?
6. What is interaction strength? Do all species have the same interaction strength?
7. Are most interactions in food webs weak or strong? Why?
8. What is ecological network analysis?
9. How do the relative sizes of consumers and prey affect the strength of interactions in food webs?
10. Do more species in a food web make it more stable?
11. Why does enriching a food web with nutrients not necessarily increase the food-chain length?

11 Secondary Production

11.1 APPROACH

This chapter provides complementary approaches to the estimation of biomass production in ecosystems that we explored with mass-balance bioenergetics approaches in Chapter 8. The allure of the approaches herein is that complicated physiological models, especially those requiring specific estimates of metabolism, are not required. We end by suggesting that the trophic basis of secondary production has merit and should be applied more frequently.

11.2 INTRODUCTION

When we think of ecosystem production and productivity, we generally think green, which is primary production by plants and other autotrophs. As discussed in Chapter 2, primary producers are the foundation of ecosystem productivity, but energy must be moved up other trophic levels in the food web. Most texts focus on primary production and provide little discussion on concepts, methods, and patterns associated with secondary production, because measuring and interpreting primary production is relatively straightforward. Secondary production, on the other hand, is more difficult to conceptualize and quantify. Here, we provide some background and an overview of how secondary production can be estimated, and the utility of such estimates in studies of trophic ecology.

Secondary production is the formation of heterotrophic biomass through time and across space, regardless of fate (e.g., biomass may die, decompose, or be eaten by a predator, but it was produced at some point). As a simple example, let us start with a terrarium with 100 *Lumbricus* worms called night crawlers weighing 1 g each. This gives us a starting point of 100 g of biomass in our closed terrarium system. Over time, the worms grow and multiply, so we might pull them all out of the terrarium and weigh them after 1 month and find that we now have 120 g of worm biomass, yielding a simple production estimate of 20 g of biomass per terrarium per month. This yields a crude estimate of secondary production for the first part of the definition (formation of heterotrophic biomass), but things get a bit trickier when we consider the second part of the definition (regardless of fate). The issue here is that we do not know if any of the worms died and decomposed over the course of the month. Accordingly, models used to estimate animal production follow *cohorts*, individuals born or hatched at the same time, through time, or estimate cohort structure based on demographic analyses of sampled populations.

While secondary production is often depicted as energy associated with trophic levels of heterotrophic organisms (e.g., primary consumers), actual measurements of secondary production are generally based on single cohorts or populations, rather than entire trophic levels (Golley 1968; Waters 1977). Population-based methods for estimating secondary production have been around for ~40 years (e.g., Waters 1977), but they are not widely applied in ecological studies and are generally absent from most general ecology texts. Given the importance of consumers in all ecosystems on the planet, and the utility of secondary production estimates for addressing a wide variety of basic and applied ecological concepts and questions, it is puzzling that there is not more emphasis placed on the topic. In fisheries and some other fields, models of biomass production have been developed using mass-balance bioenergetics approaches, based on individual-based expectations for organisms (Chapter 8). This chapter provides a more general, ecosystem-based approach to those tools developed in that chapter.

11.3 WHY MEASURE SECONDARY PRODUCTION?

Secondary production is a rate, generally expressed as some unit of mass (e.g., grams) per unit area or volume (square meter) per unit time (e.g., year). Abundance, density, and biomass, which are more common metrics by which we measure animals or other consumers (e.g., fungi), are static measures. We can estimate the average abundance or biomass of an animal over the course of a month or year, but these values are still not rates. While static measures such as these can be useful in a variety of ways, their use for examining the ecological roles of consumers is limited.

Ecologists are increasingly interested in the ecological roles of consumers, such as how they influence overall energy flow and ecological processes like nutrient cycling and decomposition. Ecosystem functions and processes are rates, not static measures, and thus linking consumer organisms to them using static measure like abundance and biomass is not always appropriate or informative. On the other hand, production, as a rate, is considered the best measure of the relationships of animals to energy flow and can be directly linked to processes such as nutrient cycling (Odum 1957; Huryn 1996). As a simple example, if we estimate that grasshopper production in a prairie is 5 g m^{-2} year^{-1}, and grasshopper tissues are about 10% nitrogen, then we can quickly assess the contribution of grasshoppers to the cycling of an important limiting nutrient in the prairie; 10% of the 5 g m^{-2} year^{-1} is N, so 0.5 g of N cycles through the grasshoppers each year. This rate can then be interpreted in the context of inputs, internal cycling, and outputs of N in the system, all of which are also rates.

Raymond Lindeman's highly influential paper published in 1942, "The Trophic Dynamic Aspect of Ecology," set the stage for development and expansion of secondary production studies (Lindeman 1942). Among other things, Lindeman suggested that ecologists could compress the wide array of interactions among community components into a common currency such as energy flow to quantify them. Lindeman also suggested that an organism's overall success could be a function of its ability to acquire and retain energy. Both of these concepts are central to studies of secondary production, which quantitatively link organisms to energy flows among community and ecosystem components.

One of the first comprehensive studies of energy flow through an entire ecosystem, primary and secondary production estimates, took place in a spring system in Florida. H. T. Odum's classic study (Odum 1957) of Silver Springs utilized a variety of methods, including growth studies of snails in cages and respiration estimates of fishes, to examine growth and tissue turnover rates of major animal groups as part of developing a complete energy budget for this ecosystem. This system was ideal for studying trophic dynamics and associated energy flow because of its very stable temperature, chemical, and flow conditions associated with large springs. The Silver Springs study has served as a model for examining energy flow patterns in ecosystems.

Secondary production has been referred to as the *ultimate dynamic variable* because it integrates abundance, biomass, growth, survival, and other important population measures into one value (Benke 1993). Secondary production studies draw upon elements of population biology (population growth and factors affecting it) and ecosystem ecology (energy flow and nutrient cycling) (Benke and Whiles 2011). Along with their value for linking animals to ecosystem processes, production estimates have been used to examine animal responses to a variety of natural (e.g., drought, disease) and anthropogenic (e.g., logging, pollution) perturbations. As an example, exclusion of leaf litter inputs from a riparian forest to a headwater stream had cascading effects through the food web, resulting in lower secondary production of multiple trophic levels (Figure 11.1; Box 11.1).

Secondary production is also an accurate way to assess energy available integrated through time, in the form of prey, to predators. This aspect of secondary production can apply to conservation issues regarding predatory species ranging from large carnivores to amphibians. Here again, abundance and biomass can be informative to some degree, but they do not reflect the biomass produced, and thus available

FIGURE 11.1 Section of the mesh canopy that was constructed over an entire 100 m reach of headwater stream in the southern Appalachian Mountains. The canopy and riparian fencing eliminated almost all litter inputs from the forest to the stream, resulting in lower growth rates and secondary production of detritivorous invertebrates and predatory invertebrates and salamanders.

BOX 11.1   RIPARIAN FOREST SUBSIDIES AND
SECONDARY PRODUCTION OF STREAM ANIMALS

Energy and material exchanges between habitats and ecosystems are of great interest to ecologists because they are central to understanding ecological processes and connections at large spatial scales (e.g., landscapes). These so-called *ecological subsidies* can be obvious and even spectacular events, such as the annual salmon migration runs that move significant quantities of marine-derived nutrients from oceans to streams (Naiman et al. 2002). Others may be subtler, such as annual inputs of autumn-shed leaves from riparian forests to streams and wetlands (Baxter et al. 2005). Human disturbances that interrupt or alter subsidies can have significant impacts on receiving ecosystems.

Researchers at the Coweeta Hydrologic Laboratory in the Blue Ridge Mountains of western North Carolina performed a large-scale experimental manipulation using a coarse mesh canopy and riparian fencing to exclude leaf litter inputs to a headwater stream for multiple years (Wallace et al. 1997; Figure 11.1). The manipulation greatly reduced inputs and standing stocks of detritus, and resulted in lower production of many stream animals, including detritivores that fed directly on leaf inputs from the riparian forest and the predators that fed on them (Wallace et al. 1997). Reduced secondary produc-tion during the litter exclusion experiment was a function of reduced popula-tion sizes, biomass, and individual growth rates, and these negative responses cascaded through invertebrates up to the predatory salamanders in the stream, which also grew slower and had reduced production (Johnson et al. 2003; Johnson and Wallace 2005).

The results of this manipulation have important applications. In particular, they indicate that human activities such as logging that reduce litter inputs to streams can have bottom-up effects, resulting in reduced secondary production of multiple trophic levels including top predators. This study also demonstrates the utility of using secondary production as a response variable that integrates responses at multiple ecological scales (e.g., individual and population growth rates).

to the next trophic level, by a prey population. In this regard, secondary production has practical applications; humans manage populations of many predatory species (e.g., predatory fishes, waterfowl), and maximizing energy available to them is of obvious interest. One of the more well-known secondary production studies was designed to assess energy available to a valuable fishery in New Zealand streams (Box 11.2).

While some studies focus on prey production for fish and other animals that humans harvest and consume, others focus directly on the production of the man-aged or harvested species. Such information is useful for managing populations of interest and establishing harvest regulations such as bag limits. Methods for estimat-ing fish production generally employ modeling approaches that are very similar to

BOX 11.2   ALLEN PARADOX

One of the more well-known secondary production studies was designed to assess energy available to a valuable fishery in New Zealand streams. New Zealand is an increasingly popular destination for anglers because of its scenic streams with abundant large trout; these streams and the trout that inhabit them fuel tourism, a major part of the New Zealand economy. Allen (1951) examined trout and invertebrate prey production in a New Zealand trout stream and found that the amount of invertebrate prey production in the stream was much less than would be required to sustain the level of trout production observed. This pattern has since been observed in trout streams in other regions and was dubbed the Allen paradox (Hynes 1970; Waters 1988).

A comprehensive production study designed to investigate the Allen paradox in a New Zealand stream solved some of the apparent mystery by accounting for invertebrate production deeper into the stream bottom (hyporheic zone), terrestrial invertebrate inputs, and cannibalism by trout (Figure 11.2; Huryn 1996). However, this study still found that trout required almost all prey production that was available in the system; invertebrate production on surface substrata of the stream accounted for ~80% of trout demand, followed by hyporheic invertebrates (~13%) and terrestrial inputs (~8%); cannibalism contributed very little.

Ecologist's views of the Allen paradox have shifted from primarily methodological issues in developing production estimates to accepting that trout in productive trout streams (i.e., those producing at least 100 kg WM trout ha^{-1} year^{-1}) consume an extremely large proportion of benthic prey production (e.g., >80%); margins of error associated with production estimates could therefore encompass estimates indicating deficits to small surpluses in prey production. These patterns also indicate that trout exert strong top-down controls in streams where they are productive, an important consideration in regions such as New Zealand where they are not native; remaining native galaxiid fish populations exert much less energetic demands on these streams.

the methods used for invertebrates, accounting for the production of fish biomass and the rate by which fish biomass is removed by natural mortality and harvest (Ricker 1975; Box 11.3). These models have moved on to more sophisticated, physiologically based approaches outlined in Chapter 8.

The secondary production literature reflects a strong bias toward aquatic habitats (Benke and Whiles 2011), and this is likely linked to long-term human interests in fish production. Along with estimating fish production, fisheries managers are also interested in prey production for fish. As such, many of the methods and early studies of animal production were based in aquatic habitats, particularly fishes and the invertebrates many of them feed on (Waters 1977). While secondary production studies are becoming increasingly common in studies of aquatic ecosystems, terrestrial secondary production studies remain relatively scarce in the literature; studies

BOX 11.3   CONCEPT OF YIELD IN FISHERIES

Fisheries biology has long grappled with the concept of production of fish populations. The concept of maximum sustainable yield (MSY) was developed based
on models of secondary production (Ricker 1975). The basic idea behind these
yield models is that the catch rate (C) in terms of biomass of fish per unit of fishing effort (E) is maximized at some level below maximum E. Effort can be any
kind of harvest mortality including netting, angling, or trawling. Yield or C is the
production of the fishery and should be maximized when the population is reduced
below its carrying capacity. The reason this should happen is that the surviving fish
grow faster and have higher reproductive rates, increasing production and allowing the fishery to reach the MSY or, in the context of this chapter, the maximized
secondary production of the fish population. This was considered a rule of thumb
in fisheries management, although identifying the MSY is problematic for a variety of reasons (Larkin 1977). Finding the MSY is difficult, if it actually exists in
the population, and usually requires overfishing (and recovery) to discover. More
sophisticated models of fish production have arisen as a result (Chapter 9).

of terrestrial game species tend to focus more on population analyses such as life
tables and related approaches, which actually draw on the same data and information
that are used for estimating secondary production (Benke and Whiles 2011).

11.4   PRODUCTION EFFICIENCIES

Secondary production is ultimately the end result of a series of ecological inefficiencies (Chapter 2). The obvious distinction between primary producers and consumers is that while primary producers create their tissues using energy from the sun
or chemicals, consumers are converting materials that they ingest into their own
tissues. Not all that is ingested is assimilated, setting up the first inefficiency. The
difference between what is ingested and what is egested is often expressed as an
assimilation efficiency (AE), which is the percent of ingested material that is actually assimilated. Of that actually assimilated, some is used for maintenance (e.g.,
respiration) and lost through excretion of nitrogenous wastes; what is left, which
is secondary production, is used for somatic growth, reproduction (gamete growth
and related), or storage (think fat). Thus, a simplified individual energy budget for a
consumer can be expressed as

$$P = A - (R + U + S)$$

where P is production, A is assimilation, R is respiration, and U is excretion. *Specific
dynamic action*, which is the energy used by an animal in digestion, transport, and
deposition of assimilated energy (Beamish 1974), is represented as S; simplified
approaches sometimes lump S in with R. The percent of assimilated material that
is actually converted into production is the *net production efficiency (NPE)*. The

AE and NPE can be multiplied together to estimate the *gross production efficiency* *(GPE)*, which is the percent of ingested material that becomes actual production.

Assimilation efficiencies vary greatly across animal species and with type of food; animals that feed on energy-rich, easily assimilated materials (e.g., carnivores) tend to have higher AEs, whereas herbivores and detritivores have generally lower AEs. Net production efficiency values for invertebrates and ectothermic vertebrates are often around 50% (Pough 1980; Benke 1996). Endothermic groups have much higher metabolic demands than ectotherms, and thus lower NPE because a higher proportion of assimilated energy is used to fuel metabolic activity. Lower NPE in endotherms is compensated for to some degree through higher AE associated with sophisticated digestive machinery and relatively constant internal body conditions for digestion. Higher energy demands of endotherms also translate into higher overall food ingestion rates.

Assimilation and production efficiencies can be related to important functional roles in ecosystems. For example, detritivores such as earthworms and isopods have low GPE because the materials they feed on, like decaying leaves on the forest floor, are of relatively low nutritional quality and can be hard to digest. To compensate for this, they ingest large quantities of material to meet their energetic needs. High rates of ingestion translate into rapid decomposition rates and recycling of materials and energy. Thus, species with relatively low secondary production rates are not necessarily less important at the ecosystem scale; they may not produce as much heterotrophic biomass as more productive species, but they can contribute significantly to energy flow and nutrient cycling because of their inefficiencies.

11.5   HOW DO WE ESTIMATE SECONDARY PRODUCTION?

There are a variety of ways to estimate secondary production, most of which involve following changes in density and individual mass in a cohort (Ricker 1946; Waters 1977; Benke and Huryn 2006). A cohort is a group of individuals that started life at about the same time. The relationship between cohort density and individual mass over time can be depicted with an *Allen curve* (Allen 1951; Figure 11.2). An Allen curve shows the production of a cohort for a given time interval (the trapezoid underneath the Allen curve defined by rectangle A + triangle B in Figure 11.2), which is the biomass produced by those that survived (A) and those that died (B) in the time interval. Total production by the cohort is the total area under the Allen curve. The Allen curve is also useful for visualizing secondary production; if there were no mortality, secondary production would be the entire area of the plot.

The trapezoid for a time interval can be calculated from densities and individual mass by multiplying the average number of individuals during the time interval by the change in individual mass:

$$\frac{N_t + N_{t+1}}{2}(W_{t+1} - W_t)$$

where N_t is the number of individuals at the beginning of the time interval, and N_{t+1} is the number at the end of the interval; likewise, W_t is the average individual weight

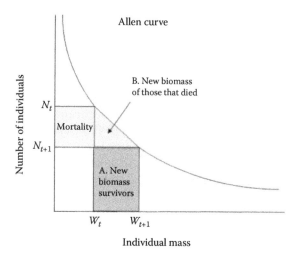

FIGURE 11.2 An Allen curve showing the relationship between the number of individuals in a cohort and individual mass over time. N is number and W is individual weight or mass; subscripts t and $t+1$ indicate points in time. As time proceeds, individuals become larger (moving right on the X-axis) and numbers decline (moving down the Y-axis; the difference between N_t and N_{t+1} is mortality). Rectangle A and triangle B together represent production for the time interval depicted.

at the beginning of the interval, and W_{t+1} is the average individual weight at the end. This yields the production for a given time interval; total production is the sum of the interval values. This method for estimating production is called the *increment summation method* (Waters 1977; Benke 1996). The increment summation method is considered fairly accurate, but can only be used when a cohort can be followed through time. Table 11.1 shows procedures for estimating the production of the caddisfly, *Brachycentrus spinae*, an aquatic insect that can be an important food item for insectivorous fish. Note that the table is set up so that the first sampling date is when the caddisflies have hatched as larvae; the density of larvae (N) clearly declines with time as they grow in size (W); if these two columns were plotted on a graph, they would yield an Allen curve. If this pattern is not evident (density decreasing and individual mass increasing with time since hatching or birth of the cohort), then a cohort method like the increment summation method cannot be used, and noncohort methods should be explored (see below).

The production estimate for the caddisfly in Table 11.1 of ~2.6 g m^{-2} year^{-1} is modest for an invertebrate primary consumer. Estimates for single species of invertebrates range from less than 1 mg m^{-2} year^{-1} up to extremes of ~8 g m^{-2} day for highly productive species (e.g., those that have both high biomass and rapid growth rates; Huryn and Wallace 2000). Production can be expressed as annual production (the sum of the interval production estimates), or interval estimates can be divided by the number of days in the interval to express daily production, as in the last column of Table 11.1.

The two major components of production are population biomass (which in turn is a function of density and individual size) and individual growth rates. Daily

TABLE 11.1
Data and Production Calculations for *Brachycentrus spinae*, a Stream-Dwelling Caddisfly with a 1-Year Life Cycle

Date	Interval Days D	Density N No. m^{-2}	Individual Weight W mg	Biomass B mg m^{-2}	Weight Change (ΔW) $(W_{t+1} - W_t)$	Mean Density Mean N No. m^{-2}	Interval Production Mean $N \times \Delta W$ (P) mg m^{-2}	Daily Production P/D mg m^{-2} Day^{-1}
18-May	14	283	0.021	5.94	0.036	255.0	9.2	0.656
1-Jun	12	227	0.057	12.94	0.031	204.5	6.3	0.528
13-Jun	16	182	0.088	16.02	0.085	160.5	13.6	0.853
29-Jun	14	139	0.173	24.05	0.179	124.0	22.2	1.585
13-Jul	13	109	0.352	38.37	0.588	98.5	57.9	4.455
26-Jul	22	88	0.940	82.72	0.266	74.0	19.7	0.895
17-Aug	13	60	1.206	72.36	0.590	54.0	31.9	2.451
30-Aug	16	48	1.796	86.21	0.025	42.5	1.1	0.066
15-Sep	18	37	1.821	67.38	1.378	32.0	44.1	2.450
3-Oct	42	27	3.199	86.37	0.358	20.0	7.2	0.170
14-Nov	23	13	3.557	46.24	1.074	11.0	11.8	0.514
7-Dec	51	9	4.631	41.68	2.222	6.5	14.4	0.283
27-Jan	21	4	6.853	27.41	1.624	3.5	5.7	0.271
17-Feb	24	3	8.477	25.43	3.071	2.5	7.7	0.320
13-Mar	45	2	11.548	23.10	3.252	1.0	3.3	0.072
27-Apr		0	14.800	0.00				
Average annual B				41.01				
Total production							256.0	
Production/biomass =		6.2						

Source: Data are from Ross, D., and J. B. Wallace. 1981. *Environmental Entomology* 10:240–246.

Note: Dates are days when samples were collected from the stream. May is used as a start date because this is when newly hatched larvae first appear. Production is calculated using the increment summation method.

production estimates standardize across sampling intervals of different lengths and can be used to examine production dynamics on a finer scale and assess the influence of the individual components of production. One can compare daily production values with density, biomass, and growth (changes in weight) during each interval. As evident for the caddisfly production patterns in Table 11.1, interval production is often highest at some intermediate point on the Allen curve, when numbers are still fairly high, individuals are larger than when they were born or hatched, and growth is still fairly rapid.

11.6 NONCOHORT METHODS

For many smaller organisms with rapid life cycles and overlapping generations, cohorts often cannot be followed and cohort procedures such as the increment summation method cannot be used. In these cases, the *instantaneous growth* method is sometimes used. The instantaneous growth method requires only an estimate of average biomass and an estimate of growth for a given interval. Biomass estimates are easily obtained from quantitative field samples, but estimates of growth can be more difficult because of overlapping generations and/or rapid generation times. In these cases, growth may be estimated by laboratory rearing studies, caging individuals in the field, or marking and recapturing individuals. Changes in weight over time are then used to estimate instantaneous growth as follows:

$$G = \ln\left(\frac{W_{t+1}}{W_t}\right)$$

where G = instantaneous growth rate, W_t = weight at the beginning of the interval, and W_{t+1} = weight at the end of the interval. The instantaneous growth rate is often standardized to a daily rate by dividing by the number of days in the interval. Once G and average biomass for the interval are obtained, interval production is estimated as the product of the two.

The data in Table 11.1 can be used to estimate production with the instantaneous growth method by calculating G from the change in individual biomass during sampling intervals. For example, for the first interval of May 18–June 1:

$$G = \ln\left(\frac{0.057}{0.021}\right) = \ln(2.714) = 0.998$$

Using biomass estimates for the same interval in Table 11.1 (5.94 and 12.94 mg), the average biomass was 9.44 mg/m² and thus production for the interval was

$$P = 0.988 \times 9.44 = 9.42 \text{ mg/m}^2 \text{ for the May 18–June 1 interval}$$

While they require less detailed information and are thus easier to calculate, production estimates using the instantaneous growth method are generally considered less precise than those obtained with cohort procedures such as the increment summation method. In this case, the estimate from the instantaneous growth (9.4 mg/m^2) and increment summation (9.2 mg/m^2) methods are close, but this is not always the case.

One issue with growth rate estimates is that they can vary with conditions such as temperature and with the life stage of the animal of interest (e.g., newly hatched individuals vs. those close to maximum size). For this reason, instantaneous growth procedures are often incorporated into temperature- and size-dependent models (e.g., Walther et al. 2006). Growth models are often developed from controlled laboratory studies or field studies using growth chambers deployed across environmental gradients of interest.

There is increasing use of the *size-frequency method* for estimating production because it can be applied fairly accurately in situations where clear cohorts can be followed and when cohort structure is not clear. This method involves constructing an average cohort from sampled size classes, much like constructing a life table. Subsequent calculations are then similar to the increment summation method, except that interval production estimates are multiplied by the number of size classes or developmental stages present because they are not actually being followed through time. The interval production estimates are then summed to estimate annual production for species with 1 year life cycles. Those with shorter or longer life cycles can be expressed as production over the length of the life cycle, or standardized to 1 year using a cohort production interval correction, which is obtained by simply dividing 1 year by the length of the life cycle. For example, a species with an 8 month life cycle would have a CPI correction of 12/8 = 1.5; the production estimate for one cohort of this species would be multiplied by 1.5 to estimate annual production. Details on the size frequency method and other production methods can be found in Benke and Huryn (2006).

11.7 PRODUCTION/BIOMASS RATIOS

Annual production (P) divided by average annual standing stock biomass (B) yields an estimate of growth and turnover, or P/B (year^{-1}). Production-to-biomass ratios vary tremendously across animal taxa as a function of individual growth rates and life histories. In general, small, rapidly developing species will have high P/B values (e.g., >10), whereas slower growing species may have values less than 1. Production and biomass values for *B. spinae* presented in Table 11.1 yield a P/B of 6.2, which is moderate among insects; P/B estimates for some fast-growing insect tax such as midges can exceed 100, whereas some large slower-growing insects such as some dragonflies and stoneflies can have P/B values ranging from 1 to 2.

Actual tissue turnover time for the animal of interest can be estimated as the inverse of the annual P/B. In the case of the *B. spinae*, tissue turnover is 365/6.2 = 59 days for complete turnover of the tissues as the animal grows. As with production, estimates of tissue turnover time are valuable for linking animals directly to ecosystem rates.

These relationships reinforce that production is heavily influenced by growth rate, which, in turn can be influenced by environmental factors such as temperature, pH, or salinity. While there have been some attempts to model production based on environmental variables, meta-analyses of production studies conducted in freshwater (Benke 1993) and marine (Cusson and Buorget 2005) habitats indicate that biotic variables such as life history characteristics are more important than environmental variables for explaining variability in production and P/B estimates.

11.8 TROPHIC BASIS OF PRODUCTION

Production and P/B estimates themselves can be very valuable for ecological studies, and they can also be used in conjunction with other information to quantitatively assess material and energy flows through animals at relatively fine scales. Studies of the trophic basis of production use information on diets, production efficiencies, and secondary production to quantify the actual contributions of different food items to observed production (Benke and Wallace 1980). Such examinations can accurately reveal the functional roles (e.g., grazing algae) and trophic status of an animal, sometimes with surprising results.

Food demand, or consumption required to sustain the level of production observed by an animal, can be estimated using ecological efficiencies:

$$C = \frac{P}{AE * NPE}$$

where C = consumption, P = production, AE = assimilation efficiency, and NPE = net production efficiency. For example, if an animal that has 10 g of annual production feeds on tree leaves and it assimilates 50% of what it consumes, and 60% of what it assimilates is converted to actual production, then

$$C = \frac{10}{0.5 * 0.6} = 12$$

This indicates that 12 g of leaf material are required per year to maintain the 10 g of production by the animal. These are the same basic methods used by Allen in the New Zealand trout streams (Box 11.2). Of course, accurate estimates of assimilation and net production efficiencies are needed. In some cases, these can be gleaned from the literature, or laboratory experiments may need to be performed.

These types of analyses can be taken a step further to examine the contribution of individual food types to secondary production, or the trophic basis of production. Trophic basis of production studies generally start with diet information converted into percentages (e.g., 80% detritus, 20% animal material on average in guts of a species of interest). Assimilation and net production efficiency estimates are then applied to each food type to estimate relative amounts to production. Relative amounts to production are then converted to percentages of production attributed to each food type, which are then multiplied by total production to estimate actual contribution of each food type to production, or the trophic basis of production (Table 11.2).

TABLE 11.2
Procedure for Calculating the Contribution of Different Food Types to Central Stoneroller (*Campostoma anomalum*) Production in a Kansas Stream

Food Type	Amount in Gut (%)		AE		NPE		Relative Amount to P	% P Attributed to Food Type	P Attributed to Food Type mg m^{-2} Year^{-1}
Misc. detritus	55.7	X	0.10	X	0.5	=	2.79	30	78
Diatoms	25.3	X	0.18	X	0.5	=	2.28	24	64
Filamentous algae	10.4	X	0.41	X	0.5	=	2.13	23	59
Animal	5.6	X	0.70	X	0.5	=	1.96	21	55
Leaves	3.0	X	0.10	X	0.5	=	0.15	2	4
Total							9.30	100	260

Source: Data are from Evans-White, M. A. et al. 2003. *Journal of the North American Benthological Society* 22:423–441.

Note: AE is assimilation efficiency, NPE is net production efficiency, and P is production. The percentage of production attributed to each food type column is calculated by dividing each individual value in the relative amount to production column by the sum of all those values. The annual production (260 mg m^{-2} year^{-1}) is then multiplied by the percentage of production attributed to each food type to estimate the amount of production attributed to each food type. Note that large differences in assimilation efficiencies for different food types can override amounts consumed in determining contributions to production.

These rather simple calculations yield basic information on consumption and the relative contributions of food items to production; more sophisticated modeling procedures that incorporate variables likely to affect these relationships, such as size and age of the consumer and temperature regimes, have been used for more detailed studies of consumption patterns and the trophic basis of production (Box 11.3). Along with their utility for understanding energy flow patterns into consumers, trophic basis of production estimates can reveal the true trophic status of a consumer, which may or may not be the same as its functional role. For example, a consumer may ingest mostly plant material, and thus function as a grazer, but because of differential assimilation, it may derive most of its energy from animal material such that it is not classified as a herbivore even though it functions as a grazer.

BOX 11.4   TROPHIC BASIS OF RIVER FISH PRODUCTION

A study of the trophic basis of production of three fish species (smallmouth bass, rock bass, and flathead catfish) in a Virginia river was undertaken to examine the potential impacts of human harvest of crayfish and hellgrammites (aquatic larvae of the dobsonfly) for bait on fish production. Roell and Orth (1993) used simulations of fish energetics to estimate annual consumption of prey by cohorts of each fish species. Their consumption estimation procedure, which was based on methods detailed by Hewett and Johnson (1987), incorporated fish abundance and mortality rates, average fish growth history for each cohort, energy densities of fishes and their prey, fish diet composition, annual thermal history of the fishes, and a mass balance equation, algorithms, and parameter estimates for fish energetics and physiology.

Their analyses demonstrated that aquatic insects were the dominant contributors to production of age 0 and age 1 individuals of all three fish species, whereas crayfish contributed most to production of older individuals. The importance of crayfish to all three fish species resulted in high diet overlap among them. Hellgramites contributed little to fish production because they were rarely consumed. The three fish species combined consumed a high proportion of crayfish production (~75% of age 1 and age 2 annual crayfish production). Bait harvest by humans removed an estimated additional 5% of crayfish production from the river.

This study demonstrated how concepts and procedures often used in basic ecology could have important applications. The use of sophisticated modeling techniques to examine the trophic basis of production of popular sport fishes demonstrated the importance of crayfish to this valuable fishery, and how increased human harvest of crayfish for the bait industry could have deleterious effects. In contrast, given their low contribution to fish production, hellgrammite harvest was not predicted to influence the fishery.

11.9   CONCLUSIONS

Quantifying the trophic basis of production based on gross estimates of growth within ecosystems is a robust technique and complements the complicated bioenergetics-based models described in early chapters. A comprehensive examination of the utility of these various approaches to predicting production within trophic ecology will yield new vistas for research (Box 11.4).

Particularly useful for this approach to secondary production is the ability to expand upon estimates of autotrophic production (Chapter 2) and derive standard metrics by which ecosystems can be compared. Ratios such as P/B can be used to estimate the growth potential and responses of populations within ecosystems to human effects such as harvest and climate change.

QUESTIONS AND ASSIGNMENTS

1. Why is quantifying secondary production more difficult than primary production?
2. What is meant by secondary production being the ultimate dynamic variable in trophic ecology?
3. What is assimilation efficiency in the context of secondary production?
4. Describe what an Allen curve is and how it works.
5. How is the instantaneous growth method used? How does it differ from the increment summation method?
6. How do P/B ratios differ among taxa with different life histories?
7. What is the trophic basis of production? How can it be used to estimate the relative trophic position of consumers in ecosystems?

Section V

Quantifying Material
Flux and Synthesis

12 Nutrient Dynamics and Stoichiometry

12.1 APPROACH

At this juncture, we have explored the role that nutrients play in the growth and trophic dynamics of organisms in ecosystems. Chemical reactions must balance both within organisms and in the environment. However, the resulting ratios of nutrients stemming from stoichiometric processes may lead to mismatches between organisms and their environment. Autotrophs often have stoichiometric ratios that match the environment, and these nutrient concentrations have a strong impact on the growth, survival, and food web role of consumers. We will explore the basic processes involved and determine implications for trophic ecology.

12.2 INTRODUCTION

Trophic interactions are driven by many processes. As we have shown throughout this book, natural selection underlies all foraging events. As the physical environment changes, organisms with sufficient genetic diversity express phenotypic characteristics that allow them to consume the necessary food to persist and reproduce. Those individuals that cannot adjust their trophic roles will go extinct. As the past chapters have illustrated, selection for characteristics that maximize energetic intake has been the focus of much ecological research. Behaviors and chemicals that affect energy intake have been studied theoretically and empirically. However, this is not the whole story. In Chapter 9, we introduced the concept that factors not directly related to energy intake affect trophic interactions. Essential materials in the environment must be consumed in sufficient quantities to ensure that organisms grow, survive, and reproduce. The presence or absence of the simplest essential materials—critical elements—may be the most important factors affecting the fate of species and perhaps ecosystems. Justus Freiherr von Liebig predicted that a single material would limit chemical reactions, leading to Leibig's law of the minimum. The relative concentration of a single nutrient can determine the survival of organisms and the function of ecosystems.

Unlike energetics, where the currency (e.g., heat) is fairly easy to conceptualize and quantify, tracking the flux of elements within and through organisms can be tricky. Elements are usually consumed by organisms as molecules, which have predictable properties. Molecules are constrained by physical rules. Because of biological and physical constraints, the elements consumed, absorbed, and lost by organisms should be predictable. Evolution does not dabble in fission and fusion, meaning that organisms are incapable of transforming elements from one type to another. However, elements can be rearranged biochemically, resulting in an energetic cost or gain (Sterner and Elser 2002). Although the configuration of the elements in molecules changes, the elements entering and exiting the organism can be followed using a mass-balance approach. The mass of elements entering an organism through consumption or cellular transport must equal those incorporated into tissues or lost as waste (Figure 12.1).

Whereas organisms and their food usually contain molecules that conform to rules, many of the elements in the environment combine in a variety of complex ways. Reactions form strict ratios in the environment and are considered *stoichiometric*. Two hydrogen atoms and one oxygen atom form a stable molecule of water. Biological reactions between nitrogen and oxygen atoms may form many different, relatively stable molecules such as nitrate, nitrite, and nitrous oxide, each having different properties, again as per predictable stoichiometric accounting (Figure 12.2). These stoichiometric reactions and the molecules they generate in the environment greatly affect the successes and failures of organisms, trophic interactions, ecosystem processes, and ultimately biogeochemical cycling.

Because organisms must process elements in a stoichiometric manner (Sterner and Elser 2002), ratios of elements within organisms tend to be predictable, although those in the environment might not because of biogeochemical, climatic, human-induced, and other factors. In this chapter, we will explore the stoichiometric relationships that define the composition of organisms. We will discover that micro-organisms, plants, and animals vary greatly in their stoichiometry. Understanding the differences in stoichiometric relationships among these organisms provides insight into their trophic relationships. In a stoichiometric world, autotrophs and heterotrophs are both consumers of elements. In this context, we consider nutrients taken up directly from the environment as *food* to be consumed. Elements accumulate in the autotrophs and move through the food web, typically in ratios that differ from the environment. Consumers accumulate elements from the autotrophs in their

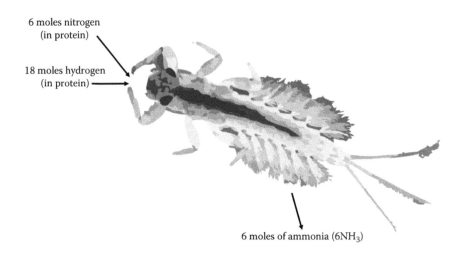

6 moles nitrogen
(in protein)

18 moles hydrogen
(in protein)

6 moles of ammonia ($6NH_3$)

FIGURE 12.1 Stoichiometry of molecules consumed and excreted from organisms.

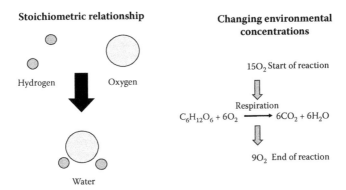

FIGURE 12.2 Stoichiometric reactions use exact quantities of elements. These reactions (such as respiration) will change the concentration of oxygen in the environment. Similar changes in environmental concentrations of nutrients occur as those that are needed are used, while those that are not consumed at high rates become proportionally more concentrated in the environment.

own stoichiometric ratios. We will find that the ratios of elements in the world do not match those of the biota, meaning imbalances in the nutrients in the living and abiotic pools. The dietary requirements of organisms and the elemental content of their food are often quite different, leading to a landscape with different combinations of elements.

As Chapter 3 demonstrated, plants and animals are inefficient at using materials, often leaking or excreting unneeded or potentially harmful elements or molecules (e.g., ammonia) back into the environment. Organisms rapidly release their accumulated nutrients at death. These recycled elements are either quickly reabsorbed by other organisms or lost geologically depending on their ecological importance, their chemical properties and resulting reactions with environmental chemicals, and ambient conditions such as temperature and pH. Because life is pervasive on the planet's surface, the movement of elements scales up from individuals, to ecosystems, and eventually to the entire globe. Trophic ecology and global biogeochemistry are linked because of stoichiometric processes (Schrama et al. 2013). These processes have important effects on conservation, climate, and ultimately the ability for Earth to support humans.

12.3   ELEMENTAL STOICHIOMETRY

Many of us recall high school chemistry and cringe at the word *stoichiometry*, if we remember the term at all. Recall that elements combine by sharing electrons orbiting their nuclei. Elements differ in their oxidative states, which means that they have different numbers of electrons (Figure 12.3). Electronegative elements like oxygen and chlorine have negative oxidation states and *steal* electrons from metals, like magnesium and calcium, which have positive oxidation states. The difference in oxidative numbers between elements predicts how strongly they will react. Elements receiving electrons are reduced while those donating electrons are oxidized. In the

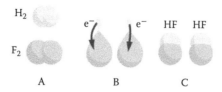

FIGURE 12.3 Reduction–oxidation reaction between hydrogen and fluoride. Fluoride atoms gain electrons and are reduced; hydrogen atoms contribute electrons and are oxidized.

stoichiometric reaction between hydrogen (H^+) and oxygen (O^{-2}), two hydrogen atoms are oxidized for each oxygen atom reduced to make water. The reaction is stoichiometric because the ratios are predictable and neither oxygen nor hydrogen remains after the reaction is complete. Recall that elements and molecules are measured as moles, a way of standardizing the amount of these materials. A mole is equivalent to the weight generated by 6.023×10^{23} chemical units (i.e., Avogadro's number) of the element or molecule. The atomic weight for one mole of a hydrogen atom is 1.00794 g. For oxygen, it is 15.994 g. When these elements combine to make one mole of water, the weights $1.00794 + 1.00794 + 15.994 = 18.010$ g, which is the weight of 6.023×10^{23} molecules of water. No free atoms remain in this stoichiometric summation (Figure 12.2).

The efficiencies created through natural selection should ensure that biochemical reactions within organisms are stoichiometric (Agren 2004). Reactions that do not use all the reactants available are energetically wasteful and should be selected against. The problem facing most organisms is that the environment has a variable elemental composition. Most organisms must take the seemingly random elements occurring in the environment and maintain them at precise internal ratios. The energy cost of maintaining biochemical reactions can be substantial. Hypothetically, the closer the environmental stoichiometric composition is to that of the organism, the less energy cost and the greater scope for growth and reproduction. Organisms with elemental requirements matching their environment should have a competitive advantage over those having *unbalanced* relationships with the reactive chemicals in the environment.

12.4 ENVIRONMENTAL CHEMISTRY

The composition of Earth affects trophic interactions. The planet's crust and seawater differ considerably in composition (Figure 12.4). The crust contains many metals including aluminum, calcium, iron, potassium, sodium, and silica. Oxygen and carbon are also abundant. Seawater, being mostly water, is predominately hydrogen and oxygen. It is salty because it contains many dissolved elements and molecules, with sodium, chloride, and carbon being common. An analysis of the rank order of these elements in crust and seawater from Sterner and Elser (2002) shows no relationship (Pearson's correlation, $r = 0.19$, $p = 0.35$). All organisms, including humans, have ancestors that originated in seawater. Humans contain mostly hydrogen, oxygen, sodium, and chloride. Not surprisingly, humans and most organisms

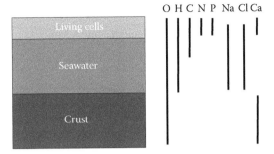

FIGURE 12.4 Ranks of common elements in organisms, seawater, and the Earth's crust. Elements sharing the same bars have similar ranks for an element. Elements only sharing half a bar are less closely related. Those sharing no bars are not similarly ranked.

have elemental compositions that correlate in rank order with seawater (Pearson's correlation, $r = 0.67$, $p = 0.0002$). It may be cliché to say, but seawater does run through the bodies of organisms both aquatic and terrestrial.

The rankings of elements in the ocean and in organisms may be related, but the stoichiometric ratios and concentrations still differ (Sterner and Elser 2002). Both nitrogen and phosphorus exist in high concentrations in animals and plants (Figure 12.4), although they are rare in the crust and seawater. Nitrogen is the most abundant gas in the atmosphere. However, it persists in elemental form with three covalent bonds that require immense energy to break. Phosphorus is highly reactive with oxygen and is often insoluble. It is testament to the importance of these elements to the function of all living organisms that they are concentrated in organisms relative to the environment in both water and on land. As we will see, the ratio of nitrogen atoms to phosphorus atoms is the primary focus of stoichiometric approaches in trophic ecology. How organisms accumulate and maintain these two elements in strict ratios is critical for processes at all ecological scales.

12.5 ORGANISMAL CHEMISTRY

In plant and animal physiology, a few organisms are *conformers* whereas the vast majority are *regulators*. A conformer has an internal chemistry that is similar to that of the environment. *Scenedesmus*, a genus of freshwater algae, has an elemental chemistry that conforms well to its surroundings (Sterner and Elser 2002). Conformers are not confined to single-celled organisms. Hagfish, a vertebrate, is an osmoconformer, meaning that its internal water concentration is the same as that of its environment. Most organisms are regulators, maintaining internal elemental and water concentrations that differ from the environment. Regulators have homeostatic set points in both their cells and in the extracellular environment (Schmidt-Nielsen 1997). No organism is perfectly homeostatic, with internal pH, solute content, temperature, and water concentrations varying around the set points. If the internal environment deviates too much from the species' set point, then it dies. Homeostatic organisms range from being strict regulators to tolerating broad changes in the internal environment (Figure 12.5).

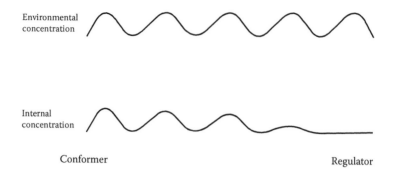

FIGURE 12.5 Relationship between variation in the external and internal concentration of an element for organisms that are conformers or regulators.

Particularly important to trophic ecology is the ability for plants and animals to store materials internally that do not contribute immediately to biochemical activity. Storage includes vacuoles in cells, particularly in plants that may contain high concentrations of nitrogen, phosphorus, and sugars. Animals store fats within adipose cells. If these components are included as part of the total concentration of elements within organisms, it would appear that internal concentrations of some materials vary more than they do in the bioactive space in cells including the cytoplasm and organelles.

In addition to water, the cytoplasm of all organisms is dominated by carbon, nitrogen, and phosphorus. Carbon makes sense. The atmosphere had as much as 10 atm of carbon dioxide before life arose on Earth (Walker 1985; Kasting 1993). Although atmospheric concentrations are much lower now, carbon is still generally unlimited in ecosystems. We will show that this is not always true. Nitrogen and phosphorus are often limited in specific ecosystems unaffected by humans because of their rarity in water and land. Because nitrogen is often limited in soils and does not occur in a reactive form (recall that it forms trivalent bonds between nitrogen atoms to form N_2), it is rapidly absorbed by plants when in ionic form with oxygen or hydrogen. When concentrations exceed the demand of plants or microbes, nitrate, nitrite, and ammonia are water-soluble and rapidly flush off of the landscape into streams and lakes, making nitrogen an ephemeral nutrient on land. Many freshwater systems have apparently unlimited concentrations of nitrogen for plants and animals because of runoff from the watershed or effluent from urban areas. Nitrogen is consumed by freshwater organisms and, before humans began altering the globe's nitrogen cycle, this element was again limited when it exited the rivers into the oceans (Vitousek et al. 1997b). As we will see, the concept of nitrogen being limited or unlimited in terrestrial and aquatic ecosystems depends on its stoichiometric relationship with phosphorus. Plotting total nitrogen or phosphorus concentrations against primary production may not be a good predictor of energy fixation in an ecosystem.

Phosphorus behaves very differently than nitrogen, making its availability to organisms differ. Because it readily forms covalent bonds with other atoms and molecules, it *sticks* to particles in the environment. In terrestrial systems, phosphorus binds to the soil and does not move. It can become freely available to plants as phosphate ions, which are obtained through roots. In freshwater, phosphorus is often unavailable to organisms (Figure 12.6). If free phosphorus becomes available through chemical activity or the decomposition of an organism, it rapidly reacts with oxygen to become phosphate, which is rapidly absorbed by plants. Phosphorus can bind with iron in the presence of oxygen and becomes insoluble, causing it to sink to the bottom of lakes, wetlands, and the ocean. Thus, there is a race between the time it takes for plants to absorb phosphorus and for it to settle out. As plants and other organisms die and sink to the bottom, they take their accumulated organically bound phosphorus with them. The bottom of some lakes, streams, and wetlands become anoxic. Under those conditions, phosphorus often loses its bond with iron and can become soluble (Figure 12.6). If the water upwells, the nutrients will promote primary production. In marine systems, phosphorus is typically not limited for autotrophs because of the scarcity of

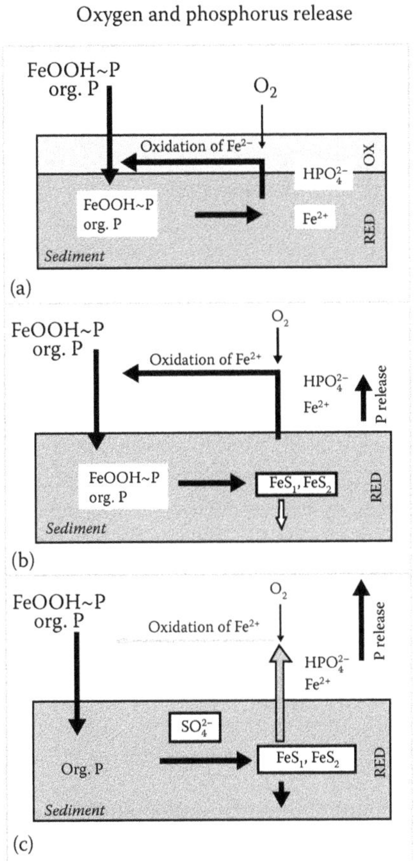

FIGURE 12.6 Fate of phosphorus at the interface between the sediment and the water. In aerobic conditions at the sediment interface (a), iron is oxidized in the sediment and oxidized iron is available to hold phosphate. In the absence of oxygen at the sediment layer (b), iron is reduced and bound with sulfur, allowing phosphorus to be released. A third condition (c) occurs where iron is limiting due to high sulfate concentrations. (From Hupfer, M., and J. Lewandowski. 2008. *International Review of Hydrobiology* 93 (4–5):415–432. With permission.)

iron. Determining whether phosphorus is limited depends on its stoichiometric relationship with nitrogen and other elements like oxygen and iron.

12.6 ECOLOGICAL STOICHIOMETRY

Leibig's law of the minimum is a pervasive ecological concept that uses the stoichiometric approach (Chapter 9). It states that once any needed resource is abundant, then another resource becomes limited. Focusing on a single element like carbon, nitrogen, and phosphorus tells us nothing about the availability of the resource. Rather, the stoichiometric relationships (i.e., ratios) among them tell us whether one is limited or not.

Redfield (1934) discovered that the seston (i.e., suspended detrital matter) in the ocean had a consistent composition of carbon, nitrogen, and phosphorus. When he processed his samples, the ratio was 106C:16N:1P. When nitrogen and phosphorus concentrations were plotted against each other, the intercept overlapped zero. This means that the elements become colimited when the concentration of phosphorus is zero. When either of these elements is regressed against carbon, the intercept is always positive, suggesting that carbon is in excess in the environment. Although other ecosystems have detritus that deviates markedly from this metric, which is now called the *Redfield ratio*, the elemental needs of autotrophs and consumers are remarkably consistent with the Redfield ratio across the globe.

The reason this ratio is ubiquitous within oceans has been the source of much conjecture. These elements are important building blocks of many biological components (see Chapter 9). Proteins are predominately nitrogen, comprising tissues and enzymes. Collagen is an important connective material in animals. RuBisCo, critical for photosynthesis, is one of the most abundant enzymes on the planet. Phosphorus forms the backbone of nucleic acids used to produce genes and to store energy. This element is also the primary component of bones and teeth. Carbohydrates, lipids, and pigments are dominated by carbon. When comparing the whole body or cellular C:N:P of most organisms, they do not match the Redfield ratio. But the Redfield ratio does consistently determine the outcome of biological interactions.

The C:N:P ratios of the environment relative to those of organisms are the ecologically relevant benchmark for determining how they should fare. Both the Earth's crust and bone have low N:P (Figure 12.7). Agricultural runoff and protein have

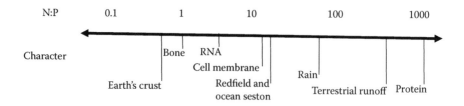

FIGURE 12.7 Nitrogen-to-phosphorus ratios of several common items on Earth. (Modified from Sterner, R. W. and J. J. Elser. 2002. *Ecological Stoichiometry: The Biology of Elements from Molecules to the Biosphere*. Princeton University Press, New Jersey.)

high N:P (Figure 12.7). With the exception of marine seston, most materials in the environment do not occur at Redfield ratios, meaning that organisms must regulate intake of carbon, nitrogen, and phosphorus to be successful. The Redfield ratio occurs because of the cellular machinery used to manufacture proteins (Sterner and Elser 2002). Later, we will explore more specifically how this ratio arises so consistently among living organisms and the seston they leave in the oceans.

12.7 STOICHIOMETRY OF ORGANISMS

We already determined that organisms vary in their elemental composition. However, the ratio of elements determines the requirements of organisms and availability in the environment. Organisms differ greatly in their C:N:P. These ratios also differ among life stages within species. The interaction between body size and environment affects C:N:P. Plants and animals in terrestrial systems must invest in structures (e.g., wood and bone) that are rigid and dense, using carbon and other materials like phosphorus and calcium. Conversely, organisms in water typically do not need to invest as much in these structures because of buoyancy. Elemental differences in structures, particularly in large, heavy organisms will affect whole-body C:N:P.

Although autotrophs make their own energy, they are no different than heterotrophs in their dependence on external sources of elements, including carbon, nitrogen, and phosphorus. The cells of autotrophs have cell walls that are composed of silica, carbon, and some nitrogen, contributing to their rigidity. The cytoplasm of these cells contains nitrogen and phosphorus, depending on the amount of protein synthesis. Algae are among the simplest autotrophs, existing as single cells or colonies. Investment in carbon is low because structural needs are limited. Thus, the C:N:P is low, with the cells having a high turnover rate of nutrients. Vascular plants, especially those on land, have a high C:N:P. Because these organisms have large, complex structures, the turnover of elements from tissues is much lower. Plants carry materials in the large vacuoles of their cells, with these structures varying widely in their capacity and content (Martinoia et al. 2007). A vacuole may contain high concentrations of nitrogen and phosphorus, making the whole-organism stoichiometry and total nutrient content vary widely.

Many animals grow through many developmental stages in life, with body size increasing markedly (Figure 12.8). As with plants, size and location (e.g., terrestrial versus aquatic) influence the structural and thereby the chemical characteristics of the organism. Small animals usually are either growing rapidly or have high metabolic activity. Protein synthesis for enzymes and tissue anabolism are elevated, leading to a high N:P. Animals that grow large typically have to invest in support structures such as an exoskeleton or bone. As maximum size is reached, protein synthesis declines. If bones are present, then N:P will be low. Because most animals have considerably less carbon in their tissues than plants, the C:N:P of animals is typically lower than that of plants.

Many other types of organisms exist. Bacteria are quite important, influencing processes ranging from decomposition in ecosystems to internal digestion of organisms. These microbes have high concentrations of nitrogen and phosphorus relative to other life, with C:N < 7 and C:P < 70. In contrast, another group of decomposers, fungi,

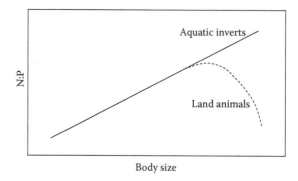

Body size

FIGURE 12.8 Body size of organisms versus their N:P stoichiometry. Aquatic organisms should continue to increase in nitrogen content, whereas terrestrial organisms require phosphorus in bones and other structure to overcome gravity. (Modified from Sterner, R. W. and J. J. Elser. 2002. *Ecological Stoichiometry: The Biology of Elements from Molecules to the Biosphere.* Princeton University Press, New Jersey.)

have relatively low concentrations of nitrogen and phosphorus, often being composed of high concentrations of minerals. Reef-building corals are well known for accumulating high concentrations of calcium carbonate to build their skeletons. If the entirety of the calcium and other minerals incorporated into the skeletal matrix in reefs was incorporated into the stoichiometry of the animals, nitrogen and phosphorus would be very low. This highlights how important the stoichiometry of elements other than carbon, nitrogen, and phosphorus is to the materials budget of many ecosystems and how they need to be considered in trophic ecology (Showalter et al. 2016).

12.8 STOICHIOMETRY AND GROWTH RATE

Might the Redfield ratio be an important predictor of the outcome of all trophic interactions? Loladze and Elser (2011) concluded that this ratio is inherent to cellular activity in all organisms on Earth, relating to the ratio of protein relative to ribosomal RNA (rRNA). Genes encode rRNA, which combine with proteins to build *ribosomes*. Ribosomes are the structures that convert codons in messenger RNA (mRNA) into proteins (Figure 12.9). Whereas the proteins that are translated within cells are largely composed of nitrogen, ribosomes are a combination of phosphorus-containing nucleotides and protein. Growing organisms increase the number of ribosomes, rRNA concentration, and protein in cells. This generates the stoichiometric relationship of 16N:1P.

Because of its ubiquity among all life, the Redfield ratio may be an important predictor of growth in all organisms. The match or mismatch between the stoichiometry of nitrogen and phosphorus in the environment and how efficiently organisms can obtain the elements in the proper ratio in their cells to grow while minimizing energy costs should be the evolutionarily relevant issue for trophic ecology. This idea is not a new one. Tilman et al. (1982) developed the concept of R^*, which is the concentration of a nutrient necessary to keep a plant at its carrying capacity. If R^* falls

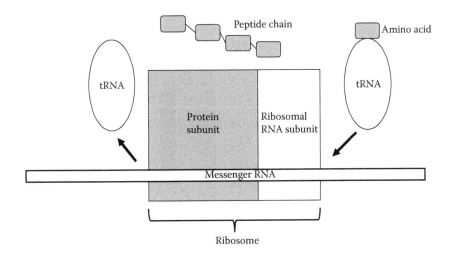

FIGURE 12.9 Structure of ribosomes and how they are related to peptide chains (protein precursors). Ribosomes are constructed of a ribosomal RNA and protein subunit. Messenger RNA contains codons to translate codes in DNA to amino acids. Transfer RNA (tRNA) delivers each amino acid to be added to the peptide chain (right) and leaves the ribosome empty (left).

below this threshold, then the nutrient is limited. Above this level, the $R*$ of another nutrient or other factor will become limiting. Tilman et al. did not incorporate the stoichiometry of the nutrients into this approach.

The *threshold element ratio* (TER) is similar to Tilman's approach but treats elements stoichiometrically (Urabe and Watanabe 1992). Each organism has a TER for an element. The threshold value of a nutrient is determined relative to carbon. If the C:nutrient in the environment exceeds the organism's TER, then the nutrient is limited. If the C:nutrient is less than the TER, then the nutrient is unlimited. Carbon is generally unlimited in the environment (but perhaps not in some locations such as soil), serving as a constant concentration in the organism. The TER for an organism is determined by evaluating its growth and reproductive rate along a continuum of nutrient ratios (Figure 12.10). The TER does not tell the whole story for an organism. Becoming an efficient consumer of a nutrient in short supply (high C:nutrient) in the environment may incur high energetic costs. When the nutrient is available (low C:nutrient), the consumer might be at some physical or energetic disadvantage. Consider two plants (Figure 12.11). One has shallow

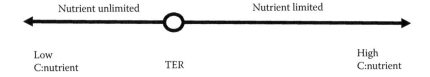

FIGURE 12.10 TER as a function of the stoichiometry of carbon and an essential nutrient.

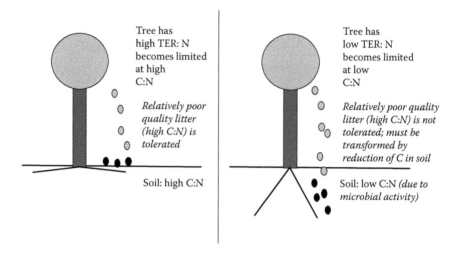

FIGURE 12.11 How tree species with different TERs for C:nitrogen respond to their ecosystem. Trees in ecosystems with limited nitrogen (high C:N) have a high TER, meaning that nitrogen only becomes limited when it is scarce relative to carbon (high C:N). Shallow roots and an ability to rapidly mineralize nitrogen are necessary via rapid decomposition. Trees in locations with relative high nitrogen (or low C:N) should have low TERs, meaning that nitrogen will become quickly limited if the C:N rises. These species may develop deep roots and relationships with symbiotic nitrogen fixers.

roots that allow it to quickly extract nitrogen from the environment (high TER: high C:N). This strategy commonly occurs in tropical forests where nitrogen is extremely limited on the forest floor. A plant with a low TER (low C:N) would not be able to survive in this system. However, in a system with high nitrogen concentrations in its soil, a high TER for nitrogen would probably not be advantageous. Plants that have a high TER for other limiting nutrients such as phosphorus or iron might dominate. Growth rate is negatively related to the TER for C:P, presumably because growing organisms have a greater demand for phosphorus in RNA synthesis (Frost et al. 2006).

Using foraging theory (Chapter 4), the environment might be viewed as a mosaic of patches with different stoichiometric characteristics (Leroux et al. 2012). The time spent foraging in patches should depend on the TERs of the foragers (Figure 12.12). Ideally, the TERs of the organisms across patches should match the local stoichiometry, similar to the ideal free distribution (Chapter 4). As with foraging theory, the nutrient concentrations of the patches will change as foraging occurs. Models of patch use and giving-up times (see Chapter 4) should incorporate not only the relative energy intake of patches but also the stoichiometric characteristics of the nutrients in the patch. If the energy needed to meet the TER of the organism is too high, then the organism should leave the patch. To our knowledge, patch dynamics in this context have not been explored in trophic ecology. Determining the relative contribution of energy intake and materials influx may be an important new development in predicting the distribution and abundance of organisms.

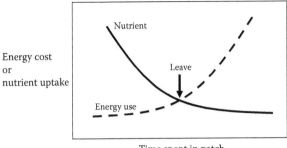

FIGURE 12.12 Hypothetical relationship between time spent foraging in a patch and the relative energy cost (dashed line) and nutrient decline (solid) of food.

12.9 RECYCLING

Organisms are sloppy eaters, and no plant or animal is perfect at consuming and assimilating materials, even if its TER matches that of the food in the environment. Predicting how inefficiencies in uptake of nutrients affect ecosystems is important for determining patterns of nutrient cycling and ultimately the trajectory for ecosystem change.

For most of history, large organisms were viewed by ecologists as inconsequential in nutrient cycling. In aquatic ecology textbooks, the role of fish and other vertebrates in physical processing of materials is often downplayed. However, in many freshwater ecosystems, the most reliable source of phosphorus for planktonic autotrophs is not through physical processes, but rather from fish, which contain the majority of phosphorus in the open water (Schindler and Eby 1997). Whether fish and other organisms are sinks or sources of phosphorus, nitrogen, and other materials depends on mechanisms that are still not well understood.

The stoichiometry of the food available in the ecosystem will affect patterns of nutrient recycling. Using carbon as the benchmark in the environment, as C:nutrient increases in the environment, the nutrient excreted by the organisms should decline (Figure 12.13). In other words, the scarcer the nutrient is (again assuming that carbon concentration is fixed), the more efficient organisms in the environment will be at

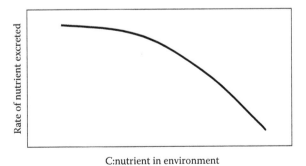

FIGURE 12.13 Hypothetical relationship between nutrient stoichiometry and rate of excretion. As a nutrient becomes scarce in the environment, it will be excreted less by consumers.

keeping it incorporated in their biomass. In contrast, a low C:nutrient in the environment leads to inefficiencies in uptake as the organisms meet their TERs and excrete the remainder as waste. In the case of nutrients like nitrogen and phosphorus, these elements begin to accumulate, leading to a condition called *eutrophication*.

The assemblage of organisms in the ecosystem will affect how efficiently nutrients are retained in an ecosystem (Figure 12.14). The relative TERs of organisms should play a role. If assemblages are dominated by species with high TERs (i.e., meaning that the organisms have a large threshold for an element and a minor need), then nutrients will be recycled rapidly. If the TERs are low, then the organisms should have a high need to maintain the nutrient internally, causing them to be sinks for that element or molecule (Figure 12.14).

The impact of TERs on nutrient cycling in the environment depends on the physical characteristics of the organisms (Figure 12.15). Plants are able to store many

Organism:	Low TER			High TER
Nutrient uptake:	Slow uptake			Rapid uptake
Example:	Wetland	Temperate forest	Tropical forest	Coral reef
	⟵━━━━━━━━━━━━━━━━━━━━━━━━━━━━━━━━━━━⟶			
Recycling:	Low recycling rate	Intermediate recycling rate		High recycling rate
Environment:	C:nutrient low	C:nutrient intermediate		C:nutrient high

FIGURE 12.14   Rate of nutrient recycling as a function of organism TER, rate of nutrient uptake (i.e., efficiency), and the stoichiometry of the nutrient in the environment.

FIGURE 12.15   Cyanobacteria floating on freshwater.

nutrients as they become available. Alkaline phosphatase is produced by many auto-trophs (Chapter 2), allowing them to absorb phosphorus when it is scarce. Thus, the effects of autotrophs on nutrient recycling may not be about their rates of excretion. Rather, the availability of nutrients from plants will depend on their turnover rates—how rapidly they live, reproduce, die, and decompose (Figure 12.15). Most heterotrophs do not possess a large vacuole in their cells for storing spare nutrients. Rather, they rely on the content of their prey, including autotrophs, to meet their TERs. If the prey have stoichiometric characteristics that do not match those of the consumer, the ratio of nutrients excreted will be altered. Organisms with a high N:P should excrete relatively high concentrations of phosphorus, because they have high nitrogen demand. Conversely, animals with a low internal N:P should excrete high concentrations of nitrogen and concentrate phosphorus in their bodies.

Ontogeny and body size plus architecture should affect patterns of nutrient recycling. Because small organisms have less need for carbon in their structure, they rapidly assimilate nitrogen and phosphorus from the environment. The stoichiometry of growing organisms is more complex, depending on the accumulation of mineralized structures like chitin and bone. A TER for a young animal might be quite different than that of an adult (Figure 12.7). Accumulating bones requires high concentrations of phosphorus, meaning that the TER for growing animals might be low, creating a sink for this nutrient. Once adulthood is reached, the internal phosphorus pool within individuals becomes relatively inert. Even though the C:P of adults may be low, the TER for phosphorus may be high. Thus, phosphorus is recycled at higher rates by adults. Nitrogen is important for both growth and maintenance, meaning that its TER in organisms may not change through life. As populations change demographically, their impact on nutrient cycling will change as well.

12.10   ECOSYSTEM CONSEQUENCES

We have shown that the physical characteristics of ecosystems and the organisms within them interact to affect stoichiometry. Stoichiometry is a dynamic aspect of ecosystems and will depend on a host of interacting factors including the physical environment, basal trophic level, species composition, size structure, and age distribution. The TER concept should help predict competitive ability, biodiversity, ecosystem productivity, and biogeochemical feedback that may scale to the entire planet.

A classic example of how TERs affect ecosystems revolves around algae in freshwater. Although phosphorus is limited in most natural freshwater systems, human activities have modified this condition. Phosphorus is a major component of many products, specifically industrial and household detergents. The ability for phosphate to *stick* to many substances as a surfactant makes it an ideal cleaner. Unfortunately, phosphate from soap and other applications is washed into water and is rapidly sequestered by algae as mineral food.

The process by which phosphorus enriches a stream or lake is called *eutrophication*. Generally, total phosphorus concentration is positively related to algal production. However, the composition of algae depends on the stoichiometry of nitrogen and phosphorus. The algal composition, in turn, affects the physical characteristics of the ecosystem. Freshwater typically has a N:P > 16, meaning that phosphorus is limited based on Redfield predictions. Loading of phosphorus from external sources skews the ratio to <16, causing nitrogen to become limited, even if the relative concentrations of nitrogen are apparently high. Species of autotrophs called cyanobacteria or blue-green algae that are specialists at obtaining limiting nitrogen are favored at these low N:P ratios. These organisms float near the surface of the water and contain oxygen-free heterocysts that can fix atmospheric nitrogen gas into organic nitrogen (Figure 12.15). The heterocysts must be thick-walled and can only convert nitrogen to useable forms in the absence of oxygen, an energy-intensive strategy that is only competitive when nitrogen is limiting. Cyanobacteria is typically unpalatable, and for some species, it produces toxins.

Blooms of cyanobacteria modify the physical structure of lakes and streams. Many of these species form large mats on the surface (Figure 12.15). Light is unable to penetrate, and other algal species and aquatic plants die. Dying vegetation and organisms that consume them promote microbial activity (see Chapter 3) that quickly consumes oxygen in the water. This ensures that blue-green algae remain dominant in the ecosystem. By the early 1970s in the United States, the influence of nutrients on aquatic ecosystems was not well understood. However, blue-green blooms were prevalent, largely as a function of the use of phosphorus in detergents and untreated wastewater dumped into streams. By understanding the role of phosphorus in the stoichiometry of freshwater, phosphorus use was curbed, leading to a reversal in eutrophication, most notably in the North American Great Lakes (Ludsin et al. 2001). Increased nitrogen deposition from agricultural runoff and atmospheric deposition from burned fossil fuels make phosphorus more limited in freshwater.

Nitrogen deposition is a global problem for terrestrial and marine systems. In many terrestrial ecosystems, higher nitrogen rather than phosphorus stimulates overall primary production. Like aquatic ecosystems, not all species respond similarly. Increasing the N:P of the soil favors species that do not require mechanisms to capture nitrogen. When N:P is low in terrestrial systems, plants that have complex relationships with root symbionts (mycorrhizal fungi and bacteria; Chapter 3) are able to capitalize on limited nitrogen. These species lose their competitive edge when N:P is high. In marine systems, runoff with high N:P shifts stimulates algal production, causing high densities of algae to accumulate, die, and create zones of low oxygen concentration (Rabalais et al. 2002). In this case, eutrophication is caused by high N:P rather than low N:P, as occurs in freshwater.

Humans are altering more than the stoichiometry of nitrogen and phosphorus in the biosphere. Cycling of carbon is intimately linked to the transport of other nutrients. For example, as carbon dioxide increases in the atmosphere with the combustion of fossil fuels (Chapter 3), the C:nutrient of autotrophs increases. From an energetic perspective, this change in food quality increases the amount of effort that herbivores must expend to capture nutrients, potentially reducing growth, survival, and ultimately reproductive fitness. Assemblages of organisms should change to those that are more efficient at extracting nutrients from autotrophs. Food webs should become simpler with shorter chain lengths.

The stoichiometry of calcium and other minerals changes as a function of the environment and may affect trophic interactions. Carbonate and calcium are tightly linked in the oceans (Figure 12.16). Their stoichiometric relationship depends largely on the pH of the water. As pH declines, carbonate becomes dissociated from the calcium ions; carbonate further transforms to carbonic acid and dissolved carbon dioxide. As humans increase the amount of carbon dioxide in the atmosphere and in the oceans, they are pushing the stoichiometric relationship to one where pH is low, carbonate ions are scarce, and calcium ions are in solution. Without available carbonate ions, organisms that rely on the stoichiometric relationship between carbonate and calcium to build skeletons or shells, like corals and mussels, will be unable to create these structures. Reduced calcium carbonate causes thin shells, making mussels and snails more vulnerable to predation. Changes in the formation of coral reefs will affect trophic interactions in complex ways.

In Chapter 3, we explored how temperature affects biological rates, particularly decomposition. Increased temperature, like that caused by predicted global warming, should increase biological rates and favor smaller organisms. Because smaller organisms have a low C:N:P, the concentration of nitrogen and phosphorus captured in this pool will increase, leading to increased dominance of smaller taxa (Chapter 8). The stoichiometric change in ecosystems should result in higher C:N:P for large organisms and poor food quality. This coupled with higher sequestration of carbon into autotrophs should reduce the prevalence of large-bodied organisms in ecosystems (Figure 12.17).

While human-induced climatic shifts are predicted to reduce the overall body size in ecosystems, humans are already systematically shifting the mean body size of organisms in ecosystems to smaller statures. We cover this topic in Chapter 9, showing that humans are simplifying food webs by selectively removing large-bodied, top consumers. Because body size influences patterns of nutrient cycling, the stoichiometry of ecosystems is likely changing as humans harvest the megafauna

Calcium ion solubility as a function of CO_2 partial pressure at 25°C ($K_{sp} = 4.47 \times 10^{-9}$)		
(atm)	pH	$[Ca^{2+}]$ (mol/L)
10^{-12}	12.0	5.19×10^{-3}
10^{-10}	11.3	1.12×10^{-3}
10^{-8}	10.7	2.55×10^{-4}
10^{-6}	9.83	1.20×10^{-4}
10^{-4}	8.62	3.16×10^{-4}
3.5×10^{-4}	8.27	4.70×10^{-4}
10^{-3}	7.96	6.62×10^{-4}
10^{-2}	7.30	1.42×10^{-3}
10^{-1}	6.63	3.05×10^{-3}
1	5.96	6.58×10^{-3}
10	5.30	1.42×10^{-2}

FIGURE 12.16 Calcium as unconventional food for aquatic organisms such as gastropods and corals. Increased carbon dioxide dissolved in water reduces pH and increases the solubility of calcium. Thus, high carbon dioxide concentrations can *starve* them of calcium carbonate.

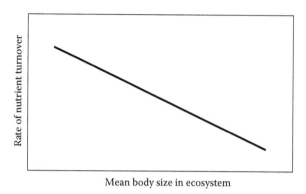

FIGURE 12.17 Predicted nutrient turnover as a function of the size of organisms in an ecosystem.

of the world. The retention time of N:P in biota will decline as it is captured in small organisms. Because these taxa have shorter lifespans, concentrations of nutrients in the environment will likely vary more as organisms die in large numbers and then sequester the nutrients as they recover (Figure 12.17). Differences in relative rates of uptake will render the stoichiometry of the environment more variable and less predictable. Organisms with the ability to function across a broad range of stoichiometric conditions will be favored (Figure 12.17).

12.11 CONCLUSIONS

Trophic ecology typically focuses on energetic transactions. But the uptake of essential nutrients is equally important (Chapter 9). In this sense, autotrophs are consumers as well as heterotrophs. Natural selection leads to organisms that must sequester nutrients to meet the stoichiometric requirements of their cells. Sequestration must occur in ways that minimize energetic and survival costs or organisms will go extinct.

Assessing the nutrients in the environment and organisms is more complex than quantifying the quantity of energy. Nutrients affect each other. When one becomes unlimited, another likely becomes limited. Stoichiometric relationships may influence trophic interactions more than the individual concentrations of the nutrients. The relative concentrations of nitrogen, phosphorus, and carbon seem to be the most important factors affecting the growth, survival, and reproduction of organisms in ecosystems. The biological importance of nitrogen and phosphorus is highlighted by the lack of natural bioavailability of these elements, at least before humans altered their cycles.

How the stoichiometry of nitrogen and phosphorus affects the growth of organisms appears to be consistent on Earth, following the Redfield ratio of 16 atoms of nitrogen for every atom of phosphorus. The basic machinery of ribosomal and protein production is responsible for this universal pattern. Although the Redfield ratio predicts which nutrient is limiting growth, the whole-body composition of organisms typically deviates markedly from this ratio, as nutrients are stored in different body parts such as vacuoles of plant cells, chitin of invertebrates, and bones of animals.

With the exception of the seston of oceans, environmental stoichiometry of nutrients in the environment differs considerably from the Redfield ratio, influencing biological interactions.

Plants and animals differ in how well they can hold nutrients. Plants sequester carbon, nitrogen, and phosphorus. Bacteria harbor large pools of nitrogen and phosphorus, but not carbon. The relative rate of nutrient turnover by these organisms increases with increased death rate and small size. Most metazoans have guts, and these vary in efficiency of nutrient uptake. The nutrient stoichiometry of food is typically not perfectly matched with the needs of the organism. Thus, waste occurs either through excretion or egestion. This waste is usually highly mineralized and immediately available for autotrophs and decomposers. Thus, animals can contribute greatly to the recycling of nutrients and alter environmental stoichiometry.

Humans are greatly altering environmental stoichiometry, affecting trophic interactions. The N:P of the surrounding water or soil changes species composition with potentially detrimental feedback to ecosystem function. Declining N:P causes eutrophic conditions in lakes and streams. Conversely, high N:P modifies the function of marine and terrestrial ecosystems. Carbon cycling is changing as humans increase carbon dioxide concentrations in the atmosphere and seawater. Increased C:nutrient of autotrophs decreases their quality for herbivores. The ultimate outcomes of most human-induced stoichiometric change are loss of biodiversity and degraded ecosystem function.

A challenge for the future of trophic ecology is to merge traditional energetic-based foraging models with nutrients as income (Allgeier et al. 2015). The influence of fish on aquatic ecosystems through excretion has been modeled. However, how organisms distribute across the landscape as a function of both the energy and nutrients in patches has not been a focus of trophic ecology. Quantifying patterns of food quantity and quality and understanding how they change should predict how ecosystems respond to alterations in the physical environment including energy availability and nutrient stoichiometry.

QUESTIONS AND ASSIGNMENTS

1. Why must reactions between molecules and elements balance in a stoichiometric manner?
2. How do stoichiometric reactions change the relative concentrations of nutrients in the environment?
3. Do organisms vary dramatically in their ratios of elements as does the environment? Why is there a difference?
4. What is the difference between a conformer and a regulator? What are the implications for trophic ecology?
5. Why is the Redfield ratio significant for trophic ecology?
6. Describe ways that elemental ratios differ among various ecosystems.
7. Relate rRNA and the ratio of N:P in organisms.
8. How does TER affect the competitive ability of an organism for limiting nutrients?
9. How should low N:P ratios affect algal dynamics in freshwater ecosystems?

13 Elements and Isotopes as Tracers

13.1 APPROACH

In previous chapters, we have identified ways that trophic transactions may be quantified from bioenergetics, diets, consumption models, food-web topologies, and secondary production estimates. The fate of materials and energy can also be quantified using the physical variation among elements that commonly occur within ecosystems. Although elemental analysis can be used to answer questions in biogeochemistry, we focus on how isotopes of atoms can be used to tease apart feeding relationships in trophic ecology. This chapter is not intended to provide an exhaustive review of the rapidly expanding use of these techniques in trophic ecology, but to familiarize the many of us who barely remember our chemistry and need access to basic and more sophisticated tools. We end the chapter by identifying some of the potential pitfalls of relying solely on elemental tracers for assessing trophic interactions.

13.2 INTRODUCTION

Quantifying the movement of materials through food webs is accomplished using tracers. A tracer in a food web is usually first accumulated at the base of the trophic pyramid by autotrophic production and is identifiable as it is either assimilated into the consumers or egested or excreted back into the environment. In this chapter, we will explore how many human-made and naturally occurring materials in the food of plants and animals may be followed through complex chains within ecosystems. In addition to identifying the source of materials, tracers can be used to determine the length of food chains. Over the past two decades, our ability to detect very low concentrations of materials in food and consumers has improved markedly while the cost of processing samples has decreased. The use of tracers to *map* food webs is standard practice in trophic ecology and will increase in the future.

In this chapter, we first provide a brief discussion on the use of tracers in trophic ecology from those that are neutral (i.e., do not have much impact on metabolism or nutrition) to those that are biologically active. Many biological tracers are radioactive, which makes them easy to detect but creates concerns about health and disposal. Modern techniques in the field rely on nonradioactive elements, which are rare in the field and fall below detection limits of most instruments. These rare tracer elements often arise from different food sources as well as the environment itself (e.g., water). Once the elements enter organisms, they are metabolized and distributed differentially in tissues. We will review quantitative techniques to sort out these processes of mixing and fractionation.

13.3   TRACERS

Many materials can serve as *tracers* in organisms in ecosystems. They are divided into those that are inert and those that are assimilated slowly or rapidly into organisms. Tracers may target organisms including microbes, plants, and terminal consumers. Inert tracers are not particularly useful for food-web studies. Adding materials like powdered charcoal to food is cheap and easy to quantify in feces relative to the biomass assimilated (Chapter 9). However, these materials only provide information about assimilation at the level of the individual, because they will not propagate through the food web.

Isotopes of elements are the most frequently used tracers in trophic ecology. To review basic chemistry, each element is defined in the periodic table by the number of protons in its nucleus and the number of electrons in its shell. The atomic mass, also included in the periodic table, is actually an estimate of the weight of the most common isotope of the element, which is driven by the number of neutrons plus protons in its nucleus and shown by a raised number to the left of the element's symbol, such as ^{12}C, which has 6 protons and 6 neutrons (Figure 13.1). Each element may have many different isotopes, depending on the number of neutrons it has in its nucleus. Neutrons have no electric charge, so they do not affect the net charge of each

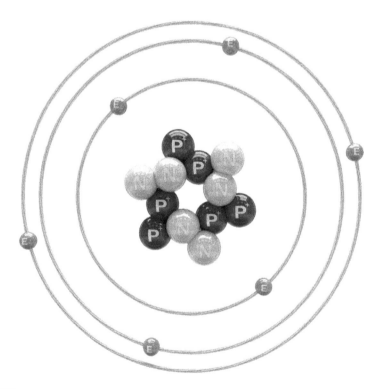

FIGURE 13.1   Conceptual diagram of a carbon atom ^{12}C, showing the number of protons and neutrons in the nucleus. Other isotopes of carbon have the same number of protons but different numbers of neutrons. (Courtesy of Shutterstock.)

atom. However, neutrons can affect the stability of the isotope. Stable isotopes are ones that will retain their neutrons for long times. Unstable isotopes are ones that are unable to maintain their arrangement of protons, electrons, and neutrons.

Radioactive isotopes of elements (i.e., radioisotopes) are unstable, meaning that they decay (i.e., by losing particles) into stable isotopes of other elements. The resulting decay causes radioactivity, which is typically measured as energy lost by the atoms. This energy is easy to quantify but also potentially harmful to organisms. As we note in Chapter 12, phosphorus (^{31}P) is one of the most important nutrients in ecosystems. The radioactive isotope of this element is ^{32}P, which decays into stable sulfur, ^{32}S, while emitting 1.79 MeV. The half-life of ^{32}P is 14.3 days, so it does not linger for long in the environment. Radioisotope studies were conducted in ecosystems in the past to determine the fate of phosphorus, which was considered a limiting nutrient. Newbold et al. (1983) released ^{32}P as phosphate into a stream in Tennessee. The phosphorus was incorporated into the detritus being transported by the water, but only 3% was found in the stream organisms. Thirty percent of this small fraction accumulated in predators within the stream. This study and others showed that other nutrients, for example, nitrogen and organic carbon, may play a greater role in trophic dynamics in streams. While radioactive ^{14}C can be used, nitrogen does not have a reasonably long-lasting radioisotope for use in the environment.

Many biologically relevant elements have nonradioactive, stable isotopes (Table 13.1; Chapter 8). Neutrons also make bonds between atoms harder to break (Figure 13.2). Rather than merely being tracers in the environment, stable isotopes interact with physical and biological chemistry in ways that allow us to determine their origin and residence in food webs. The elements with stable isotopes used to explore food-web relationships include hydrogen, carbon, nitrogen, oxygen, and sulfur (Table 13.1). Stable isotopes exist at low, albeit detectable, concentrations in the environment. Sources include the environment and the diets. The assumption is that consumers are feeding in an open system where changes in the environment

TABLE 13.1
Some Common Isotopes Used in Ecology

Element	Symbol	Number Protons (Atomic Number)	Number Neutrons	Proportion in Environment
Hydrogen	^{1}H	1	0	0.99984
	^{2}H	1	1	0.0016
Carbon	^{12}C	6	6	0.9889
	^{13}C	6	7	0.111
Nitrogen	^{14}N	7	7	0.9964
	^{15}N	7	8	0.0036
Oxygen	^{16}O	8	8	0.9976
	^{18}O	8	10	0.002
Sulfur	^{32}S	16	16	0.9502
	^{34}S	16	18	0.0421

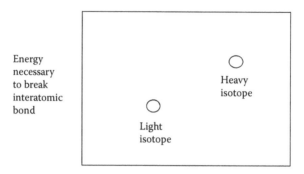

Interatomic distance

FIGURE 13.2 Illustration of why isotopes with less neutrons are more likely to be used in biological process than heavier ones with more neutrons. Heavier isotopes have a larger distance between bonds in molecules and thus are more energetically costly to break. Organisms have evolved enzyme pathways that minimize energy intake, thus preferentially using molecules with lighter atoms.

and activities of the organisms do not change the ratio of stable isotopes available (Boecklen et al. 2011). The concentrations of common and rare isotopes vary widely among ecosystems. Thus, several standards have been adopted for each tracer element. The ratio of the common, light element to the rare, heavy isotope is constant in these standards (Table 13.2). This leads to the common notation for heavy isotopes:

$$\delta = [R_{sample}/R_{standard} - 1]*1000$$

where

$$R = {}^{H}F/{}^{L}F,$$

with F being the relative amount of the heavy (H) or light (L) isotope in the sample or standard.

TABLE 13.2

Standard Ratios of Heavy (Rare) to Light (Common) Elements Used in Trophic Ecology

Source	Ratio (H/L)	Value of Ratio
Ocean water	${}^{2}H/{}^{1}H$	0.00015576
PeeDee Belemnite (Vienna)	${}^{18}O/{}^{16}O$	0.0020052
Air	${}^{13}C/{}^{12}C$	0.011180
Canyon Diablo Troilite	${}^{18}O/{}^{16}O$	0.0020672
	${}^{15}N/{}^{14}N$	0.0036765
	${}^{34}S/{}^{32}S$	0.0441626

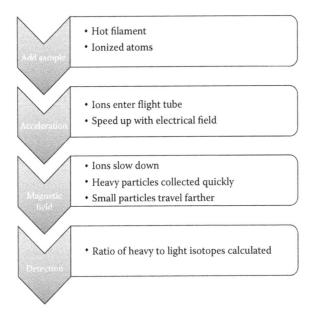

FIGURE 13.3 Pathway of quantification of isotopes in an isotope-ratio mass spectrometer. Element is added to the instrument and heated with a filament that boils electrons off the atoms. Ionized (positively charged) atoms are accelerated through a flight tube with an electrical field. These atoms enter a detection chamber with a magnet; ions slow down based on their mass and inertia. Heavier atoms are attracted to detectors first. Lighter isotopes have a longer flight path and do not drop onto detectors until later. A computer calculates the relative counts of each isotope and generates a ratio of heavy to light isotopes.

Isotopes of elements can be detected using many devices, most typically a mass spectrometer. One of these devices is an isotope ratio mass spectrometer, which uses magnets to separate out ionized (i.e., charged) isotopes of different masses (Figure 13.3). Locations of the collection points on detectors in the instrument are correlated with the mass of the isotopes. The result is the ratio (R) of the counts of atoms collected for the standards and samples. From this, the δ-value of the sample is calculated.

It often is difficult to conceptualize what a ratio of ratios like δ means. We view this an index for comparing values within an ecosystem. The multiplication factor of 1000 is used to convert the small proportional differences that arise to parts per thousand (or per mil, ‰). A value of $\delta = 0$ from an ecological sample out of context is relatively meaningless. In this case, the sample has a ratio of heavy to light isotopes that is identical to the standard (and you likely had someone mistakenly load a standard rather than a sample in the mass spectrometer).

13.4 MIXING MODELS

Values of δ allow us to do detective work to determine how heavy isotopes are mixing, fractionating, diluting, and turning over in organisms and food webs. Any baker understands the concept of mixing. A large quantity of dry flour only needs a small pinch of salt and baking powder. Once these ingredients are added, they quickly disappear into the bowl and are impossible to detect by eye. But, if forgotten, they will be missed in the flat, tasteless bread after it is baked. *Mixing models* allow us to *trace* the small amount of salt and baking powder in all that flour, in this case, the rare heavy isotopes of atoms swimming in a sea of light ones. If we have a sample and know that there are two food sources with different heavy isotope ratios of one elemental tracer (e.g., two diet items, *a* and *b*), then we can develop a mixing model to determine the relative contribution of each diet source to the total heavy isotope composition of the consumer (δ_{sample}).

$$\delta_{sample} = \delta^*_{source\ a} f_{source\ a} + \delta^*_{source\ b} f_{source\ b}, \text{ where}$$

$$f_{source\ a} = (\delta_{sample} - \delta_{source\ b})/(\delta_{source\ a} - \delta_{source\ b}) \text{ and}$$

$$f_{source\ b} = 1 - f_{source\ a}$$

If you have a sample with $\delta = 9‰$ and food item *a* having $2‰$ and *b* having $11‰$, the mixing model can determine the relative contributions of each item to the consumer. The proportion of heavy isotope coming from source $a = (9–11)/(2–11) = 0.22$, or 22%. Source *b* then contributes $1–0.22$, or 78% to the sample. This makes sense given that the isotopic ratio of *b* is much closer to that of the sample than *a*.

Mixing models with more than two food sources and multiple elements can be developed, using the same logic as above (Figure 13.4; Phillips et al. 2014). The ratios of two common elemental isotopes on which we will focus in later sections are $\delta^{13}C$ and $\delta^{15}N$. Consider a mixing model with three food sources (*a*, *b*, *c*) and incorporating both $\delta^{13}C$ and $\delta^{15}N$.

$$\delta^{13}C_{sample} = \delta^{13}C^*_{source\ a} f_{source\ a} + \delta^{13}C^*_{source\ b} f_{source\ b} + \delta^{13}C^*_{source\ c} f_{source\ c}$$

$$\delta^{15}N_{sample} = \delta^{15}N^*_{source\ a} f_{source\ a} + \delta^{15}N^*_{source\ b} f_{source\ b} + \delta^{15}N^*_{source\ c} f_{source\ c}$$

$$f_{source\ a} + f_{source\ b} + f_{source\ c} = 1$$

This approach is depicted graphically as the biplot of $\delta^{13}C$ and $\delta^{15}N$ of the food sources and the consumers, with the consumer isotopic ratios falling in the centroid (Figure 13.4). Expanding mixing models can become computationally complicated, requiring computer packages such as Isosource, available at http://www.epa.gov /eco-research/stable-isotope-mixing-models-estimating-source-proportions.

Mixing is a simple concept, but not the only process occurring. The different chemical properties of stable isotopes cause them to *fractionate* within organisms as they move through food webs. When an organism, which is typically an autotroph, is fixing an element from the environment, the lighter isotope will be favored in the

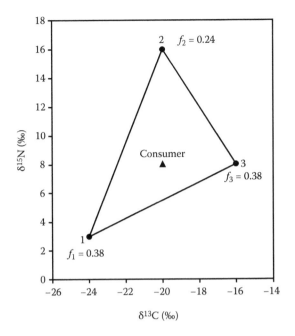

FIGURE 13.4 Predicted proportional (*f*) contribution to consumer stable isotope composition from three food sources (1, 2, 3). These proportions are determined from mixing models. (From Phillips D. L. 2012. *Journal of Mammalogy* 93 (2):342–352.)

uptake. When an organism respires, excretes, or egests some elements (but not all), the heavier isotope is lost at a slower rate than the lighter one. So, fractionation can either be negative or positive for a heavy isotope, depending on the biological process and organism. Positive fractionation is often called *trophic enrichment* or *bioaccumulation*. Fractionation in chemical reactions is fairly well understood, because lighter elements will react at faster rates leaving the heavier ones behind in the autotroph or consumer. Fractionation in organisms depends on the elements involved and will be the focus of each section about each biologically relevant element.

If an organism consumes an item that contains a unique δ and then moves to eating other food sources, the concentrations of the assimilated isotopes will change due to dilution and tissue turnover. *Dilution* happens as organisms grow and accumulate new tissues. The proportion of the isotopes consumed and present will decline as new ones from other food sources are accumulated. It is particularly important to account for dilution in young organisms, because they are accumulating mass rapidly (e.g., often exponentially; Figure 13.5). Turnover is critical as well. The isotopic composition of tissues changes via catabolic and anabolic mechanisms for all ages of organisms. The isotopic composition of bodies of organisms takes time to *catch up* with the food in the environment. The expectation is that after this lag occurs and accounting for potential dilution, tissues and food reach some equilibrium between intake and tissue turnover (Hesslein et al. 1993; Figure 13.6).

Different tissues have different mixing, fractionation, dilution, elemental concentrations, and turnover patterns. Bony or carboniferous (e.g., chitin, wood) parts may

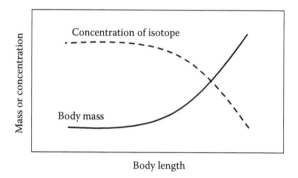

FIGURE 13.5 Potential dilution of an isotope or rare element in an organism as it grows. The isotope will comprise a high concentration in the body. Body mass increases exponentially, and if the isotope is not consumed after initial consumption, then it becomes less concentrated in the body. Note that this does not account for loss of the isotope due to catabolism and excretion/egestion.

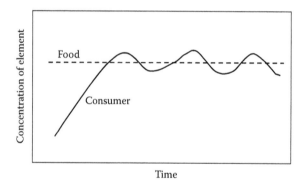

FIGURE 13.6 Hypothetical concentration of a rare element or stable isotope in a consumer that is consuming these materials constantly in the environment. The concentration in the tissues will reach some equilibrium with the food item and then fluctuate as a function of growth dilution and excretion/egestion of the material.

be relatively inert. Many animals and plants accumulate these hard tissues at different rates through time, depending on photoperiod, temperature, resources, and other factors. Mining these parts for isotopes or unique elements may provide a long-term history of trophic interactions (Box 13.1). Muscle tissue drives the rates of whole-body tissue turnover in many organisms (Hesslein et al. 1993), making it the most likely indicator of recent trophic history. The concentration of isotopes in blood plasma of organisms is another option, reflecting recent meals (Hobson and Clark 1993). Hobson and Clark (1993) found that the half-life for ^{13}C in bird blood plasma was 2.9 days. A variety of tissues may be sampled to develop a differential chronology of trophic history (Lourenco et al. 2015).

A recent modeling approach, Stable Isotope Analysis in R (SIAR) (Parnell et al. 2013; https://cran.r-project.org/web/packages/siar/index.html), incorporates

BOX 13.1   FINDING ISOTOPES IN LAYERS

The elements accumulated by organisms can be incorporated into relatively stable components. Bone, for example, can incorporate elements such as calcium and phosphorus that may persist in an organism long after it is dead. Some organisms like trees and fish lay down rings of elements in their bodies that correspond to time, leading to the ability of ecologists to track changes in elemental composition, including the ratios of stable isotopes, through time.

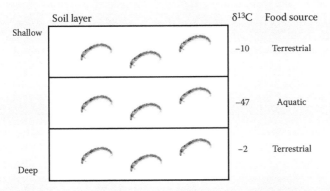

An ecologist interested in the history of invertebrate feeding in a wetland may take a core of sediment. If the sediment is relatively anoxic, insect bodies, especially the carbon-dominated shed exoskeletons, may be relatively well preserved and their composition may reflect hydrology. A wetland may vary from receiving food (i.e., carbon) from terrestrial sources during flooding and from local sources (i.e., algae) during dry periods. The $\delta^{13}C$ in the exoskeletons may reflect that of the food sources through time and provide a chronology of feeding of the insects. In the above figure, we show wetland clay soil that has accumulated the remains of midge larvae over several months. Midge larvae consume dead leaves and other vegetation. Each layer of clay corresponds to either a flooding period or drying period, with the top and bottom being formed during a flood and the middle formed during a drought. The exoskeletons are processed with a ratio-isotope mass spectrometer, yielding $\delta^{13}C$ values of −10, −47, and −2 in the top, middle, and lower layers. Terrestrial plants are less likely to have low $\delta^{13}C$ than aquatic counterparts (see text). Thus, the $\delta^{13}C$ values in the surface and the bottom layer suggest that midge larvae were consuming terrestrial-derived matter during flooding. The low $\delta^{13}C$ during the drought suggests that autochthonous production from algae in the water was providing energy during that time.

Any approach, in which layers—whether they be within organisms or the environment—are *mined* for elemental composition must account for cause and effect. Ecologists can speculate that the elemental signatures are due to environmental effects and past feeding. However, experiments need to be conducted where the food sources and other environmental factors are quantified to determine whether they indeed translate to patterns of elemental composition in the organisms. In the case of the midge larval exoskeletons, without knowledge of the elemental composition of the terrestrial and aquatic sources of food and a direct experimental link, conclusions about the relationships between foraging and flooding are speculative at best.

the most important factors affecting the source contributions to consumers. It is a hierarchical Bayesian model, which uses as much data as possible to generate a distribution of actual estimates of overlap values between the food and the consumer (i.e., unlike parametric statistical models that provide error estimates around the parameter values). The model accounts for the contribution of each food source based on the ratios of heavy isotopes in the food items and the consumers, the concentrations of the elements in each diet item, and the fractionation of the isotopes (i.e., trophic enrichment factor) from food to consumer trophic levels. The model also incorporates the density distribution of possible observations among samples, allowing the results to be generated as a distribution of estimates of source and consumer overlap. One example of the utility of this approach is that it can show iso-space plots of predicted consumer and source overlap. Parnell et al. (2013) showed that this approach can provide temporal patterns of overlap between swallows and their prey (Figure 13.7). Box 13.2 provides a more detailed example of how to use this and related modeling approaches in packages like R.

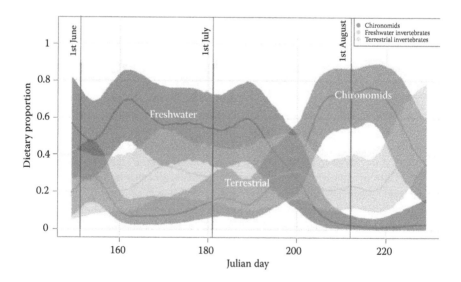

FIGURE 13.7 Dietary composition as predicted by a hierarchical Bayesian mixing model developed by Parnell et al. (2013). The model provides predicted distributions of the proportional contribution of each diet item to the consumer (swallows) through time.

BOX 13.2   USING A STATISTICAL PACKAGE

Models such as Isosource and SIAR allow ecologists to load in sources and consumer δ values of heavy isotopes, account for the differences in concentrations of these isotopes in the sources, and incorporate variation among individual estimates. Conventional mixing techniques such as those used in Isosource provide estimates of overlap that correspond to the statistical variation around the parameters measured. Hierarchical Bayesian techniques, such as those used by SIAR and its successors, follow use estimates of known values of consumer and source isotopic ratios to provide distributions of the actual parameter estimates, allowing for more robust analyses of diet overlap.

SIAR is a package in the statistical R platform, which is becoming a standard tool in ecologists' toolboxes (Parnell et al. 2013). The standard example in this package is for the blood plasma values of $\delta^{13}C$ and $\delta^{15}N$ in geese and the ratios found in the four aquatic plants on which they graze. The plasma values are indicative of recent meals of the birds. The first step is to load in the mean values and standard deviations for the plant sources:

```
> library("siar", lib.loc="~/Library/R/3.2/library")
> sources = read.table("source.txt", header=TRUE)
> sources
        Sources  Meand15N    SDd15N   Meand13C    SDd13C
1       Zostera  6.488984 1.4594632 -11.17023 1.2149562
2         Grass  4.432160 2.2680709 -30.87984 0.6413182
3     U.lactuca 11.192613 1.1124385 -11.17090 1.9593306
4   Enteromorpha  9.816280 0.8271039 -14.05701 1.1724677
```

The figure includes the code used in R to load the program and read a table of data for the four plant types. The sources of food, their mean $\delta^{13}C$ and $\delta^{15}N$ values, and the standard deviations around the means are in the columns with the four observations. The second step is for the program to load in data for the plasma of the geese from two collection groups:

```
> data = read.table("target.txt", header=TRUE)
> data
  Code d15NPl d13CPl
1    1  10.22 -11.36
2    1  10.37 -11.88
3    1  10.44 -10.60
4    1  10.52 -11.25
5    2  10.19 -11.66
6    2  10.45 -10.41
7    2   9.91 -10.88
8    2  11.27 -14.73
9    2   9.34 -11.52
```

Each observation in the figure above is for an individual bird. The trophic efficiency factors (i.e., fractionation) of $\delta^{13}C$ and $\delta^{15}N$ are loaded in as:

```
> tef
         Source Mean15N Sd15N Mean13C Sd13C
1       Zostera    3.54  0.74    1.63  0.63
2         Grass    3.54  0.74    1.63  0.63
3     U.lactuca    3.54  0.74    1.63  0.63
4  Enteromorpha    3.54  0.74    1.63  0.63
```

The means and standard deviations from the above figure were obtained from the literature by Parnell et al. (2013).

The program runs a Bayesian model that uses prior distributions of the source data via what is called a Gibbs sampler to find the best fits of mixtures (i.e., proportions) of each plant diet source to the diets of the two groups of geese over 500,000 iterations.

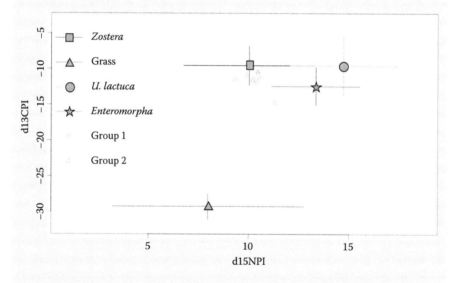

The figure from SIAR produces what looks to be a standard biplot of the relative $\delta^{13}C$ and $\delta^{15}N$ values for the two groups of geese and the resulting distributions of the food sources. Importantly, the bars represent the range of actual parameter estimates that were produced by the model. Areas of overlap between the bars and the groups show the relative importance of each diet type. These results suggest very little consumption of grass and much consumption of *Zostera*.

Another way to look at the results is by generating frequency histograms of a subset of the predicted proportion of overlap between ratios in plants versus the geese plasma. The data show considerable spread in potential overlap of diets containing *Zostera*, although on average they make up 60%. Conversely, grass is consistently less than 10% of the diets. The degree of overlap between the food groups demonstrates how well the model differentiates the relative contribution of each diet type. Here, the high overlap among the non-*Zostera* plants suggests that their importance to foraging geese is less clear.

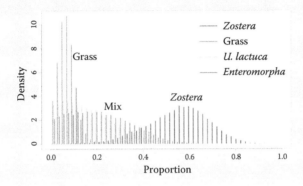

13.5 CARBON

Carbon is one of the most common elements in organisms and exists in two stable isotopes: the abundant form, ^{12}C, and the rare, heavy isotope, ^{13}C (Table 13.1). Vienna PeeDee Belemnite, which is formulated based on fossilized *Belemnitella americana* carbonate found in a Cretaceous geological layer about 70–145 million years old, is the international standard for the ratio of ^{13}C to ^{12}C (Table 13.2). The ^{13}C in Earth's atmosphere has been negatively fractionated (i.e., reduced) relative to this standard by about 1‰ because of the combustion of fossil fuels over the past century (Fry 1991). Incomplete combustion of carbon fuels preferentially releases ^{12}C as gas.

Carbon is fixed similarly among all autotrophs through the action of a highly conserved enzyme called RuBisCo (Figure 13.8). This enzyme preferentially converts ^{12}C over ^{13}C in carbon dioxide to carbohydrates, reducing the $\delta^{13}C$ in plants. In terrestrial C3 plants, which do not need to conserve water, stomates are open during photosynthesis, and carbon dioxide is converted to carbohydrates in one reactive step, leading to a fractionation of ^{12}C by 20‰ and a $\delta^{13}C = -28‰$ (Figure 13.8). In dry areas, C4 plants have evolved a second step to photosynthesis, where carbon dioxide gas captured during cool nights through stomates is stored during hot, drier days as malate in mesophyll cells. Photosynthesis by RuBisCo is then completed on the malate, leading to less fractionation of ^{12}C and a $\delta^{13}C = -13‰$ in C4 plant tissues (Figure 13.8). The use of this to trophic ecology should be fairly obvious. Mixing models can identify the relative contribution of C3 and C4 plants to the diets of herbivores. In aquatic ecosystems, the need for a C4 system is absent, because water balance is not an issue. Aquatic plants, especially algae, have higher turnover times

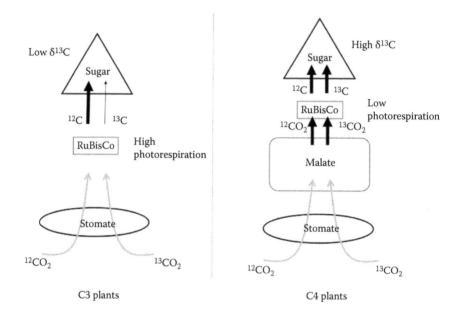

FIGURE 13.8 Pathways and fate of ^{12}C and ^{13}C from carbon dioxide fixed by plants that use the C3 and C4 pathways of photorespiration. The enzyme RuBisCo is involved in both pathways. In C3 plants, photorespiration is high, involving the use oxygen in the process of fixing carbon. The reaction is not efficient and leads to the fractionation of the lighter carbon isotope. In contrast, C4 plants store carbon as malate, reducing photorespiration. Presumably, the extra step of storing carbon as malate in this process does not lead to fractionation of ^{12}C, meaning that the δ^{13}C of these plants is higher and more similar to that of the atmosphere.

than many terrestrial plants. This leads to greater variation in accumulation of light carbon, ranging from 19‰ to 45‰. Organic matter in freshwater and oceans, which is largely derived from autotrophic production in most ecosystems, may have δ^{13}C = −35‰. Dissolved carbon dioxide in the ocean and freshwater deviates only slightly from atmospheric values, with δ^{13}C varying between 1‰ and −15‰. The organic matter fraction of soil has a similar fractionation of δ^{13}C (about −26‰) of organic matter in water.

Because aquatic autotrophs such as algae and vascular plants differ in fractionation of ^{13}C, estimating the δ^{13}C of these potential sources is useful for estimating the source of autotrophic production in aquatic ecosystems. Terrestrial vegetation may have a sufficiently different signature to identify the source of aquatic- versus land-derived autotrophic sources. Mixing models can then be used to assess the relative contribution of these sources to the consumer. Heavy carbon does not accumulate in consumers. Net fractionation of δ^{13}C, with the value becoming more positive, is uncommon in ecosystems (Figure 13.9). This ensures that carbon isotopes provide fairly good tracers of energy source for all organisms within a food web. An herbivore and an apex predator should share similar δ^{13}C values if they are obtaining the same basal resources (Figure 13.9).

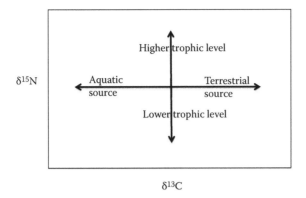

FIGURE 13.9 Typical biplot of fractionation values for heavy carbon and nitrogen in a food web.

13.6 NITROGEN

Nitrogen has historically been a key limiting nutrient in many ecosystems (Chapters 9 and 12). Nitrogen gas is common in the atmosphere, but energy is necessary to convert it into organic forms. Most nitrogen on Earth is light as ^{14}N, with only about 0.4% being heavy ^{15}N (Table 13.1). Human activities are altering δ^{15}N in various ways. Animal sewage has high δ^{15}N, likely because of the differential uptake of the lighter isotope in the guts of vertebrates. Nitrogen is a major component of the combustion of fossil fuels. As with carbon, the lighter isotope is released in the form of either ammonia or as an oxide. Because nitrogen gas in the atmosphere is inert, it is used for developing the international standard ratio of ^{15}N to ^{14}N (Table 13.2).

Nitrogen varies widely among ecosystems. It is water-soluble and travels readily through the pores in soil as well as through the currents in waterways. Nitrogen is usually quickly sequestered by plants and microbes (Chapter 12). It also is a waste product that must be eliminated by consumers. Because of its importance as food and waste, this element is moved around rapidly in ecosystems. In streams, a nitrogen atom can *spiral* as it moves downstream (Figure 13.10). The atom is taken up by primary producers or bacteria, consumed, and then excreted again where it flows a short distance to cycle again. Unlike carbon, the probability that the isotope will be excreted depends on its mass in both terrestrial and aquatic ecosystems. Nitrogen excretion is costly (Chapter 8), causing organisms to fractionate ^{15}N. Thus, light isotopes will be more likely excreted while heavy ones accumulate causing δ^{15}N to increase in consumer tissues.

The net fractionation of ^{15}N in organisms provides a convenient tool for trophic ecology. Unlike heavy carbon, heavy nitrogen increases as it is transferred between consumers. Both Vander Zanden and Rasmussen (2001) and Post (2002) found that δ^{15}N consistently increased by 3.4‰ between trophic levels (Figure 13.8). Trophic position (TP) of a consumer can be calculated as

$$TP = \lambda + (\delta^{15}N_{consumer} - \delta^{15}N_{primary\ consumer})/3.4‰.$$

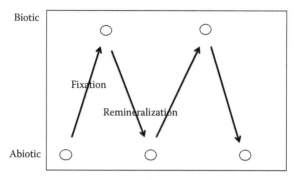

Distance from location of fixation

FIGURE 13.10 Nutrient *spiraling* in an ecosystem. Elements and their isotopes are mineralized from inorganic locations. Elements are incorporated into biotic sources (i.e., fixation). As organisms excrete, egest, lose tissues, or die, the elements are remineralized into the abiotic environment. These nutrients are then fixed again. As elements are fixed and remineralized, their concentration and perhaps fractionation of isotopes change along an environmental gradient extending away from the source of mineralization.

Critical for this equation is λ, which is the trophic position of the primary consumer (Figure 13.9). Note that $\delta^{15}N_{\text{primary producer}}$ is not used to determine the TP of the consumer of interest. Recall in Chapter 12 that plants and algae vary widely in nitrogen content and may have different isotopic signatures. Using a primary consumer to *anchor* the TP model allows the herbivores to integrate the nitrogen content of the primary producers. In these models, the source of autotrophic production is known, typically from diets and carbon isotopes of the primary consumers. In a terrestrial ecosystem, whether the $\delta^{13}C$ of the consumers overlaps that of a C3 or C4 plant allows trophic ecologists to pinpoint the base of the trophic pyramid.

The TP of most organisms is not an integer, like 3 or 4. This occurs because organisms are not strictly consuming one item in the field and may consume prey at multiple trophic levels (Chapter 2). This fact makes using heavy nitrogen as a tracer in food webs so compelling as a way to bridge individual-based consumer dynamics and the ecosystem consequences of materials and energy flux (Post 2002). The primary assumption revolves around the value of λ for the primary consumer, where it is assumed to be an obligatory herbivore. This is not always true. Many herbivores consume other prey such as insects, bacteria, and other items in addition to plants. Still, λ is assumed to equal 2 for the primary consumer, representing the second trophic level in most TP models. In a lake, the primary herbivore is a freshwater mussel, with a $\delta^{15}N_{\text{primary consumer}} = 2$. We are interested in the TP of yellow perch, an omnivorous fish, in the lake. After collecting mussel and perch muscle tissue, samples are processed and analyzed using an isotope ratio mass spectrometer. Mussel samples generate a mean of $\delta^{15}N_{\text{primary consumer}} = 2.2$, while fish samples average at $\delta^{15}N_{\text{consumer}} = 4.1$. Thus, $TP_{\text{consumer}} = 2 + (5.1-1.2)/3.4 = 3.1$. Yellow perch are expected to fall somewhere between the third and fourth trophic level in the lake ecosystem.

Different herbivores reflect different autotrophic sources. If we know that the yellow perch not only eat food from the bottom of the lake (i.e., benthos), where mussels feed, but also consume prey in the open water, we need to include an herbivore that integrates algal production in this location called the *pelagic zone*. Herbivorous pelagic zooplankton like cladocerans may improve our understanding of the trophic position of yellow perch. A mixing model needs to be incorporated into the analysis, which accounts for the amount of energy coming from the benthos and the pelagic zone, where

$$TP = \lambda + \left\{ \left(\delta^{15}N_{consumer} - \left[\delta^{15}N_{primary1} * \alpha \right] \right) + \left(\delta^{15}N_{primary2} * [1 - \alpha] \right) \right\} \Big/ 3.4\text{‰}.$$

The relative source of energy from autotrophs is determined from the relative fractionation of ^{15}N of the two primary consumers (mussels = 1; cladocerans = 2). An additional piece is needed to determine their relative contribution to the consumer. Because heavy carbon provides source and contributes little to trophic position, it can be quantified and incorporated into the equation as

$$\alpha = (\delta^{13}C_{consumer} - \delta^{13}C_{primary2})/(\delta^{13}C_{primary1} - \delta^{13}C_{primary2}).$$

As in the previous example, $\delta^{15}N_{primary\ 1} = 2.2$ and $\delta^{15}N_{consumer} = 4.1$. In this case, mean $\delta^{15}N_{primary\ 2} = 1.3$. Mass spectrometry of carbon isotopes revealed that $\delta^{13}C_{consumer} = -17.1$, $\delta^{13}C_{primary1} = -19$, and $\delta^{13}C_{primary2} = -10.3$. Thus, $\alpha = (-17.1 - [-10.3])/(-19-[10.3]) = 0.78$. Although the values are typically negative, the resulting value of α will always be a positive proportion. This changes the trophic level estimate for the yellow perch to $TP = 2.8$. The above examples highlight the importance of knowing the sources of nitrogen within food webs when evaluating trophic levels.

13.7 OTHER ISOTOPES

Several additional biologically relevant isotopes provide information about food-web dynamics. Heavy sulfur ($\delta^{34}S$) is highly fractionated by 21‰ in the sulfate reservoir found in seawater. Sulfur's standard is based on that in a meteorite in Arizona (Table 13.2), which is assumed to have sulfur that reflects that of the solar system. Sulfur is another pollutant that is released into the atmosphere by burning fossil fuels and precipitated out as sulfate, with a modified $\delta^{34}S$ that is less than that of seawater. Terrestrial organisms typically have lower fractionation of ^{34}S than ocean counterparts, which often have similar $\delta^{34}S$ as their surroundings. This difference between terrestrial and marine fractionation values can be used in mixing models to determine the relative sources of food items, particularly in coastal and estuarine ecosystems.

Two other elements, hydrogen and oxygen, are useful in trophic ecology, although they typically do not reflect diet composition, just location. They are largely consumed through water intake in food. Aquatic organisms in freshwater experience a constant influx of water from the environment. Stable isotopes of hydrogen exist as ^1H and ^2H (i.e., deuterium), with the lighter isotope being common (Table 13.1). Hydrogen is most common in water; seawater is fractionated with ^2H by up to +20 δ^2H, relative to freshwater. Evaporating water tends to be depleted in deuterium, with precipitation and freshwater in the continents having lower δ^2H. Although δ^2H is not a marker of feeding, it can provide information about the spatial habits of organisms, inferring that they fed in those areas. In rivers with interconnected shallow lakes, the deuterium content in the lake water may be increased through evaporation (Figure 13.11). Whether an organism such as a fish or waterbird spent its time foraging in the river or the lake might be assessed from the δ^2H in its tissues. Oxygen has three stable isotopes, with ^{16}O being common, ^{17}O being rarest, and ^{18}O being rare (Figure 13.2). In addition to being present in water and food, its role in respiration makes this element difficult to use as a tracer of food sources for trophic ecology.

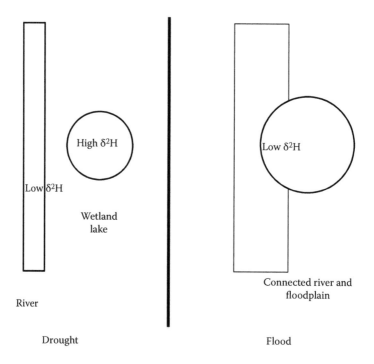

FIGURE 13.11 Differences between hydrogen fractionation as a function of location. The hydrogen stable isotope fractionation within a wetland lake and river may differ, depending on the evaporation that occurs in the lake, as long as they are separated (e.g., during years of drought). If an ecologist is interested in foraging behavior of a waterbird during this time, the δ^2H of the bird may reflect its primary habitat. Of course, flooding will erase this signature as the two systems become connected. Also, the ecologist must assume that the areas in which the bird drinks water is the same as where it forages.

13.8   HETEROGENEOUS AND RARE ELEMENTS

Many elements are not uniformly distributed across the surface of Earth. Common biologically relevant elements like barium, lithium, manganese, magnesium, lead, strontium, and zinc can provide unique geographic-specific markers, because of their heterogeneity, even at small spatial scales. Interestingly, a group of materials called *rare earth elements* are not necessarily heterogeneous. Rather, they are just found in most geologic samples at low concentrations. *Rare earth materials* include exotic names like scandium, cerium, and holmium. Because of their scarcity, they are rarely biologically significant and do not typically serve as good markers (but see Lawrence et al. 2006).

Some elements have similar biological properties and can be substituted biochemically. Calcium and strontium behave this way in the hard structures of some organisms. Calcium is most abundant but strontium can be accumulated as well. The ratio of strontium to calcium in the bones and other structures of animals may reflect origin based on the underlying strontium concentration of the environment. The sources of these elements are food and water and may reflect past foraging. The rivers of central North America vary in their concentration of strontium, which allows biologists to assess environmental source (Zeigler and Whitledge 2010). Growing fish accumulate these minerals in their bones, fin rays, and inner ear structures (i.e., otoliths) in ways that provide location information through time. Research using this technique has shown that fishes in these rivers move thousands of kilometers over short times (Phelps et al. 2012).

13.9   EXPERIMENTAL DESIGN

Elemental tracers are clearly crucial tools for understanding food-web interactions. Simply generating correlations between environmental sources and concentrations in consumers does not reveal mechanisms, leading to criticisms about using this technique in the field (Herman et al. 2000). Experiments that determine the relative sources of the elements must be conducted. Both water and food need to contain known concentrations of the tracer isotopes or heterogeneous elements. Controls with no spikes in food, water, and neither should be included in the experiment. As we have shown in Chapter 12, the tracers will first increase in the tissues and then after they are no longer provided in treatments, tissue turnover and growth will begin to eliminate them. The patterns of accumulation that arise in the tissues of the consumers will provide insight into the duration necessary for the experiment. In some tissues where turnover is rapid, like muscle, experiment duration can be short because tissues come into equilibrium with the sources rapidly. In hard parts such as bone, detectable accumulation may take months or years, making verification with experiments time-consuming and expensive.

13.10   CONCLUSIONS

Tracers are important tools for trophic ecology. Using rarely occurring materials and quantifying their pathways through organisms and ultimately through ecosystems provide insight for developing sophisticated models and testing hypotheses.

Elements that rarely occur in the environment and that are distributed heteroge-neously across terrestrial and aquatic landscapes are useful because they can be traced back to a source, whether it be a food item or foraging location. A common or widespread material cannot be used to differentiate source because it is impossible to determine its origin. Rare tracer elements are often heavy isotopes of commonly occurring atoms like carbon and nitrogen. These accumulate in food webs in unique ways. The challenge is to understand how patterns of mixing and fractionation are related to trophic interactions. The models presented in this chapter are simplistic. Future research will require developing more sophisticated ways to trace materi-als as they are accumulated and transformed in ecosystems as with mixing models being developed with hierarchical Bayesian techniques (Box 13.2) and other related approaches such as niche overlap techniques (Swanson et al. 2015).

Detecting miniscule concentrations of rare stable isotopes and elements in the environment and in experiments is becoming easier and cheaper. Quantifying isoto-pic concentrations in the field is already possible, although high concentrations (i.e., above ambient) are required to meet detection limits. Sample processing and analysis will continue to depend on laboratories with high quality control. It is important for ecologists interacting with laboratories that use mass spectrometers to quantify data to understand the process and accept both the promises and limitations of the technology.

It is critical to note that tracer analysis is no substitute for other approaches in trophic ecology including experimentation and diet analysis. Without identifying the source materials in the environment, it is impossible to differentiate how materials move through food webs. Enrichment studies, where tracers like ^{15}N are released into the field and followed through all apparent components of ecosystems, will con-tinue to be illuminating. A recent multibiome experiment called the Lotic Intersite Nitrogen Experiment (LINX) was conducted across 11 streams in arctic, tropic, and temperate biomes with different levels of precipitation (Webster et al. 2003). Uptake of nitrogen as ammonium into biotic (as organic matter) and abiotic (as ammonia, nitrite, nitrate) pools was quantified in each of these systems after the release of ^{15}N upstream. No clear patterns in uptake of nitrogen among streams were found, although biotic metabolic demand was an important predictor of uptake. This study was limited to quantifying uptake into living and nonliving compartments, because it is impossible to collect every food-web component in these studies. The risk that important pathways might be missed is always real and must be confronted with other approaches such as mechanistic foraging experiments.

Ultimately, multiple tracers should be used to gain a comprehensive picture of food-web relationships. Producing biplots of δ^{13}C and δ^{15}N is the most common mul-tivariate approach, because it likely gives insight into energy source and trophic position of potentially interacting organisms (Box 13.3). Including other stable iso-topes like δ^2H and δ^{34}N plus potentially heterogeneous elements like strontium may provide more resolution about overlap among prey and their consumers in ecosys-tems. Tracer studies are opportunities to compare food-web structure between eco-systems. They also may provide baselines by which changes in food webs through time may be quantified, for example, when new species invade, climate changes, or restoration is implemented. Humans are altering the elemental composition of soil, air, and water. This must be accounted for in any long-term monitoring program.

BOX 13.3   STABLE ISOTOPES SHOW ECOSYSTEM CHANGE

Biplots of stable isotope ratios ($\delta^{13}C$ and $\delta^{15}N$) reflect trophic interactions in ecosystems.

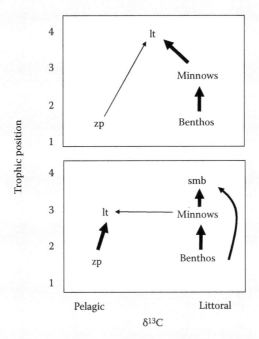

Vander Zanden et al. (2004) used these relationships to categorize the vulnerability of Canadian lakes to invasion by a novel fish species. Lake trout, a cold-water fish, is native to many of these northern lakes. As climate warms, smallmouth bass, a warm-water species, is able to invade these bodies of water. The question is how do smallmouth bass introductions affect lake trout? Northern lakes have distinctive habitats, with a shallow area around the shore called the *littoral zone* that rapidly drops off to deep water, creating an open-water area called the *pelagic zone*. Only the top layer of the pelagic zone is available to most organisms because the lower layer is dark, cold, and devoid of oxygen. Prey organisms available to fish in the littoral and pelagic zone differ in quantity and quality.

Aquatic ecologists know that lake trout and smallmouth bass differ in their behaviors. Lake trout occupy both the littoral and pelagic zones, whereas smallmouth bass are littoral zone specialists. When smallmouth bass invade a lake, lake trout foraging behavior may change due to interference and exploitative competition. A way to test this hypothesis is to quantify stable isotope concentrations of carbon and nitrogen of food and fish in invaded and uninvaded ecosystems. Zooplanktons (zp) are pelagic herbivores that are important prey for fish including lake trout (lt) and minnows. Minnows are prey for both lake trout and smallmouth bass (smb). Minnows and benthic food (benthos) are restricted to the littoral zone. What Vander Zanden discovered is that $\delta^{13}C$ reflected origin of food. The littoral zone is closer to land and should have a terrestrial signature. Pelagic food webs are dominated by autotrophic production of algae, with lower $\delta^{13}C$ values. Trophic level was reflected by the equation using $\delta^{15}N$, described in the text.

The trophic signatures of the lakes were different depending on whether smallmouth bass were present or not. Without smallmouth bass, lake trout occupied a high trophic position and most of its energy (i.e., carbon) derived from littoral sources, likely via consuming minnows. In lakes with smallmouth bass, lake trout shifted to a $\delta^{13}C$ more reflective of a pelagic consumer and its trophic position declined, with zooplankton being more important to diets. Smallmouth bass replaced lake trout's position in the food web. This study illustrates the utility of using biplots of carbon and nitrogen isotopes to generate hypotheses about food-web structure in ecosystems. However, the next step would be to treat these approaches experimentally. For example, what happens to the trophic position of lake trout if smallmouth bass are removed? What do the diets of these fish look like? Do they correspond to the patterns of isotopic fractionation? Mechanistic information collected in ways outlined in previous chapters will provide insight into these correlative patterns.

QUESTIONS AND ASSIGNMENTS

1. What is an isotope of an atom?
2. Why are radioactive elements rarely used in ecological studies of trophic interactions?
3. What is the reason that ratios of the heavy and light isotopes are used in analysis rather than the actual concentrations?
4. Provide an example of a simple mixing model with one isotope and three food sources.
5. How do $\delta^{13}C$ and $\delta^{15}N$ patterns of fractionation differ between terrestrial and aquatic ecosystems?
6. In what tissue do most trophic studies collect stable isotopes? Why? Are there other options?
7. What are limitations to the use of stable isotopes in trophic ecology?

14 Use and Importance of Lipids in Trophic Ecology

14.1 APPROACH

Lipids and their component fatty acids are becoming important tools in trophic ecology. In this chapter, we will review their structure and function and explore various statistical tools for analyzing them. Lipids also are critical for the function of ecosystems, and the quantity of some essential fatty acids may well limit trophic interactions.

14.2 INTRODUCTION

Trophic ecologists continue to search for linkages between consumers and their food sources. Tools for delineating these relationships are becoming more sophisticated and available for ecologists, allowing us to continue tackling pressing questions about biodiversity, food-chain length, and the complexity of ecosystems.

In the previous chapter, we explored how elemental tracers might be used to track the flux of energy and matter among trophic levels in ecosystems. Organisms synthesize all sorts of complex molecules that may be traced through food webs. Challenges in finding robust, naturally occurring markers exist. Natural selection has led to surprisingly consistent solutions to structural and reactive molecules in organisms (e.g., enzymes, ribosomes, nucleotides). These conserved molecules are not particularly unique, are typically easily catabolized when consumed, and thus are no more useful for tracing the flux of materials than using simple common elements like carbon or nitrogen, which comprise them.

Lipids are a group of relatively environmentally stable, biosynthesized molecules that hold promise for trophic ecologists interested in linking consumers and prey. The use of these molecules for revealing feeding relationships is not new (Lovern 1934). We will see that fatty acids, which are the primary components of lipids, have structural characteristics and metabolic properties that make them useful biotracers in trophic ecology. Plants, particularly aquatic ones, and some bacteria are capable of synthesizing unique and essential fatty acids in their storage cells, cytoplasm, and cell walls. The capacity for organisms to synthesize essential fatty acids (Chapter 9) and rearrange them biochemically appears to decline with increasing phylogenetic complexity, with many vertebrates requiring dietary sources of essential fatty acids to survive and reproduce. Recent research that we will explore later shows that essential fatty acids may limit the length of food chains in ecosystems (Wilder et al. 2013).

In this chapter, we first review the biochemistry of lipids and their component fatty acids, making sure that the basic building blocks and structures are understood. Fatty acids are quite diverse in their structure, and their *fingerprint* in the food and the consumer can be used to track down trophic relationships. We will find that these data are inherently multivariate; statistical procedures can be used to quantify trophic linkages. We will then explore how fundamental changes to food webs, especially at lower trophic levels, may well influence bottom-up processes via fatty acids synthesized by the primary producers and in the microbial food web.

14.3 LIPIDS

Lipids are ubiquitous in life because of their ideal properties for supporting body structure (e.g., cushioning), cellular materials transport, and energy storage. Chapter 9 provided a short primer on lipids in nutrition and their value to trophic ecology. When a consumer eats another organism, it ingests a variety of lipid types that have different uses and fates. Cell membranes are composed of partially charged phospholipids (Figure 14.1). The charged end of the phospholipid head is composed of a negative and positive charge, which is attracted to water molecules (hydrophilic). The fatty acid ends of phospholipids are uncharged and repelled by water (hydrophobic). Some of the simplest multimolecular structures are called *micelles* (e.g., soap bubbles), which provide some insight into how simple cell membranes are formed (Figure 14.2). Life likely arose in part because of this self-organizing behavior of some lipids in the environment. The fatty acid tails of phospholipids in cell membranes affect how they interact with the environment, particularly with temperature. We will review the impact of temperature on fatty acid shape and behavior in a later section. Triacylglycerols (TAGs) and wax esters (WEs) are neutral lipids that do not carry a charge (Figure 14.3). These serve primarily as storage for energy, but also provide structural support and insulation in some organisms.

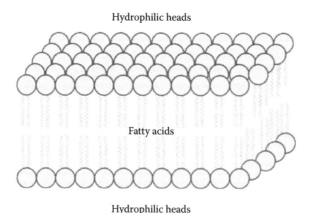

Hydrophilic heads

Fatty acids

Hydrophilic heads

FIGURE 14.1 Basic structure of cell membranes in all living things on Earth. Phospholipids have hydrophilic heads that are attracted to the water in the surrounding medium or within cells. Fatty acid tails turn inward because they are hydrophobic. This arrangement leads to a relatively impermeable barrier that can be structured with various structures made of proteins and lipids that transport materials.

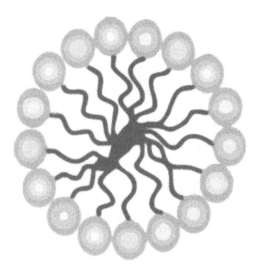

FIGURE 14.2   Simple micelle in nature. Charged parts of lipids face the water, while fatty acid tails face inward. It is not difficult to see how micelles might arise into biologically meaningful structures.

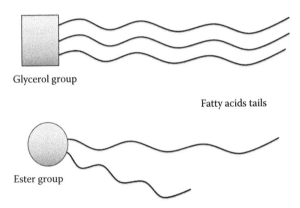

Glycerol group

Fatty acids tails

Ester group

FIGURE 14.3   TAGs (above) and WEs (below). Whereas WEs are typically not catabolized, TAGs are rapidly hydrolyzed and the fatty acids mobilized for many uses.

In some superficial ways, fatty acids are similar to DNA in their basic simplicity but ability to be arranged in complex ways (see Chapter 7). As DNA can be identified by the pattern of base-pair combinations of nucleotides, lipids and their component fatty acids can be categorized by some simple chemical characteristics associated with their carbon and hydrogen bonds. Both phospholipids and lipids have *heads* to which the fatty acids attach (see Chapter 9). Hydrolysis is the process that the acidic digestive tract of organisms uses to remove the fatty acids from the glycol or nitrogenous head of the lipid or phospholipid. Hydrolysis is a simple electron transfer between the surrounding acidic medium and the weakest *point* of the lipid molecule at the junction between the head and the fatty acid chains.

Each fatty acid is composed of several basic chemical structures (Figure 14.4). A carbonyl group (CO) is at one end, attached to an oxygen atom to form an ester (COO). Esters are responsible for smells we humans recognize as fruity and are the precursors to many acids. The addition of a hydrogen atom forms a carbonyl (COOH) group that carries a slightly positive charge. Attached to the carbonyl is a carbon skeleton, or aliphatic tail, composed of only carbon and hydrogen atoms, ending with a methyl group (CH₃; Figure 14.4). This basic structure may sound simple, but the configuration of the hydrogen atoms and resulting double bonds between carbon atoms in the aliphatic tail profoundly affects the chemical properties, biological characteristics, and environmental behaviors of the fatty acid molecules.

Even given the ubiquity of fatty acids in ecosystems, most that have complex structure are biologically synthesized and have some general characters based on the organisms and enzymes used to make them. Fatty acids rarely occur in the open environment because they combine with other molecules to form micelles, such as the foam of bubbles that occur on ocean beaches. They also are quickly consumed, stored, or transformed by plants, algae, and microorganisms. On Earth, most naturally synthesized fatty acids have an even number of carbon atoms due to the propensity for enzymes in organisms to add, otherwise known as elongate, fatty acids in carbon pairs. Fatty acids appear to have a length limit in ecology, with chains of greater than 28 carbon atoms rarely occurring. Six or fewer double C–C bonds exist in nature. The structural limits are likely due to the energetic limits of certain enzymes to build fatty acid molecules. The energetic benefit of synthesizing enzymes to build long chains of relatively repeating carbon molecules must exceed the value of storing or locating these materials (Figure 14.5).

Fatty-acid aliphatic chains hanging off of lipids, phospholipids, or WEs have different shapes depending on length, the position of the C–C double bonds, and environmental conditions such as the presence of water and temperature. We will specifically define nomenclature later, but those fatty acids with no double bonds are called saturated and typically exist as highly packable *sticks* in a lattice (Figure 14.6). At environmental temperatures (i.e., homeotherm temperature), these

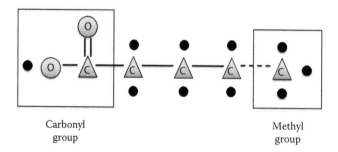

Carbonyl
group

Methyl
group

FIGURE 14.4 Basic structure of a fatty acid. Triangles are carbon atoms. Large and small circles are oxygen and hydrogen atoms, respectively. Solid lines are covalent bonds. Dotted lines are the potential carbon–carbon double bonds in the aliphatic chain.

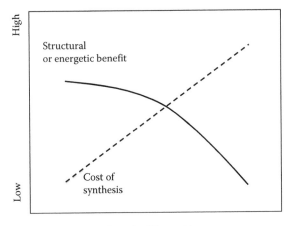

FIGURE 14.5 Hypothetical relationship between size of fatty acids and the costs (dotted lines) and benefits (solid line) of producing them. Fatty acids are energetically costly to build, requiring structural enzymes to reduce saturation and increase length. At some point (intersection), the cost of maintaining this synthetic machinery is greater than its benefit to body structure or energy availability.

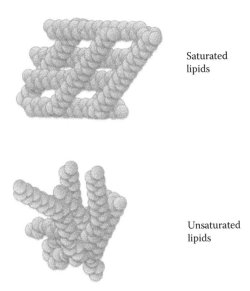

FIGURE 14.6 Differences in behavior of saturated and unsaturated fatty acids in lipids. Saturated lipids are straight and form a lattice structure at environmentally relevant temperatures. Unsaturated lipids have bends in their aliphatic tails that make them less able to pack into a solid, decreasing their melting point and allowing them to be more fluid at environmental temperatures.

fatty acids are solid because they have a high melting temperature (i.e., point). Unsaturated fatty acids have at least one double bond in the aliphatic chain, making them bend in unique ways and less apt to pack into a tight lattice (Figure 14.6). They tend to remain liquid at environmental temperatures because of their lower melting point.

Patterns of bending in unsaturated fatty acids make them behave in unique and useful or deleterious ways in organisms. These molecules have two possible orientations, otherwise known as *isomers* (Figure 14.7). The *cis* isomers of fatty acid and other molecules mean that the bends created by the double bonds of carbon orient them in the same direction, similar to a spring (Figure 14.7). These isomers have lower melting points and are dominant in organisms, likely because they are suppler and less likely to *seize up* with cooling temperatures. In contrast, *trans* isomers have double-bond configurations that make them bend in opposite directions, allowing them to stack up on top of each other and increasing their melting point (Figure 14.7). Certain microbes like *Pseudomonas* that must withstand environmentally high temperatures or other conditions that require rigidity in the cell membrane may preferentially synthesize *trans* isomers.

Although *trans* isomers of fatty acids are relatively rare in nature, humans have successfully and intentionally created these molecules through industrial hydrogenation. A primary reason to accomplish this is to make vegetable oils that are often liquid at room temperature in their *cis* isomer form to remain solid. Some animals that undergo intensive fermentation in their guts by microbes (e.g., ruminants) have elevated levels of trans fatty acids in their digestive tracts (Wahle et al. 2004). Tracing trans fatty acids through food webs might be a useful way to assess the contribution of ruminants to trophic interactions.

Fatty acids are largely tied up in the biota of ecosystems. They are not water-soluble and degrade rapidly in the environment. Thus, the source of fatty acids in trophic levels is largely through de novo synthesis by the primary producers, and transfer occurs mainly via consumption. Environmental factors do not play a role in uptake as they do for elemental concentrations in organisms (Chapter 13). As many

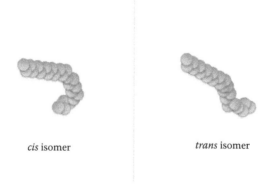

cis isomer   *trans* isomer

FIGURE 14.7   Two fatty acids with different carbon double-bond configurations. The chemical formulas are the same, but the spatial orientation is different.

as 70 fatty acids may be isolated in tissues of organisms, meaning that there is tremendous variability to quantify and compare across trophic levels.

Because of their importance to human nutrition and especially the food industry, tremendous research has been conducted on the chemistry of fatty acids. This also means that there is quite a bit of differences in descriptive language and nomenclature depending on discipline. We follow Iverson et al. (2002). Short-chain (SC) fatty acids have less than six carbons in their aliphatic tails. Medium-chain (MC) fatty acids are characterized by 6–12 carbon atoms; long-chain (LC) fatty acids have 13–21, and very long chains (VLC) have greater than 21 carbon atoms in their aliphatic skeletons. A monounsaturated fatty acid has only one C–C double bond. The terms polyunsaturated (PUFA) and highly unsaturated (HUFA) fatty acids are used by some ecologists. A PUFA has more than one double bond, whereas HUFAs typically have more than one double bond and greater than 19 carbon atoms.

We described the nomenclature system used typically by ecologists in Chapter 9. Recall that fatty acids are described by the convention $C{:}Dn{-}x$, where C is the number of carbon atoms in the molecule, D is the number of double C–C bonds, and x is the location of the first double bond from the methyl end of the molecule. Using the conventions, we can easily identify the differences between fatty acids. Saturated fatty acids, like the common stearic acid, with 18 carbon atoms and no double bonds is 18:0, with no n-x included because of the absence of a double bond. Docosahexaenoic acid (DHA) is unsaturated with 22 carbon atoms and 6 double bonds, beginning at the third carbon atom from the methyl end. The standard nomenclature for DHA in trophic ecology is *cis* 22:6n-3.

14.4 SYNTHESIS AND FATE OF FATTY ACIDS

Fatty acids are deceptively simple. Long chains of carbon atoms do not occur by environmental accident and are the result of biological activity. De novo production of unsaturated fatty acids is limited to plants and algae because they possess enzymes called *desaturases* that allow them to generate them from saturated fats. Two particularly important, *essential* fatty acids that are produced by certain autotrophs are *DHA* and *eicosapentaenoic acid* (*EPA*; 20:5n-3). Not all autotrophs are similar in their propensity to generate these fatty acids (see Box 14.1), with marine diatoms being the champion at producing them (Brett and Muller-Navarra 1997). Neither marine fishes (Brett and Muller-Navarra 1997) nor terrestrial, herbivorous insects (Blomquist et al. 1982) appear to have the *desaturases* necessary to convert precursor fatty acids to EPA and DHA. In contrast, freshwater fishes and terrestrial omnivores, including humans, possess the desaturases necessary to convert plant-derived linolenic (18:3n-3) and linoleic (18:2n-6) acids to EPA and DHA. However, the rate by which young organisms can convert the precursors to EPA and DHA may be unable to meet demand, thus necessitating a dietary source to make up for short falls (Brett and Muller-Navarra 1997).

Lipogenesis of saturated fatty acids is common in organisms. Glucose is converted to fatty acids from fatty acid synthase. Many organisms have desaturases, with limited activity, that allow them to convert saturated fatty acids to monounsaturated fatty acids. Some marine mammals store fatty acids in the form of WEs (Figure 14.7). Obviously, quantifying profiles composed of saturates or monounsaturates will

BOX 14.1

The distribution of fatty acids depends on the source. Brett and Muller-Navarra (1997) showed that the fatty acid compositions from oceans, especially the essential ones like EPA and DHA and their precursors, differ among phytoplankton taxa.

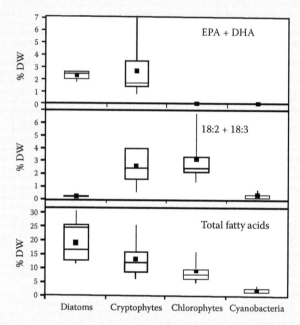

The authors showed that diatoms had the highest concentrations of fatty acids, of which the highest proportion were composed of EPA and DHA. Precursor fatty acids linoleic and linolenic acid were highest in green algae. Cyanobacteria had low concentrations of fatty acids, leading the authors to consider these organisms as a trophic dead end. The contribution of cyanobacteria to fatty acids in food webs may depend on taxa present and conditions within ecosystems.

provide little information about trophic linkages because they are easily synthesized in all organisms.

When attempting to link the profiles of the unsaturated fatty acids among trophic levels, we must understand their fate as they are processed in the digestive tract and tissues of consumers (Figure 14.8). Lipids are hydrolyzed into fatty acids and glycerol via the action of lipases and esterases in the gut. Because fatty acids are not water-soluble, they must be transported through the organism with the aid of carrier molecules called *lipoproteins*. Some fatty acids may enter cells through diffusion or active transport. In vertebrates, fatty acids with 14 or less carbons are transported to the liver to be catabolized into energy or re-esterified as lipid. Longer-chain fatty

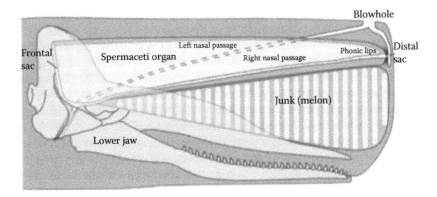

FIGURE 14.8 Head of a sperm whale, which is dominated by a spermaceti organ. This organ is filled with WEs that are used in echolocation. Other organisms like copepods accumulate WEs to enhance buoyancy. WEs are typically not stored for energy use.

acids, especially those that are unsaturated, may be transported throughout the body and to specific tissues, rather than being catabolized or stored as fat. These unsaturated fatty acids also may be stored as lipids or WEs. As we noted earlier, some consumers such as marine copepods are capable of elongating and further desaturating unsaturated fatty acids (Iverson et al. 2002). The relative influence of catabolism, elongation, and desaturation needs to be understood when linking the occurrence of HUFAs between trophic levels.

14.5 FATTY ACIDS IN TROPHIC ECOLOGY

The above primer demonstrates that prey containing lipids composed of unsaturated fatty acids are potentially traceable in their consumers. This is possible because of the conservation of these molecules. Because these fatty acids can be transformed in some organisms and catabolized, they probably are best used to quantify bitrophic interactions. In the next section, we will look at several examples of how fatty acid profiles or *fingerprints* have been used successfully to identify feeding relationships in food webs.

Long-chain and very-long-chain unsaturated fatty acids may be anabolized in the many tissues of organisms. For small organisms, the entire body may be used, losing resolution about the source and potential physiological mechanisms underlying uptake of the fatty acids. Typically, fatty acids are obtained from adipose tissue either stored in the body cavity or adjacent to other tissue layers, such as muscle. For mammal research, the fat found in milk produced by lactating females is a potentially powerful way to link prey and mammalian consumers (Brown et al. 1999; Iverson et al. 2002). Blood plasma is not a good source of fatty acids because they are only going to be present as attached to lipoproteins and may only represent a very short timescale, such as the fatty acid composition of a single meal.

Techniques used to quantify fatty acids in organisms are well developed. Lipids, phospholipids, and WEs are isolated from homogenized tissue by solvent extraction (e.g., chloroform–methanol; Bligh and Dyer 1959). Extracted nonpolar and polar lipids are separated. Before lipid fractions can be analyzed for fatty acids, the fatty acids need to be transformed into fatty acid methyl esters (FAMEs) by means of acid-catalyzed transmethylation (Christie 1982). The resulting FAMEs can then be separated using a gas chromatograph (GC). Other detectors such as a mass spectrometer also can be used. A sample of FAMEs is injected into the GC, and an inert carrier gas such as helium transports the sample up the column where it is combusted at specific temperatures. A mass detector measures the FAMEs as they are burned (each has a unique retention time), which are then compared with the retention times of a standard. Resulting peaks from the GC can be used to quantify the relative concentrations of FAMEs in the lipid fractions (e.g., g FAME/g tissue).

14.6   USES AND ADVANCES

Fatty acids can be used as tracers between trophic levels and also to quantify spatial sources of energy and matter. We explore aquatic and terrestrial examples, showing how fatty acids can be used to find linkages between these systems. Statistical tests that analyze multiple fatty acids simultaneously will be briefly described.

As we noted earlier, essential fatty acids are synthesized de novo by plants and algae in freshwater and marine ecosystems. This has been recognized in fish aquaculture, where the availability of HUFAs may limit growth and survival, ultimately affecting yield (Brett and Muller-Navarra 1997). Fatty acid profiles from blubber cores of elephant seals of the South Pacific Ocean provided a strong match to their marine prey (Bradshaw et al. 2003). Seasonal variation in foraging in marine amphipods was shown to occur using fatty acids (Legezynska et al. 2012). Laureillard et al. (2004) used fatty acid profiles to determine the relative importance of bacteria to foraging deep-sea protists. Many aquatic examples exist (Iverson et al. 2002).

A compelling use of fatty acids in aquatic systems is to determine the spatial source of energy from the autotrophs (Rude et al. 2016). Fatty acid profiles from freshwater and marine algae are often distinct due to difference in algal producers. In the San Francisco Bay estuary, HUFAs from diatoms in freshwater were a critical source of fatty acids and presumably energy (Canuel et al. 1995). The relative sources of aquatic versus terrestrial organic matter were traced with fatty acids in the surface substrate of the Mississippi River and downstream to the Gulf of Mexico (Waterson and Canuel 2008). Terrestrial sources of fatty acids were found far out in offshore areas, suggesting high importance of nonriver sources of energy.

In terrestrial ecosystems, less research appears to have been conducted on the fate of fatty acids in food webs. In our view, this is not because of a lack of utility of this technique on land. Several investigators have used fatty acids to trace trophic interactions in soil food webs (Porazinska et al. 2003; Ruess and Chamberlain 2010), although aboveground research is scarcer (see Ferlian et al. 2012). Underground, the relationships between fungi and their nematode predators were partially determined using fungal-specific fatty acid biomarkers (Ruess et al. 2002). Ruess et al. (2004) showed that long-chain fatty acids became more diverse as they moved from the soil to consumers above the ground, likely as consumers ate a greater diversity of prey items.

Fatty acid profiles are complex and may be difficult to analyze because of the sheer number of possible molecules involved in trophic transfer (Budge et al. 2002). Anyone familiar with statistics knows that data sets with as many as 70 dependent variables cannot be analyzed with simple parametric procedures. The nature of the data is important. Often, fatty acids are presented as percentages of the total lipid concentration. Percentages or proportions typically need to be transformed before being processed because these are not normally distributed. One of the most conventional ways to process multivariate fatty acid data is to use a discriminant function analysis (DFA; Iverson et al. 2002; Box 14.2). A DFA takes known profiles of fatty acids from the food sources; it then takes the fatty acid profile from the consumers and determines how well the consumer profiles *match* or assign back to the profiles of the food sources. Strong assignments allow trophic ecologists to determine the source of prey (Box 14.2). Ordination techniques like nonmetric multidimensional scaling (NMDS) with a statistical procedure called an analysis of similarity (ANOSIM) also can be used (Box 14.2). Ordination takes multiple variables and assesses what combinations explain the most variation in the data structure. The

BOX 14.2

Fatty acid data are complicated. In the example below from Iverson et al. (2002), profiles of several fatty acids from herring prey and monk seals are compared.

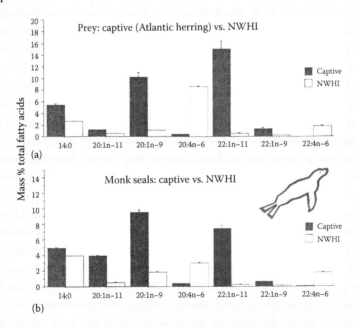

The data are derived from both captive and wild-collected animals. Although visually, the bars for prey and consumer seem to match, statistics are needed to compare them. One method is a DFA. The DFA takes samples from known data, such as assignments for herring and seals in captivity. It then assigns unknown profile data from wild caught animals and assigns them back to the known data. The better that the known and unknown data match, the stronger the overlap is in fatty acid profiles between prey and predator. A DFA for these data suggests strong dietary use of herring by seals in the field.

Many alternate pathways for data analysis exist. In the figure below, multiple fatty acid data are combined into an ordination analysis such as principal components analysis or NMDS.

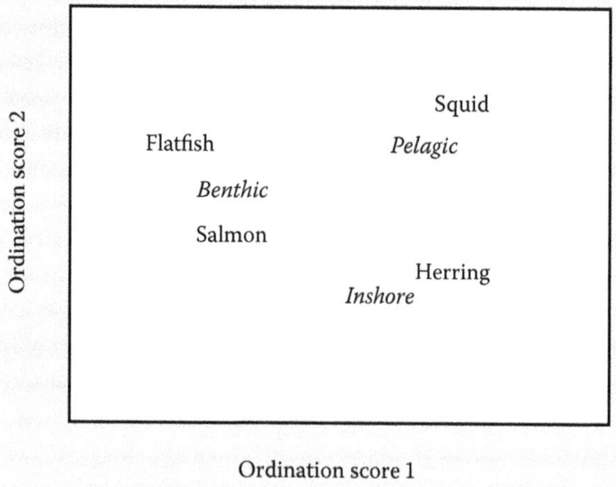

Ordination score 1

Ordination techniques reduce multiple variables into ordination scores that explain the most variation in the data set. A single score represents all the fatty acids quantified in the profile. The scores for consumers and prey are plotted together. If they overlap, then they have similar fatty acid profiles. Here fish species and squid consumers are compared to prey types from the benthos, the open water (pelagic), and the coast (inshore) of the ocean. This analysis shows that flatfish and salmon consume prey from benthic sources. Squid depend on prey with pelagic fatty acid profiles. Herring consume prey with inshore profiles.

variables that are most similar will *cluster* in the ordination. Thus, statistics from the NMDS of prey and consumers with similar fatty acid signatures will overlap in the ordination plots. Those that differ greatly will not overlap (Box 14.2). The ANOSIM of the variables in the plots can be used to determine whether the clusters are different from those that might occur randomly.

The above analyses are correlative in nature. Iverson et al. (2002) called for quantitative fatty acid signature analysis (*QFSA*). The model of QFSA weights fatty acid quantities based on their presumed catabolism or retention in the predator. Short-chain fatty acids that are easily catabolized should not be incorporated into fatty tissues of consumers and would receive low weights. Conversely, essential HUFAs that are not easily synthesized and are retained like EPA and DHA will likely be weighted strongly. These weights are determined using calibration coefficients, which are the proportions (or concentrations) of fatty acids in predators divided by those in the prey. As Iverson et al. (2002) showed with marine mammals and birds, using calibration coefficients was critical for tightening up the fingerprint analysis matching predators to their prey. Presumably a *library* of calibration coefficients could be generated for each consumer type and prey and used to improve comparisons between predator and prey. A next step in trophic ecology will be to generate novel approaches for QFSA. Refining our understanding of catabolic processes and anabolic pathways of fatty acids in dominant organisms in food webs will improve this approach. Bioenergetics models (Chapter 8) tracing the fate of fatty acids as they are consumed would be a valuable way to accomplish this.

14.7 FOOD-WEB STRUCTURE

Trophic ecology revolves around factors limiting the intake and loss of energy and materials within ecosystems. Fatty acids, like some elements and their stable isotopes, allow ecologists to trace source and fate. As with any tracers, the integrity and composition of these molecules as they pass between prey and consumers are still not well understood and need more experimental exploration.

As we have seen throughout this book, food webs are limited in many ways, including the number of species, the number of connections, the total energy available, the energetic constraints of the component organisms, and the total number of trophic levels. The ability for primary producers to synthesize essential fatty acids and to produce enough to support the consumers in the ecosystem may be key to predictions. Also, the role of fatty acid subsidies from other ecosystems must by considered.

Early research by Muller-Navarra et al. (2000, 2004) explored why enrichment or eutrophication of ecosystems did not necessarily lead to an increase in the size of food webs or elongation of the food chain. In lakes with high concentrations of phosphorus, a limited nutrient in many of these ecosystems (Chapter 9), community structure and ultimately the number of trophic levels decline with eutrophication. These researchers showed that the green algae and diatoms present in oligotrophic and mesotrophic lakes had higher concentrations of HUFAs and PUFAs than those eutrophic lakes dominated by cyanobacteria. Cyanobacteria thrive in lakes when high phosphorus concentrations alter the stoichiometric ratio of lakes, making nitrogen limited (Chapter 10). Perga et al. (2013) found that many cyanobacteria taxa do produce precursor fatty acids that could be synthesized by consumers, suggesting that cyanobacteria might not necessarily be a dead end for ecosystems. Obviously, this issue is yet to be resolved by trophic ecologists.

Fatty acids may have strong bottom-up effects on food webs. Wilder et al. (2013) conducted a review and experiments to quantify the role of proteins and lipids in arthropod food webs. Spiders were considered the top predators in this approach.

BOX 14.3

Essential nutrients may limit the size and length of food chains in ecosystems. Wilder et al. (2013) analyzed data for arthropods that were herbivorous and carnivorous and determined the relative protein and lipid content in each group and their prey.

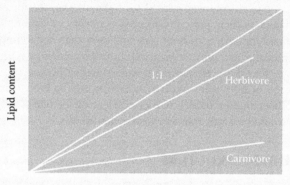

Protein content

What they found is that the protein content was consistently higher than lipids in both herbivorous and carnivorous arthropods because the slope was less than 1:1. As the protein content increases in these organisms, lipids become more limiting. The most notable finding was that the slope for carnivores was much lower than that for herbivores. This suggests that carnivores will be much more lipid-limited than herbivores. The availability of lipids may limit the number of carnivores and their predators in food webs.

Recall earlier that most herbivorous arthropods cannot synthesize essential fatty acids but must get them from plant sources. The authors indeed found that arthropod food webs were more lipid-limited at higher trophic levels than nitrogen-limited (Box 14.3). Spiders are very selective, being obligate predators and not omnivorous. Thus, specialist predators will likely become more lipid-limited because they do not have the ability to pursue prey at lower trophic levels through omnivory.

14.8 CONCLUSIONS

Fatty acids in lipids have been recognized as important to trophic ecology for nearly a century. They are deceptively simple molecules that do not occur by accident. Rather, they are synthesized by organisms and are quite diverse in their composition and geometry. Essential fatty acids are highly unsaturated, characterized by the location of their carbon–carbon double bonds.

Analyses of fatty acids have become routine in the laboratory, with time and cost involved declining. One of the great limitations to tracing fatty acids as they move

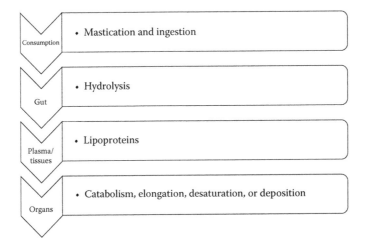

FIGURE 14.9   Process by which lipids are consumed, digested, and processed in consumers.

through food webs is analytical. Most of the techniques available are correlative but do not follow fatty acids in a quantitative manner. Techniques that allow for the catabolism and transformation (i.e., desaturation and elongation) of fatty acids as they move through consumers across multiple trophic levels will greatly improve their use for trophic ecology (Figure 14.9).

Essential fatty acids may well limit ecosystems. Determining under what circumstances fatty acid synthesis and subsidies have strong bottom-up effects will be a challenge for ecologists. The productivity of aquatic versus terrestrial or marine versus freshwater may differ based on the algal diversity, proximity of ecosystems, and environmental processes involved. General theory needs to be developed to guide research.

Fatty acids will only tell a portion of the story about trophic relationships in food webs. Multiple tracers exist in ecosystems. These combined with fatty acids and other approaches such as energetics modeling, behavioral analyses, and food-web network analysis will provide insight into trophic interactions. In the next chapter, we will determine ways that different techniques in trophic ecology can be combined to provide synthetic information about food webs.

QUESTIONS AND ASSIGNMENTS

1. Why are lipids useful markers for trophic interactions?
2. What are the differences between polar and nonpolar lipids? How does this apply to trophic ecology?
3. Why are fatty acids rarely found free in the environment?
4. What are the most common isomers of fatty acids in organisms?
5. Describe how saturated and unsaturated fatty acids differ.
6. Why are EPA and DHA limited in many vertebrates?
7. What is QFSA?
8. How might the availability of HUFAs and PUFAs limit food chain length?

15 Synthesis for Trophic Ecology

15.1 APPROACH

Trophic ecology at its simplest is the study of pairwise consumptive interactions, but it underlies all levels of ecology. We review the evolutionary history of trophic interactions and suggest that humans may be pushing back the clock of complex trophic interactions. We look toward future conceptual approaches and tools that will be used to tease apart mechanistic interactions and how these will be used to merge metabolic theory with alternative approaches such as mass-balance bioenergetics, ecological network analysis (ENA), and stoichiometry.

15.2 INTRODUCTION

Trophic ecology is deceptively complicated, with the term being used in all subdisciplines of ecology. Trophic ecology, in our view, passes through all levels of ecological organization, tying organismal (e.g., behavioral, population, and community) and ecosystem ecology (Figure 15.1). Trophic ecology applies wherever energy and matter are fixed, transported, and transformed between their physical and biological states. This subdiscipline therefore focuses on individual-scale processes—all driven by natural selection—that expand up to ecosystems, landscapes, and, ultimately, the biosphere.

Ecology, as a discipline, has long been at a crossroads, with one facet moving toward molecular techniques and the other toward macroscale ecosystem approaches. Organismal subdisciplines like population and community ecology are important but seem to have hit a conceptual wall (Lawton 1999). As Simberloff (2004) points out and Ulanowicz et al. (2014) would likely agree, community ecology is not a dead end. In our view, trophic ecology, with its embrace of intimate pairwise interactions between food and consumer, is the *missing link* leading to many new advances in ecology at all scales of inquiry. Interactions between organisms within populations and communities, as mediated by natural selection and the environment, regulate the critical contributions of individual organisms to the transfer of energy and matter through ecosystems. This, coupled with the rising roles of metabolic theory and ENA, a variety of powerful statistical and modeling tools, and the use and analysis of elemental and molecular tracers, should lead to new approaches in the future.

We briefly review the main points of this book, identifying the most salient concepts and looking for lessons from the past and present to predict the future. The geological record provides clues about how trophic interactions arose and how fragile they might be. Conceptual developments in trophic ecology can be traced back at least a century and are still useful today. As tools and techniques become more

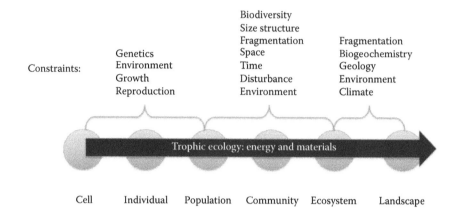

FIGURE 15.1 Trophic ecology is complicated because it is relevant to all levels of organization in ecology, ranging from the cellular/molecular to the biosphere. Key to understanding energy and matter transport through all of these levels is an appreciation for the emergent properties and constraints at each point along the continuum.

sophisticated, analyses must follow. We explore some of the challenges facing trophic ecology in terms of understanding how we deal with the complexity involved in the data that ecologists are producing. We end with some predictions for the emerging discipline of trophic ecology and its application to environmental issues.

15.3 GEOLOGIC PERSPECTIVE

Trophic interactions were likely pretty dull during most of the time life has been on Earth. Autotrophic production by prokaryotes dominated the planet 3.5 billion years ago, meaning that energy was restricted to a single trophic level. Heterotrophy of either live or dead prokaryotes began at some unknown point. It certainly occurred when eukaryotes arose about 2 billion years before present. Some researchers suggest that predation or parasitism between prokaryotes may have led to the endosymbiotic relationships that evolved into eukaryotes (Davidov and Jurkevitch 2009). This is complicated by the fact that prokaryotes are incapable of phagocytosis, meaning that the consumption or parasitism occurred via some other mechanism. Because the earliest eukaryotes only contained mitochondria, the first eukaryotes were heterotrophs or consumers. The primary mode of consumption by the first eukaryotes was either phagocytosis or absorption of materials expelled from other single-celled organisms. Autotrophy via the development of chloroplasts likely did not come until later in early eukaryotes.

The rise of complex trophic interactions that form the basis for this book was only possible with the evolution of eukaryotes and multicellularity, which is very recent in the history of Earth, beginning about 545 million years ago in ancient oceans and seas (Fedonkin 2003). The invasion of land by plants and animals 444 million years ago during the Silurian led to a host of new trophic interactions in open niches that were both above and below ground. Significant heterotrophy, leading to high trophic complexity, requires key environmental conditions including ample oxygen, relatively cool environmental temperatures (<40°C), water, essential nutrients, neutral alkalinity, and low concentrations of carbon dioxide. Many microorganisms are mixotrophs, capable of fixing energy from the environment and consuming other organisms. Mixotrophs will switch from adopting autotrophy to heterotrophy only depending on some environmental conditions (Wilken et al. 2013), suggesting that higher trophic interactions are not always optimal in the environment.

Heterotrophy and complex trophic interactions thus rely on environmental conditions that are relatively new and fragile on Earth. Heterotrophy will only persist on the planet as long as the energetic benefits outweigh the costs of procuring energy through autotrophy. If energy transfer becomes highly inefficient due to physiological, behavioral, or environmental constraints, then heterotrophy will cease to be a force on Earth that dominates some ecosystems through indirect and direct pathways (e.g., trophic cascades).

The business of obtaining energy and giving it up has led to many of the interesting evolutionary characteristics and the biological diversity that arose during the Cambrian explosion and persists to present day. Trophic interactions underlie the outcome of interactions that occur between genes and the environment, which led to the broad range of reaction norms that are responsible for transferring energy through ecosystems. Thus, the flux of energy and matter through ecosystems is not just in the realm of ecosystem ecologists. It matters to evolutionary, behavioral, population, and community ecologists as well.

In summary, trophic ecology is a discipline without a home and is relevant to our understanding of past, present, and future ecological interactions. Complex trophic interactions are not the norm for life on Earth. The rise of heterotrophy either through coevolution or autocatalytic processes is a relatively new phenomenon in geologic time and contributes to the vast diversity of life on the planet. We hypothesize that the rise of consumers is fragile and, once complex consumers go extinct, heterotrophic complexity may take millions of years of evolution to return. Or it may not arise again if conditions such as warm global temperatures that occurred in Earth's early history return.

15.4 APPROACHES TO TROPHIC ECOLOGY

Concepts in trophic ecology are evolving. This book is an attempt to draw on the genius of such forward-thinking ecologists as Forbes, Elton, Lindeman, Hutchinson, Ulanowizc, Carpenter, Power, Paine Polis, Elser, Sterner, and Odum, who recognized the importance of energy and matter flow through organisms. As we illustrated in Chapter 2, the concept of the trophic pyramid is typically the first construct taught to students as early as grade school. Although trophic pyramids might be considered an oversimplification, they are useful for both teaching and research.

Ecology, especially community ecology, has been argued to be too complex for validation (Lawton 1999). Concepts like trophic pyramids and food chains are not difficult to grasp and do provide important steps for constructing useful, testable hypotheses in communities (Simberloff 2004). Identifying important trophic levels or food-web nodes within communities from data sets allows ecologists to identify promising questions to be tested with experiments and models. These groupings, functional groups, guilds, consortia, or any of many terms used interchangeably in ecology are the conduits by which energy and matter move across the planet. Identifying these important groupings and developing a common language among subdisciplines remain a challenge in ecology.

What occurs within the nodes of trophic levels in communities and ecosystems matters. The component genes, cells, tissues, organisms, learned information, species, and communities affect the characteristics of each trophic level. The continuing struggle for ecologists is determining what processes are most important in governing the flow of energy and matter among the nodes and through each trophic level. Each species has a unique blend of behavior, morphology, physiology, population responses, community interactions, and environmental responses that ecologists must unravel. In this book, we provide tools for identifying important groupings and underlying mechanisms. The discipline of ecological network analysis (ENA) is rising and will lead to new insight into the topology and mechanisms with food webs.

Following the tenants of ENA, a useful way to organize all the complexity in trophic ecology is through hierarchy theory (Allen and Hoekstra 1992). This approach derives from information theory, where hierarchical groupings are characterized by emergent properties and constraints. In trophic ecology, each trophic level (or food web) has a series of complex components that can be divided into increasingly finer scales (e.g., community to population to individuals to genes). These components, considered separately, are messy and likely uninformative. When operating together in an intact ecosystem, these components become a functioning trophic level with unique nodes that serve as a conduit for flowing matter or energy. This is the emergent property of the trophic level. Trophic levels are constrained by factors from the bottom and top of the food web. These constraints are critical for maintaining each trophic level. If the constraints are altered, then the nature and function of the trophic level change. For example, altering the nature of autotrophic production or the selectivity of tertiary consumers may profoundly affect trophic level function (Chapter 10). The problem with hierarchy theory is that it makes intuitive sense, but applying it to ecology is difficult, without quantitative tools. The discipline of ENA, quantifying the flows of energy through nodes, provides one framework, although it has yet to quantify many of the characteristics that make certain species uniquely important in holding the puzzle in one piece. D'Alelio et al. (2016) successfully used an Ecopath/ENA type model to compare planktonic food webs during "green" (phytoplankton dominated) and "brown" (herbivore dominated) phases and understand the functional significance of the species present to maintain stability through time. Mass-balance bioenergetics modeling, secondary production models, and statistical models such as hierarchical Bayesian and ordination may hold promise.

Emergent properties of trophic levels surely occur, leading to intuitive and relevant groupings such as autotrophs, heterotrophs, scavengers, decomposers, herbivores, predators, competitors, and symbionts. Making complete sense of the underlying mechanisms is not in the realm of this book because we have not yet found a unifying answer. The future of trophic ecology is determining what parsimonious groupings lead to the most meaningful definitions of trophic levels and energy transfer in ecosystems. In this search, natural selection always needs to be considered because genes are ultimately regulating the ability for organisms to contribute to trophic interactions.

Some authors have argued that identifying underlying mechanisms is unnecessary (Peters 1980). Rather, ecological predictions are possible from observation and correlation. This falls in the realm of macroecological approaches like metabolic theory (Chapter 9). Macroecology may well help ecologists identify broad testable hypotheses for trophic ecology. However, failing to identify underlying mechanisms structuring trophic levels will cause us to miss the important structural components that lead to emergent properties in ecosystems (Allgeier et al. 2015). We will be unable to predict when trophic levels change fundamentally, altering their contribution to ecosystems. Ecologists must use both reductionist and holistic approaches to capture relevant patterns and processes in trophic ecology.

15.5   ANALYTICAL APPROACHES

The eminent ecologist, E.O. Wilson, said that the next generation of ecologists is going to be dominated by synthesizers. Ecological data sets, including those involving trophic interactions, are growing faster than ecologists can analyze them. The culprit is largely technology. The ability for ecologists to collect data from legacy data sets stored on the Internet and to collect data from various tools has increased markedly. Bioenergetics data are easily obtained with better respirometers in the laboratory and the field. Energy budgets of organisms can be quantified using all manners of remote sensing technology such as transmitters and video. Satellites and drones are now being used to track the production, movement, and foraging behavior of organisms at global scales. Ecological tracers such as unique elements, stable isotopes, and fatty acids are becoming easier to isolate and quantify. Molecular tracers such as DNA also are gaining traction for the identification of prey in diets. Statistics and models that combine tracers and look for covarying patterns should increase resolution in trophic interactions. To illustrate, a new mixing model for using both stable isotopes and fatty acids to characterize diets has been developed (Neubauer and Jensen 2015). Environmental factors that affect trophic interactions such as temperature, light, water discharge, erosion, flooding, drought, and salinity can be measured in real time at very fine scales of resolution. Combining these data sets correctly and making sense of them to predict the role of trophic processes in energy transfer and ultimately biogeochemical pathways on Earth is a daunting, exciting challenge.

Computing capacity in terms of speed and storage should keep pace with the growth in data. Mechanistic and statistical techniques such as ENA, multivariate analysis, regression trees, ordination, discriminant function analysis, information analysis, genetic algorithms, and a host of other approaches from diverse disciplines such as physics, chemistry, economics, and geostatistics need to be applied to find important emergent properties. Computer simulation modeling such as EcoPath combined with theoretical approaches will lead to further discoveries and help trophic ecologists generate hypotheses. As research budgets shrink, the days of big-scale, manipulative ecology appear to be waning. The ability to completely alter ecosystems and study changes in trophic structure has given way to large-scale observation networks and small-scale experiments. Ecology needs to continue to explore large-scale perturbations and quantify changes in trophic interactions to improve trophic ecology. Experiments like Hubbard Brook and the trophic cascade lakes (see Chapter 10) need to be repeated and replicated as much as possible.

Searching for general patterns and processes in trophic ecology using macroecological approaches should be encouraged as well. Certain processes, such as the relationship between stoichiometric ratios of elements and cellular replication, scale from the molecular to the ecosystem and perhaps to biogeochemical cycling. Similarly, almost all autotrophs rely on the same photosynthetic machinery to operate, meaning that large-scale predictions about autotrophy and trophic implications are possible. Metabolic theory is similar in that practically all organisms share the same mitochondrial equipment. Processes occurring metabolically in cells likely scale up to the entire biosphere. Macroecological approaches are useful. However, there always is error around those large-scale relationships among multiple species, driven by other constraints and emergent properties that need to be understood. Within the error around the mean predictions lies the important variation that tells us how resilient ecosystems should be to environmental change as well as internal changes that are occurring such as in biodiversity, behavior, and physiology. Allen and Gillooly (2009) took a first step toward unifying the apparently disparate concepts of ecological stoichiometry and metabolic theory by combining the energetic costs of with nutrient processing in plants and animals.

15.6 FUTURE OF TROPHIC ECOLOGY

Trophic ecology needs to be formally defined as a subdiscipline of ecology that uses complex rising approaches including metabolic theory, ENA, bioenergetics, stoichiometry, and foraging behavior to make relevant predictions about the flow of energy and matter across the entire realm of ecology as a science. Whether our definition holds is open to debate, but the organismal contribution to energetic and geochemical processes that scale from the molecular to the biosphere needs to be considered and studied. Fuzzy, broadly defined terms like trophic level, food web, food chain, autotroph, trophic pyramid, and consumer will persist. Ecologists need to work toward developing unified definitions for all of these concepts.

Trophic ecology needs to incorporate natural selection into its framework at all scales of inquiry. At the cellular level, slight changes in photosynthesis, metabolic processes, and cell repair and replication through the interaction between natural selection and environmental change may have huge effects on trophic interactions and biogeochemical processing. Theory is emerging from models and data mining. Experiments will still be needed to test and refine these concepts (Carpenter et al. 1995).

The behavior of organisms as they forage has a strong theoretical basis and much empirical research to support it. These results need to be scaled up to the level of ecosystems and landscapes. The movement of organisms has been shown to strongly impact nutrient dynamics through ecosystem subsidies and consumer impacts (Chapter 10). The roles of natural selection and the environment need to be linked to foraging and movement models to improve predictions for trophic ecology.

The role of biodiversity on the stability of ecosystems has been studied for decades. How biodiversity affects trophic interactions, ultimately influencing the stability and function of ecosystems, needs to be better understood. This is a critical role for community ecology in the future because processes such as predation, competition, and mutualisms all affect the efficiency by which energy moves from one trophic level to another, ultimately affecting the fate of materials like carbon, nitrogen, and phosphorus on Earth.

Scavenging and decomposition are common trophic processes on Earth. They are not well understood given their apparent importance to ecosystem processes (Chapter 3). These trophic interactions are misunderstood by many ecologists, believed to be a cooperative sequence in the remineralization or reuptake of energy and matter in dead organisms. Rather, there appears to be an arms race occurring between scavengers and decomposers that may well influence biogeochemical processes that scale up to major global patterns such as climate.

We could list many more examples of persistent challenges in trophic ecology as outlined in the chapters of this book. Habitat alterations, fragmentation, invasive species, pollution, harvest, and extinction are all factors affecting trophic ecology. There is clearly much research to do.

15.7 APPLICATION

No organisms on Earth are able to exist without subsidies from the environment. If energy was not continually renewed by sunlight and, in some rare ecosystems, by chemosynthesis, the whole biosphere would grind to a halt in a very short time. The last organisms to persist would be the decomposers until all of their consumed energy was converted to useless heat. Virtually every component of living matter has cycled through other organisms for billions of years. The geologic record tells us that this matter was recycled within simple food webs for most of Earth's history and that complex interactions are relatively new. As humans systematically remove top consumers and simplify landscapes, we are turning the dial back to simpler trophic interactions and less specialized consumers. Whether this simplification of food webs through the loss of biodiversity and physical complexity on Earth will lead to less efficient trophic interactions is still a matter of debate. Given that humans consume a large proportion of the Earth's primary productivity to survive (Chapter 2), loss of efficiency in trophic transfer may affect the planet's carrying capacity for our species.

Human-driven eutrophication of terrestrial and aquatic ecosystems alters trophic interactions in complex ways (Chapter 2). Autotrophic production may increase, but the nutritional quality and quantity of the autotrophs may change in ways that comprise the success of herbivores and their consumers. Eutrophication is increasing globally and should continue to compromise food webs by simplifying them to fewer species, a loss of biodiversity, and ultimately to less efficient capture of sunlight to primary production.

If predictions of metabolic theory are correct (Chapter 8), then temperature rises with increased atmospheric carbon loading should benefit decomposition over scavenging in food webs on a global scale. Although speculative at best, this change in balance should favor the rapid remineralization of nutrients and increased release of carbon dioxide and other greenhouse gases into the atmosphere. Scavenging organisms often are an important source of energy for higher-level consumers, meaning that their decline should lead to declines in trophic complexity in many ecosystems.

Both conservation and restoration ecology are disciplines that struggle with either protecting Earth's remaining natural resources or bringing them back from the dead. The problem facing both disciplines is that they might be trying to reconstitute trophic conditions (e.g., nodes, flows, emergent properties) that are no longer supported by a changing environment. Both rely on organismal approaches (i.e., conserving or restoring historical, native biodiversity). However, if the processes by which native species obtain and process energy to ensure their fitness are changed, then it will be impossible to meet conservation or restoration goals. Trophic ecology plays an important role in informing conservation and restoration.

Trophic ecologists need to be broad-minded, synthetic, and flexible in order to see the big picture despite the continued specializations occurring in ecology. Rather than focusing on one particular subdiscipline of ecology (e.g., is macroecology or ENA the unifying answer? See Figure 15.2), trophic ecologists must grasp how energy, matter, and information are transformed from the cellular to the ecosystem level and identify the important functional groups for conservation and restoration. How these groups or nodes form either through natural selection or autocatalytic, self-organization is a critical question for the future. In Figure 15.2, we generate a conceptual diagram showing how energy demand and energy availability change as trophic structure becomes more complex in ecosystems (Chapter 8). As metabolic theory predicts, the energy demands of each successive trophic level increase overall, but actually decline per unit body mass, as organisms in higher trophic levels typically have larger bodies and have lower mass-specific metabolic rates. As Brown and others have shown (see Chapter 8), there appears to be an empirical pattern of energy available to trophic levels, with a sound metabolic explanation behind it (Figure 15.2). However, variability exists among the empirical relationships among ecosystems (Figure 15.2, circles). We hypothesize that the variability among these ecosystems is due to the complex trophic relationships scaling from cells to food

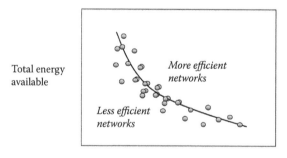

Trophic level

FIGURE 15.2 Hypothetical relationship between metabolic theory and ecological networks (i.e., food webs). As the trophic level (or depth) increases, the mass-specific metabolic demand of each trophic level declines, but the total energy available declines as well. Metabolic theory fits a predictive line through the empirical data, allowing us to make general predictions for ecosystems. Variability in the individual data from food webs (circles) is caused by individual differences in food-web efficiencies caused by differences in food-web structure and function as quantified by ecological networks.

webs to the environment that influence the efficiency of energy transfer in ecological networks. Trophic ecology may well be one of the great new frontiers of synthesis in ecology and environmental science merging complex concepts such as elemental ratios, metabolism, natural selection, genetics, energy transfer, and networks.

QUESTIONS AND ASSIGNMENTS

1. At its simplest, trophic ecology is the study of what?
2. Was heterotrophy common during most of the evolutionary history of Earth? Why might it be a fragile process?
3. Do trophic levels really exist? Are they useful?
4. Are large-scale manipulations of ecosystems to learn about trophic ecology common?
5. How will temperature rises in the environment tip the balance between scavenging and microbial activity?

Glossary

Allen curve: Production of a cohort through time.

Apex carnivores: Species at the top of the trophic pyramid that typically have no natural predators as adults.

Aposematic signals: Morphological or behavioral warnings displayed by prey to consumers.

Arbuscular mycorrhizal fungi: Fungi that penetrate the tips of roots of vascular plants.

Assimilation efficiency: The efficiency by which energy consumed by an organism or ecosystem is incorporated into the biota (but not necessarily used for growth or production).

Asymmetric competition: Two consumers equally affected negatively by competition.

Autoecological: Focusing on single organisms and their feeding relationships.

Autotrophs: Organisms able to meet almost all their energy needs through internal synthesis.

Batesian mimicry: Nonharmful species adopt warning traits similar to those of harmful species.

Bioenergetics: The study of the accounting (mass-balance equations) of energy transformations in organisms.

Biogeochemistry: The interaction between living organisms and the cycling of inorganic matter.

Bomb calorimeter: Instrument used to quantify the energy contained in food and waste products.

Carnivores: Animals that only consume other animals.

Chemosynthesis: Fixation of energy through chemical reactions not by the capture of light.

Cohorts: Groups of individuals within species or trophic levels that are produced at the same time and grow.

Compensation: Predation on prey at low densities increases prey survival (of non-eaten prey) and increases birth rates. It is also the hypothesis that plants and animals can compensate for loss of tissue or poor growth by stimulating growth.

Competitive exclusion principle: No two species can share and exist at the same niche.

Conformers: Organisms with an internal chemical environment that matches that in the surrounding environment.

Connectedness: Number of actual links among species within a food web.

Consortium: Group of consumers with loose physical and chemical associations that benefit each other (see *Symbiosis*).

Consumers: Organisms that derive energy needs from external sources.

CPOM: Coarse particulate organic matter in lakes and streams.

Crypsis: Blending in with the environment to avoid being consumed.

Cryptic species: Multiple genetically distinct taxa that have similar physical characters but different foraging characteristics.

Decomposers: Organisms that process dead plants or animals chemically or physically before consuming them.

Denitrifiers: Microbes that are able to use the oxidized nitrogen molecules as an energy source.

Depensation: Predation on prey at low densities of prey increases prey mortality and reduces birth rates.

Desaturase: Enzymes necessary to elongate fatty acids and remove hydrogen atoms, creating more double bonds.

Detritivory: Consuming decomposed animal or plant material.

DHA: Docosahexaenoic acid.

Dilution: Process by which elements or molecules that are bioaccumulated become less concentrated in tissues as organisms grow.

Ecological network analysis: Matrix-based models of food-web linkages within ecosystems used to determine general topological properties.

Ecological stoichiometry: Study of chemical reactions and how they pertain to ecological interactions.

Ecopath model: Mass-balance ecosystem model applied with bioenergetics.

Electivity: The choice of a forager for a prey item.

Electron acceptor: Molecule or element used in chemical energetic reactions (e.g., citric acid cycle) to regulate the transformation of stored fuel to work and heat.

Emergent behavior: Characteristic of an ecological category such as an ecosystem that is unique relative to its component parts.

EPA: Eicosapentaenoic acid.

Epilimnion: Top layer of a body of water, where light penetrates and physical mixing occurs.

Evacuation: The rate by which the guts of consumers empty into the lower alimentary canal.

Evolutionary stable strategy (ESS): Two or more foraging tactics exist in a population if they are beneficial to subsets of groups within the population.

FAMES: Fatty acid methyl esters; acid-catalyzed fatty acids.

Fatty acid: Building block of lipids, containing a carbon chain attached to a carboxyl group.

Food web: Interrelated linkages among consumers and their food resources.

Foraging ecology: Subdiscipline of trophic ecology that studies foraging behavior of animals.

FPOM: Fine particulate organic matter in lakes and streams.

Fractionation: The tendency (but not always the case) for heavy stable isotopes to accumulate in organisms while light ones are metabolized.

Functional response: Behavioral response of predators to changes in their prey.

Galling: An overgrown portion of stem that contains a larval insect.

Gamma diversity: The putative total number of species in an ecosystem; as opposed to alpha and beta diversity.

Ghrelin: Hormone that stimulates hunger.

Gross production efficiency: The total energy incorporated into organisms or eco-systems, of which some is lost to waste and metabolic costs (see *Net production efficiency*).

Herbivores: Animals that eat only autotrophs.

Heterotroph: Consumer of other organisms.

Hippocampus: Memory center of vertebrate brains.

Holistic: Ecological approach that quantifies broad patterns without regarding mechanism.

HUFA: Highly unsaturated fatty acid; more than one double carbon bond and greater than 19 carbon atoms.

Hypolimnion: Lower layer of a body of water, where light does not penetrate.

Ideal free distribution: Consumers should become distributed in the environment in proportion to food availability.

Increment summation method: Sum of growth accumulated among cohorts within a food web.

Instantaneous growth method: Using average biomass through time to sum secondary production.

Interaction strength: The impact a species has on a community when it is present versus when it is removed.

Intraguild: Groups of organisms with similar foraging needs (see *Trophic level*).

Isoclines: Coordinates on biplots where equilibria between predator and prey densities occur.

Keystones: Species with a strong interaction strength; density may be low but they exert strong impacts on ecosystems through consumption.

Lignocellulose: Large molecules composed of rings of carbon that resist decomposition.

Macroecology: Study of broad, general patterns across ecological systems.

Macronutrients: Nutrients that are required in large quantities.

Manly–Chesson alpha: A statistically robust index of prey selection based on proportions of prey in the environment and the diets.

Marginal value theorem: Consumers should remain in a location with food as long as the food intake is greater than the energy needed to get to another location with food.

Metabolic respiration: Consumption of energy within organisms that generates work and heat.

Metabolic theory of ecology: Theory that biomass in ecosystems and temperature predicts metabolism of trophic levels, growth, and production.

Micronutrients: Nutrients that are required in small amounts; high concentrations may be toxic.

Mineralized: Organic matter that is decomposed into chemical components.

Mixing models: Mathematical and statistical techniques for assessing the contribution of source elements or molecules (i.e., from food) to consumer biomass.

Mullerian mimicry: Harmful species have similar warning patterns.

Mushroom body: Memory center of some invertebrate brains.

Nematocyst: Pressure-filled structure of cnidarians used to pierce and inject venom in prey.

Net production efficiency: Assimilated material that is incorporated into growth or production (see *Assimilation efficiency*).

Niche breadth index: Quantifying how general a consumer is in foraging.

Nodes: Connecting points (usually trophic groups) within food webs.

Omnivores: Animals that consume resources from multiple trophic levels.

Optimality: The concept that evolution by natural selection should lead to perfectly efficient energy intake in ecosystems.

P/R: Ratio of production to respiration in an ecosystem; means to quantify ecosystem metabolism as heterotrophy.

Phospholipids: Polar lipids used primarily in structure of cells.

Phytochromes: Specialized pigments in algae and plants that change structure with light.

Poikilotherm: Organism with a body temperature the same as that of its environment.

Prey importance index: An index of foraging that accounts for both the biomass and energy density of prey.

Priority effects: The first consumer to arrive at a foraging patch has a competitive advantage over latecomers.

Prisoner's dilemma: Evolutionary argument for why cooperative behavior should not evolve among predators.

PUFA: Polyunsaturated fatty acid; more than one double carbon bond.

Q_{10} response: Change in the metabolic rate of an organism with a 10°C increase in temperature.

QFSA: Quantitative fatty acid signature analysis; weighting of fatty acids based on assumed catabolism within the consumer.

Rare earth materials: Rare elements in the environment; can be used as tracer spatial location of foraging activity.

Rarefaction: Sampling an ecosystem to estimate the total number of species present.

Reactivity: The response of food-web components to perturbations.

Recruitment: The point where mortality of offspring is variable and high to predictable in a survivorship curve.

Redfield ratio: The ratio of elements in the environment dictates limiting nutrients.

Reductionistic: Ecological approach that uses mechanisms to yield general patterns.

Regulators: Organisms that maintain a homeostatic internal environment that is invariant relative to ratios of elements in the environment.

Resilience: The ability of an ecosystem to return to its previous state after a perturbation.

Respiratory quotient: Ratio of carbon dioxide produced to oxygen consumed; estimate of the efficiency of fuel consumption for metabolism.

Resting metabolic rate: Respiration of organism that is completely still.

Ribosomes: Ribosomal production in cells drives the ratio of N:P to 16:1 (see *Redfield ratio*).

Risk sensitivity: The ability of an organism to assess the potential gain of a foraging attempt.

River continuum concept: Theory in aquatic stream ecology that predicts that patterns of autotrophy and heterotrophy depend on stream location.

Ruminant: Herbivore that contains digestive structures for fermentation of plant matter.

Runaway consumption: One consumer becomes a generalist and consumes a food web to extinction.

Scavengers: Organisms that consume nonliving animal or plant matter without processing it.

Secondary production: Generation of biomass as a rate that is not dependent on autotrophy.

Shreckstoff: Chemical cue created by conspecifics to alarm against predation.

Shredders: Stream organisms that tear plant matter into small pieces.

Size frequency method: Estimating secondary production by summing biomass within each cohort in a food web.

Specific dynamic action: Metabolic cost of digestion.

Stable isotope: Any one of a number of elements with different mass.

Standard metabolic rate: Respiration of organism that is not at rest; calculated by comparing activity to metabolic rate and estimating rate at zero activity.

Stoichiometric: Chemical reactions within organisms and in the environment must balance; elements not used in reactions remain in the environment and alter ratios.

Strauss linear index: Simplest foraging index, which compares the proportion of prey in the environment to that in the diet.

Subsidies: Energy and matter entering ecosystems from outside the location of the food web.

Survivorship: Longitudinal patterns of survival of organisms in a population.

Symbiosis: Usually coevolved mutual relationships between two coexisting species.

Symmetric competition: One consumer is a superior competitor.

Tertiary predators: Consumers at the top of the trophic pyramid.

Threshold element ratio: TER; the ratio of element to carbon that determines how readily the nutrient becomes limited in ecosystems.

TMII: Trait-mediated indirect interaction; plastic responses of prey that change their role in food-web interactions.

Topology: Configuration of the linkages within a food web.

Tracer: Element or molecule used to quantify the fate and biotransformation of energy and material in organisms and food webs.

Triacylglycerols: Nonpolar lipids used for energy storage.

Trophic: Nourishment, food intake.

Trophic cascade hypothesis: Top-down consumers at high trophic levels control primary producer biomass through effects on intermediate trophic levels.

Trophic ecology: Study of the interaction between a consumer and its food resources at all levels of ecological organization.

Trophic level: Groupings of organisms with similar food resource needs in a food web or ecosystem.

Trophic pyramids: A conceptual model of trophic levels arranged in an ecosystem.

Vitamins: Organic molecules necessary for biological function that are often not synthesized.

Wisconsin model: Mass-balance bioenergetics model for fish and other organisms, which is used to predict consumption and growth of consumers.

Zero-inflated data: Gut data often contain many empty diets and require special statistical models.

Bibliography

Abrams, P. A., and L. Rowe. 1996. "The effects of predation on the age and size of maturity of prey." *Evolution* 50 (3):1052–1061.

Abrams, P. A., O. Leimar, S. Nylin, and C. Wiklund. 1996. "The effect of flexible growth rates on optimal sizes and development times in a seasonal environment." *American Naturalist* 147 (3):381–395.

Abrams, P. A., R. Cressman, and V. Krivan. 2007. "The role of behavioral dynamics in determining the patch distributions of interacting species." *American Naturalist* 169 (4):505–518.

Adams, S. M., and D. L. DeAngelis. 1987. Indirect effects of early bass-shad interactions on predator population structure and food-web dynamics. Edited by W. C. Kerfoot and A. Sih, *Predation: Direct and Indirect Impacts on Aquatic Communities*. Hanover, NH: University Press of New England.

Agrawal, A. A., and M. Fishbein. 2006. "Plant defense syndromes." *Ecology* 87 (7):S132–S149.

Agrawal, A. A., C. Kobayashi, and J. S. Thayler. 1999. "Influence of prey availability and induced host-plant resistance on omnivory by western flower thrips." *Ecology* 80 (2):518–523.

Agrawal, A. A., G. Petschenka, R. A. Bingham, M. G. Weber, and S. Rasmann. 2012. "Toxic cardenolides: Chemical ecology and coevolution of specialized plant–herbivore interactions." *New Phytologist* 194 (1):28–45.

Agren, G. I. 2004. "The C:N:P stoichiometry of autotrophs—Theory and observations." *Ecology Letters* 7 (3):185–191. doi: 10.1111/j.1461-0248.2004.00567.x.

Aguzzi, A., and M. Polymenidou. 2004. "Mammalian prion biology: One century of evolving concepts." *Cell* 116 (2):313–327. doi: 10.1016/s0092-8674(03)01031-6.

Alatalo, R. V., and J. Mappes. 1996. "Tracking the evolution of warning signals." *Nature* 382 (6593):708–710.

Algar, A. C., J. T. Kerr, and D. J. Currie. 2007. "A test of metabolic theory as the mechanism underlying broad-scale species-richness gradients." *Global Ecology and Biogeography* 16 (2):170–178. doi: 10.1111/j.1466-8238.2006.00275.x.

Allen, A. P., J. H. Brown, and J. F. Gillooly. 2002. "Global biodiversity, biochemical kinetics, and the energetic-equivalence rule." *Science* 297 (5586):1545–1548. doi: 10.1126/science.1072380.

Allen, A. P., and J. F. Gillooly. 2009. "Towards an integration of ecological stoichiometry and the metabolic theory of ecology to better understand nutrient cycling." *Ecology Letters* 12 (5):369–384. doi: 10.1111/j.1461-0248.2009.01302.x.

Allen, A. P., J. F. Gillooly, and J. H. Brown. 2005. "Linking the global carbon cycle to individual metabolism." *Functional Ecology* 19 (2):202–213. doi: 10.1111/j.1365-2435.2005.00952.x.

Allen, K. R. 1951. "The Horokiwi stream: A study of a trout population." *New Zealand Department of Fisheries Bulletin* 10:1–238.

Allen, T. F. H., and T. W. Hoekstra. 1992. *Toward a Unified Ecology*. New York: Columbia University Press.

Allgeier, J. E., S. J. Wenger, A. D. Rosemond, D. E. Schindler, and C. A. Layman. 2015. "Metabolic theory and taxonomic identity predict nutrient recycling in a diverse food web." *Proceedings of the National Academy of Sciences of the United States of America* 112 (20):E2640–E2647. doi: 10.1073/pnas.1420819112.

Amend, J. P., and E. L. Shock. 1998. "Energetics of amino acid synthesis in hydrothermal ecosystems." *Science* 281 (5383):1659–1662. doi: 10.1126/science.281.5383.1659.

Amundsen, P. A., H. M. Gabler, and F. J. Staldvik. 1996. "A new approach to graphical analysis of feeding strategy from stomach contents data: Modification of the Costello (1990) method." *Journal of Fish Biology* 48 (4):607–614.

AOAC. 2016. Official methods of analysis of AOAC International. 20th Edition. Editor, George Latimer. AOAC International.

Arditi, R., and L. R. Ginzberg. 2012. *How Species Interact: Altering the Standard View on Trophic Ecology*. Oxford: Oxford University Press.

Armbruster, W. S. 1992. "Phylogeny of plant–animal interactions." *BioScience* 42 (1):12–20.

Arts, M. T., M. T. Brett, and M. T. Kainz. 2009. *Lipids in Aquatic Ecosystems*. Berlin: Springer.

Babcock, L. E. 1993. "Trilobite malformations and the fossil record of behavioral asymmetry." *Journal of Paleontology* 67 (2):217–229.

Bacon, I. E., T. A. Hurly, and S. Healy. 2011. "Both the past and the present affect risk-sensitive decisions of foraging rufous hummingbirds." *Behavioral Ecology* 21 (3):626–632. doi: 10.1093/beheco/arq031.

Bajer, P. G., G. W. Whitledge, and R. S. Hayward. 2004. "Widespread consumption-dependent systematic error in fish bioenergetics models and its implications." *Canadian Journal of Fisheries and Aquatic Sciences* 61 (11):2158–2167. doi: 10.1139/f04-159.

Bajer, P. G., G. W. Whitledge, R. S. Hayward, and R. D. Zweifel. 2003. "Laboratory evaluation of two bioenergetics models applied to yellow perch: Identification of a major source of systematic error." *Journal of Fish Biology* 62 (2):436–454. doi: 10.1046/j.1095-8649 .2003.00040.x.

Baracchi, D., S. Francese, and S. Turillazzi. 2011. "Beyond the antipredatory defence: Honey bee venom function as a component of social immunity." *Toxicon* 58 (6–7):550–557. doi: 10.1016/j.toxicon.2011.08.017.

Bascompte, J., C. J. Melian, and E. Sala. 2005. "Interaction strength combinations and the overfishing of a marine food web." *Proceedings of the National Academy of Sciences of the United States of America* 102:5443–5447.

Bassett, D. K., A. G. Carton, and J. C. Montgomery. 2007. "Saltatory search in a lateral line predator." *Journal of Fish Biology* 70 (4):1148–1160. doi: 10.1111/j.1095-8649 .2007.01380.x.

Baxter, C. V., K. D. Fausch, and W. C. Saunders. 2005. "Tangled webs: Reciprocal flows of invertebrate prey link streams and riparian zones." *Freshwater Biology* 50 (2):201–220.

Beamish, F. W. H. 1974. "Apparent specific dynamic action of Largemouth bass, micropterus salmoides." *Journal of the Fisheries Research Board of Canada* 31 (11):1763–1769. doi: 10.1139/f74-224.

Beard, K. H., K. A. Vogt, and A. Kulmatiski. 2002. "Top-down effects of a terrestrial frog on forest nutrient dynamics." *Oecologia* 133:583–593.

Beasley, J. C., Z. H. Olson, and T. L. DeVault. 2012. "Carrion cycling in food webs: Comparisons among terrestrial and marine ecosystems." *Oikos* 121 (7):1021–1026. doi: 10.1111/j.1600-0706.2012.20353.x.

Beatty, J. T., J. Overmann, M. T. Lince, A. K. Manske, A. S. Lang, R. E. Blankenship, C. L. Van Dover, T. A. Martinson, and F. G. Plumley. 2005. "An obligately photosynthetic bacterial anaerobe from a deep-sea hydrothermal vent." *Proceedings of the National Academy of Sciences of the United States of America* 102 (26):9306–9310. doi: 10.1073/pnas.0503674102.

Bednekoff, P. A., and S. L. Lima. 1998. "Randomness, chaos and confusion in the study of antipredator vigilance." *Trends in Ecology & Evolution* 13 (7):284–287. doi: 10.1016/s0169 -5347(98)01327-5.

Begona Santos, M., I. German, D. Correia, F. L. Read, J. Martinez Cedeira, M. Caldas, A. Lopez, F. Velasco, and G. J. Pierce. 2013. "Long-term variation in common dolphin diet in relation to prey abundance." *Marine Ecology Progress Series* 481:249–268. doi: 10.3354/meps10233.

Belgrano, A., U. M. Scharler, J. Dunne, and R. E. Ulanowicz. 2005. *Aquatic Food Webs: An Ecosystem Approach.* Oxford: Oxford University Press.

Bellwood, D. R. 2003. "Origins and escalation of herbivory in fishes: A functional perspective." *Paleobiology* 29 (1):71–83.

Belsky, A. J., W. P. Carson, C. L. Jensen, and G. A. Fox. 1993. "Overcompensation by plants–herbivore optimization or red herring." *Evolutionary Ecology* 7 (1):109–121.

Bengtson, S., and Y. Zhao. 1992. "Predatorial borings in late Precambrian mineralized exoskeletons." *Science* 257 (5068):367–369.

Benke, A. C. 1993. "Concepts and patterns of invertebrate production in running waters." *Internationalen vereinigung für theoretische und angewandte Limnologie, Verhandlungen* 25:15–38.

Benke, A. C. 1996. Secondary production of macroinvertebrates. In *Methods in Stream Ecology*, F. R. Hauer and G. A. Lamberti, editors, pp. 557–578. San Diego: Academic Press.

Benke, A. C., and J. B. Wallace. 1980. "Trophic basis of production among net-spinning caddisflies in a southern Appalachian stream." *Ecology* 61:108–118.

Benke, A. C., and A. D. Huryn. 2006. Secondary production of macroinvertebrates. In *Methods in Stream Ecology*, F. R. Hauer and G. A. Lamberti, editors, pp. 691–710. London: Academic Press.

Benke, A., and M. Whiles. 2011. "Life table vs secondary production analyses-relationships and usage in ecology." *Journal of the North American Benthological Society* 30:1024–1032.

Benke, A. C., A. D. Huryn, L. A. Smock, and J. B. Wallace. 1999. "Length–mass relationships for freshwater macroinvertebrates in North America with particular reference to the southeastern United States." *Journal of the North American Benthological Society* 18:308–343.

Berger, L. R. 2006. "Brief communication: Predatory bird damage to the Taung type-skull of *Australopithecus afficanus* Dart 1925." *American Journal of Physical Anthropology* 131 (2):166–168. doi: 10.1002/ajpa.20415.

Bergsma, G. S. 2012. "Epibiotic mutualists alter coral susceptibility and response to biotic disturbance through cascading trait-mediated indirect interactions." *Coral Reefs* 31 (2):461–469.

Berryman, A. A. 1992. "The origins and evolution of predator-prey theory." *Ecology* 73:1530–1535.

Bertram, D. F., and W. C. Leggett. 1994. "Predation risk during the early life history periods of fishes: Separating the effects of size and age." *Marine Ecology Progress Series* 109:105–114.

Biro, P. A., J. R. Post, and E. A. Parkinson. 2003. "From individuals to populations: Prey fish risk-taking mediates mortality in whole-system experiments." *Ecology* 84 (9):2419–2431.

Bligh, E. G., and W. J. Dyer. 1959. "A rapid method of total lipid extraction and purification." *Canadian Journal of Biochemistry* 37:911–917.

Blomquist, G. J., L. A. Dwyer, A. J. Chu, R. O. Ryan, and M. Derenobales. 1982. "Biosynthesis of linoleic-acid in a termite, cockroach and cricket." *Insect Biochemistry* 12 (3):349–353. doi: 10.1016/0020-1790(82)90093-2.

Blouin-Demers, G., K. A. Prior, and P. J. Weatherhead. 2000. "Patterns of variation in spring emergence by black rat snakes (*Elaphe obsoleta obsoleta*)." *Herpetologica* 56 (2):175–188.

Boecklen, W. J., C. T. Yarnes, B. A. Cook, and A. C. James. 2011. "On the use of stable isotopes in trophic ecology." *Annual Review of Ecology, Evolution, and Systematics* 42 (1):411–440. doi: 10.1146/annurev-ecolsys-102209-144726.

Boisclair, D., and W. C. Leggett. 1989. "The importance of activity in bioenergetics models applied to actively foraging fishes." *Canadian Journal of Fisheries and Aquatic Sciences* 46 (11):1859–1867. doi: 10.1139/f89-234.

Boisclair, D., and P. Sirois. 1993. "Testing assumptions of fish bioenergetics models by direct estimation of growth, consumption, and activity rates." *Transactions of the American Fisheries Society* 122 (5):784–796. doi: 10.1577/1548-8659(1993)122<0784:taofbm >2.3.co;2.

Boisseau, R. P., D. Vogel, and A. Dussutour. 2016. "Habituation in non-neural organisms: Evidence from slime moulds." *Proceedings of the Royal Society of London B: Biological Sciences* 283 (1829).

Bonter, D. N., and E. S. Bridge. 2011. "Applications of radio frequency identification (RFID) in ornithological research: A review." *Journal of Field Ornithology* 82 (1):1–10. doi: 10.1111/j.1557-9263.2010.00302.x.

Borer, E. T., E. W. Seabloom, J. B. Shurin, K. E. Anderson, C. A. Blanchette, B. Broitman, S. D. Cooper, and B. S. Halpern. 2005. "What determines the strength of a trophic cascade?" *Ecology* 86 (2):528–537.

Boudreau, P. W., and L. M. Dickie. 1992. "Biomass spectra of aquatic ecosystems in relation to fisheries yield." *Canadian Journal of Fisheries and Aquatic Sciences* 49:1528–1538.

Boulanger, M. T., and R. Lee Lyman. 2014. "Northeastern North American Pleistocene megafauna chronologically overlapped minimally with Paleoindians." *Quaternary Science Reviews* 85:35–46. doi: http://dx.doi.org/10.1016/j.quascirev.2013.11.024.

Bradshaw, C. J. A., M. A. Hindell, N. J. Best, K. L. Phillips, G. Wilson, and P. D. Nichols. 2003. "You are what you eat: Describing the foraging ecology of southern elephant seals (*Mirounga leonina*) using blubber fatty acids." *Proceedings of the Royal Society of London Series B—Biological Sciences* 270 (1521):1283–1292.

Bowerman, J., P. T. J. Johnson, and T. Bowerman. 2010. "Sublethal predators and their injured prey: Linking aquatic predators and severe limb abnormalities in amphibians." *Ecology* 91 (1):242–251. doi: 10.1890/08-1687.1.

Brain, C. K. 2000. "Do we owe our intelligence to a predatory past?" *James Arthur Lecture* 70:1–32.

Brett, M. T., and D. C. Muller-Navarra. 1997. "The role of highly unsaturated fatty acids in aquatic food web processes." *Freshwater Biology* 38 (3):483–499. doi: 10.1046/j.1365 -2427.1997.00220.x.

Brett, M. T., M. J. Kainz, S. J. Taipale, and H. Seshan. 2009. "Phytoplankton, not allochthonous carbon, sustains herbivorous zooplankton production." *Proceedings of the National Academy of Sciences of the United States of America* 106 (50):21197–21201. doi: 10.1073/pnas.0904129106.

Briand, F., and J. E. Cohen. 1984. "Community food webs have scale-invariant structure." *Nature* 307:264–267.

Briggs, D. E. G. 1999. "Molecular taphonomy of animal and plant cuticles: Selective preservation and diagenesis." *Philosophical Transactions of the Royal Society B—Biological Sciences* 354 (1379):7–16.

Briggs, C. T., and J. R. Post. 1997. "In situ activity metabolism of rainbow trout (Oncorhynchus mykiss): Estimates obtained from telemetry of axial muscle electromyograms." *Canadian Journal of Fisheries and Aquatic Sciences* 54 (4):859–866. doi: 10.1139/cjfas-54-4-859.

Brodin, A. 2000. "Why do hoarding birds gain fat in winter in the wrong way? Suggestions from a dynamic model." *Behavioral Ecology* 11 (1):27–39.

Brodin, A. 2001. "Mass-dependent predation and metabolic expenditure in wintering birds: Is there a trade-off between different forms of predation?" *Animal Behaviour* 62:993–999.

Brodin, A., and C. W. Clark. 2007. Energy storage and expenditure. In *Foraging: Behavior and Ecology*, D. W. Stephens, J. S. Brown, and R. C. Ydenberg, editors, pp. 221–269. Chicago: The University of Chicago Press.

Bronmark, C., and J. G. Miner. 1992. "Predator-induced phenotypical change in body morphology in Crucian carp." *Science* 258:1348–1350.

Brooks, J. L., and S. I. Dodson. 1965. "Predation, body size, and composition of plankton." *Science* 15:28–35.

Brown, D. J., I. L. Boyd, G. C. Cripps, and P. J. Butler. 1999. "Fatty acid signature analysis from the milk of Antarctic fur seals and Southern elephant seals from South Georgia: Implications for diet determination." *Marine Ecology—Progress Series* 187:251–263.

Brown, J. H., and J. F. Gillooly. 2003. "Ecological food webs: High-quality data facilitate theoretical unification." *Proceedings of the National Academy of Sciences of the United States of America* 100 (4):1467–1468. doi: 10.1073/pnas.0630310100.

Brown, J. H., J. F. Gillooly, G. B. West, and V. M. Savage. 2003. The next step in macroecology: From general empirical patterns to universal ecological laws. In *Macroecology: Concepts and Consequences*, T. M. Blackburn and K. J. Gaston, editors. The 43rd Annual Symposium of the British Ecological Society held at the University of Birmingham April 17–19, 2002.

Brown, J. H., J. F. Gillooly, A. P. Allen, V. M. Savage, and G. B. West. 2004. "Toward a metabolic theory of ecology." *Ecology* 85:1771–1789.

Brown, J. H., A. P. Allen, and J. F. Gillooly, eds. 2007. The metabolic theory of ecology and the role of body size in marine and freshwater ecosystems. In *Body Size: The Structure and Function of Aquatic Ecosystems*, A. Hildrew, D. Raffaelli, and R. Edmonds-Brown, editors. Ecological Reviews. New York: Cambridge University Press.

Brown, J. S. 1988. "Patch use as an indicator of habitat preference, predation risk, and competition." *Behavioral Ecology and Sociobiology* 22 (1):37–47. doi: 10.1007/bf00395696.

Brown, J. S., and B. P. Kotler. 2004. "Hazardous duty pay and the foraging cost of predation." *Ecology Letters* 7 (10):999–1014. doi: 10.1111/j.1461-0248.2004.00661.x.

Buchkremer, E. M., and K. Reinhold. 2011. "The emergence of variance-sensitivity with successful decision rules." *Behavioral Ecology* 21 (3):576–583. doi: 10.1093/beheco/arq026.

Budge, S. M., and S. J. Iverson. 2003. "Quantitative analysis of fatty acid precursors in marine samples: Direct conversion of wax ester alcohols and dimethylacetals to FAMEs." *Journal of Lipid Research* 44 (9):1802–1807.

Budge, S. M., S. J. Iverson, W. D. Bowen, and R. G. Ackman. 2002. "Among- and within-species variability in fatty acid signatures of marine fish and invertebrates on the Scotian Shelf, Georges Bank, and southern Gulf of St. Lawrence." *Canadian Journal of Fisheries and Aquatic Sciences* 59 (5):886–898.

Bull, J. J., and I. N. Wang. 2010. "Optimality models in the age of experimental evolution and genomics." *Journal of Evolutionary Biology* 23 (9):1820–1838. doi: 10.1111/j.1420-9101.2010.02054.x.

Burkepile, D. E., J. D. Parker, C. B. Woodson, H. J. Mills, J. Kubanek, P. A. Sobecky, and M. E. Hay. 2006. "Chemically mediated competition between microbes and animals: Microbes as consumers in food webs." *Ecology* 87 (11):2821–2831. doi: 10.1890/0012-9658(2006)87[2821:cmcbma]2.0.co;2.

Burnham, K. P., and D. R. Anderson. 2002. *Model Selection and Multimodel Inference*. New York: Springer.

Burns, C. E., S. L. Collins, and M. D. Smith. 2009. "Plant community response to loss of large herbivores: Comparing consequences in a South African and a North American grassland." *Biodiversity and Conservation* 18 (9):2327–2342. doi: 10.1007/s10531-009-9590-x.

Burns, J. G., and F. H. Rodd. 2008. "Hastiness, brain size and predation regime affect the performance of wild guppies in a spatial memory task." *Animal Behaviour* 76:911–922. doi: 10.1016/j.anbehav.2008.02.017.

Cáceres, C. E., and A. J. Tessier. 2004. "Incidence of diapause varies among populations of *Daphnia pulicaria*." *Oecologia* 141 (3):425–431.

Calkins, H. A., S. J. Tripp, and J. E. Garvey. 2012. "Linking silver carp habitat selection to flow and phytoplankton in the Mississippi River." *Biological Invasions* 14 (5):949–958. doi: 10.1007/s10530-011-0128-2.

Callaghan, T. P., and R. H. Karlson. 2002. "Summer dormancy as a refuge from mortality in the freshwater bryozoan *Plumatella emarginata*." *Oecologia* 132 (1):51–59.

Canuel, E. A., J. E. Cloern, D. B. Ringelberg, J. B. Guckert, and G. H. Rau. 1995. "Molecular and isotopic tracers used to examine sources of organic-matter and its incorporation into the food webs of San Francisco bay." *Limnology and Oceanography* 40 (1):67–81.

Carpenter, S. R., and P. R. Leavitt. 1991. "Temporal variation in a paleolimnological record arising from a trophic cascade." *Ecology* 72 (1):277–285.

Caraco, T., S. Martindale, and T. S. Whittam. 1980. "An empirical demonstration of risk-sensitive foraging preferences." *Animal Behaviour* 28 (Aug):820–830.

Carpenter, S. R., J. F. Kitchell, J. R. Hodgson, P. A. Cochran, J. J. Elser, M. M. Elser, D. M. Lodge, D. Kretchmer, X. He, and C. N. Von Ende. 1987. "Regulation of lake primary productivity by food web structure." *Ecology* 68:1863–1876.

Carpenter, S. R., P. Cunningham, S. Gafny, A. Munoz Del Rio, N. Nibbelink, M. Olson, T. Pellett, C. Storlie, and A. Trebitz. 1995. "Responses of bluegill to habitat manipulations: Power to detect effects." *North American Journal of Fisheries Management* 15 (3):519–527.

Carpenter, S. R., J. J. Cole, J. R. Hodgson, J. F. Kitchell, M. L. Pace, D. Bade, K. L. Cottingham, T. E. Essington, J. N. Houser, and D. E. Schindler. 2001. "Trophic cascades, nutrients, and lake productivity: Whole-lake experiments." *Ecological Monographs* 71 (2):163–186.

Carr, W. E. S., J. C. Netherton, R. A. Gleeson, and C. D. Derby. 1996. "Stimulants of feeding behavior in fish: Analyses of tissues of diverse marine organisms." *Biological Bulletin* 190 (2):149–160.

Carrascal, L. M., J. A. Diaz, D. L. Huertas, and I. Mozetich. 2001. "Behavioral thermoregulation by treecreepers: Trade-off between saving energy and reducing crypsis." *Ecology* 82 (6):1642–1654.

Carreon-Martinez, L., T. B. Johnson, S. A. Ludsin, and D. D. Heath. 2011. "Utilization of stomach content DNA to determine diet diversity in piscivorous fishes." *Journal of Fish Biology* 78 (4):1170–1182. doi: 10.1111/j.1095-8649.2011.02925.x.

Carter, D. O., D. Yellowlees, and M. Tibbett. 2007. "Cadaver decomposition in terrestrial ecosystems." *Naturwissenschaften* 94 (1):12–24. doi: 10.1007/s00114-006-0159-1.

Cary, S. C. 1994. "Vertical transmission of a chemoautotrophic symbiont in the protobranch bivalve, *Solemya reidi*." *Molecular Marine Biology and Biotechnology* 3 (3):121–130.

Cattin, M. F., L. F. Bersier, C. Banasek-Richter, R. Baltensperger, and J. P. Gabriel. 2004. "Phylogenetic constraints and adaptation explain food-web structure." *Nature* 427 (6977):835–839.

Charnov, E. L. 1976. "Optimal foraging, marginal value theorem." *Theoretical Population Biology* 9 (2):129–136.

Chen, C. Y., and C. Y. Tsai. 2012. "Ghrelin and motilin in the gastrointestinal system." *Current Pharmaceutical Design* 18 (31):4755–4765.

Chen, X., and J. E. Cohen. 2001. "Transient dynamics and food-web complexity in the Lotka–Volterra cascade model." *Proceedings of the Royal Society B-Biological Sciences* 268 (1469):869–877.

Chesson, J. 1978. "Measuring preference in selective predation." *Ecology* 59:211–215.

Chesson, J. 1983. "The estimation and analysis of preference and its relationship to foraging models." *Ecology* 64 (5):1297–1304.

Chiao, C. C., J. K. Wickiser, J. J. Allen, B. Genter, and R. T. Hanlon. 2011. "Hyperspectral imaging of cuttlefish camouflage indicates good color match in the eyes of fish predators." *Proceedings of the National Academy of Sciences of the United States of America* 108 (22):9148–9153. doi: 10.1073/pnas.1019090108.

Chipps, S. R., and D. H. Bennett. 2002. "Evaluation of a Mysis bioenergetics model." *Journal of Plankton Research* 24 (1):77–82.

Chipps, S. R., and D. H. Wahl. 2008. "Bioenergetics modeling in the 21st century: Reviewing new insights and revisiting old constraints." *Transactions of the American Fisheries Society* 137 (1):298–313. doi: 10.1577/t05-236.1.

Chipps, S. R., and J. E. Garvey. 2007. Chapter 11: Assessment of food habits and feeding patterns. In *Analysis and Interpretation of Freshwater Fisheries Data*, M. L. Brown and C. S. Guy, editors. pp. 473–514. Bethesda, Maryland: American Fisheries Society.

Christensen, V. 1995. "Ecosystem maturity—Towards quantification." *Ecological Modelling* 77:3–32.

Christie, W. W. 1982. Lipid Analysis, Second Edition. Oxford, UK: Pergamon Press.

Clarke, A. 2004. "Is there a universal temperature dependence of metabolism?" *Functional Ecology* 18 (2):252–256. doi: 10.1111/j.0269-8463.2004.00842.x.

Clements, F. E. 1936. "The structure of the climax." *The Journal of Ecology* 24:252–284.

Coe, M. 1978. "The decomposition of elephant carcasses in the Tsavo (East) National Park, Kenya." *Journal of Arid Environments* 1 (1):71–86.

Cohen, J. E. 1994. Lorenzo Camerano's contribution to early food web theory. In *Frontiers in Mathematical Biology*, S. A. Levin, editor, pp. 351–359. Berlin, Heidelberg: Springer Verlag.

Cohen, J. E., and F. Briand. 1984. "Trophic links of community food webs." *Proceedings of the National Academy of Sciences of the United States of America-Biological Sciences* 81:4105–4109.

Cohen, J. E., T. Luczak, C. M. Newman, and Z. M. Zhou. 1990. "Stochastic structure and non-linear dynamics of food webs—Qualitative stability in a Lotka Volterra cascade model." *Proceedings of the Royal Society Series B-Biological Sciences* 240:607–627.

Cole, L. C. 1954. "The population consequences of life history phenomena." *Quarterly Review of biology* 29:103–137.

Connell, J. H. 1978. "Diversity in tropical rainforests and coral reefs." *Science* 199:1302–1310.

Conner, W. E., and A. J. Corcoran. 2012. Sound Strategies: The 65-million-year-old battle between bats and insects. In *Annual Review of Entomology*, Vol. 57, M. R. Berenbaum, editor, pp. 21–39. California: Palo Alto.

Conover, D. O. 1998. "Local adaptation in marine fishes: Evidence and implications for stock enhancement." *Bulletin of Marine Science* 62 (2):477–493.

Conover, D. O., and E. T. Schultz. 1995. "Genotypic similarity and the evolutionary significance of countergradient variation." *Trends in Ecology and Evolution* 10 (6):248–252.

Conover, D. O., and S. B. Munch. 2002. "Sustaining fisheries yields over evolutionary time scales." *Science* 297 (5578):94–96.

Conover, D. O., S. A. Arnott, M. R. Walsh, and S. B. Munch. 2005. "Darwinian fishery science: Lessons from the Atlantic silverside (*Menidia menidia*)." *Canadian Journal of Fisheries and Aquatic Sciences* 62 (4):730–737.

Conrad, J. M. 2005. "Open access and extinction of the passenger pigeon in North America." *Natural Resource Modeling* 18 (4):501–519.

Corliss, J. B., J. Dymond, L. I. Gordon, J. M. Edmond, R. P. V. Herzen, R. D. Ballard, K. Green et al. 1979. "Submarine thermal springs on the Galapagos rift." *Science* 203 (4385):1073–1083. doi: 10.1126/science.203.4385.1073.

Cortes, E. 1997. "A critical review of methods of studying fish feeding based on analysis of stomach contents: Application to elasmobranch fishes." *Canadian Journal of Fisheries and Aquatic Sciences* 54 (03):726–738.

Costello, M. J. 1990. "Predator feeding strategy and prey importance—A new graphical analysis." *Journal of Fish Biology* 36 (2):261–263. doi: 10.1111/j.1095-8649.1990.tb05601.x.

Cothran, R. D., F. Radarian, and R. A. Relyea. 2012. "Altering aquatic food webs with a global insecticide: Arthropod-amphibian links in mesocosms that simulate pond communities." *Journal of the North American Benthological Society* 30 (4):893–912.

Craig, T. P. 2010. "The resource regulation hypothesis and positive feedback loops in plant-herbivore interactions." *Population Ecology* 52 (4):461–473. doi: 10.1007/s10144-010 -0210-0.

Crawley, M. J. 1989. "Insect herbivores and plant-population dynamics." *Annual Review of Entomology* 34:531–564.

Crowder, L. B., J. A. Rice, T. J. Miller, and E. A. Marschall. 1992. Empirical and theoretical approaches to size-based interactions and recruitment variability in fishes. In *Individual-Based Models and Approaches in Ecology*, D. L. DeAngelis and L. J. Gross, editors. London: Chapman and Hall.

Cummings, D. E., J. Q. Purnell, R. S. Frayo, K. Schmidova, B. E. Wisse, and D. S. Weigle. 2001. "A preprandial rise in plasma ghrelin levels suggests a role in meal initiation in humans." *Diabetes* 50 (8):1714–1719. doi: 10.2337/diabetes.50.8.1714.

Currie, C. R., J. A. Scott, R. C. Summerbell, and D. Malloch. 1999a. "Fungus-growing ants use antibiotic-producing bacteria to control garden parasites." *Nature* 398 (6729): 701–704.

Currie, C. R., U. G. Mueller, and D. Malloch. 1999b. "The agricultural pathology of ant fungus gardens." *Proceedings of the National Academy of Sciences of the United States of America* 96 (14):7998–8002.

Currie, D. J., G. G. Mittelbach, H. V. Cornell, R. Field, J. F. Guegan, B. A. Hawkins, D. M. Kaufman et al. 2004. "Predictions and tests of climate-based hypotheses of broad-scale variation in taxonomic richness." *Ecology Letters* 7 (12):1121–1134.

Cushing, D. H., ed. 1974. The possible density-dependence of larval mortality and adult mortality in fishes. In *The Early Life History of Fish*, J. H. S. Blaxter, editor. Berlin: Springer-Verlag.

Cusson, M., and E. Bourget. 2005. "Global patterns of macroinvertebrate production in marine benthic habitats." *Marine Ecology Progress Series* 297:1–14.

Czaczkes, T. J., A. K. Salmane, F. A. M. Klampfleuthner, and J. Heinze. 2016. "Private information alone can trigger trapping of ant colonies in local feeding optima." *Journal of Experimental Biology*.

D'Amore, D. C., and R. J. Blumenschine. 2012. "Using striated tooth marks on bone to predict body size in theropod dinosaurs: A model based on feeding observations of *Varanus komodoensis*, the Komodo monitor." *Paleobiology* 38 (1):79–100. doi: 10.5061/dryad .99qj3.

Dal Sasso, C., S. Maganuco, E. Buffetaut, and M. A. Mendez. 2005. "New information on the skull of the enigmatic theropod Spinosaurus, with remarks on its size and affinities." *Journal of Vertebrate Paleontology* 25 (4):888–896. doi: 10.1671/0272-4634(2005) 025[0888:niotso]2.0.co;2.

Darst, C. R., P. A. Menendez-Guerrero, L. A. Coloma, and D. C. Cannatella. 2005. "Evolution of dietary specialization and chemical defense in poison frogs (Dendrobatidae): A comparative analysis." *American Naturalist* 165 (1):56–69.

Darwin, C. 1859. *On the Origin of Species by Means of Natural Selection, or the Preservation of Favoured Races in the Struggle for Life*. London: John Murray.

Davic, R. D., and H. H. Welsh. 2004. "On the ecological roles of salamanders." *Annual Review of Ecology Evolution and Systematics* 35:405–434.

Davidov, Y., and E. Jurkevitch. 2009. "Predation between prokaryotes and the origin of eukaryotes." *BioEssays* 31 (7):748–757. doi: 10.1002/bies.200900018.

Davidson, J., and H. G. Andrewartha. 1948. "Annual trends in a natural population of *Thrips imaginis* (Thysanoptera)." *Journal of Animal Ecology* 17 (2):193–199.

DeAngelis, D. L., and W. M. Mooij. 2005. "Individual-based modeling of ecological and evolutionary processes." *Annual Review of Ecology Evolution and Systematics* 36:147–168. doi: 10.1146/annurev.ecolsys.36.102003.152644.

DeAngelis, D. L., B. J. Shuter, M. S. Ridgway, and M. Scheffer. 1993. "Modeling growth and survival in an age-0 fish cohort." *Transactions of the American Fisheries Society* 122:927–941.

Dettmers, J. M., and R. A. Stein. 1992. "Food consumption by larval gizzard shad: Zooplankton effects and implications for reservoir communities." *Transactions of the American Fisheries Society* 121 (4):494–507.

Dietl, G. R. 2003. "Coevolution of a marine gastropod predator and its dangerous bivalve prey." *Biological Journal of the Linnean Society* 80 (3):409–436.

Dingemanse, N. J., I. Barber, J. Wright, and J. E. Brommer. 2012. "Quantitative genetics of behavioural reaction norms: Genetic correlations between personality and behavioural plasticity vary across stickleback populations." *Journal of Evolutionary Biology* 25 (3):485–496. doi: 10.1111/j.1420-9101.2011.02439.x.

Dirzo, R., H. S. Young, M. Galetti, G. Ceballos, N. J. B. Isaac, and B. Collen. 2014. "Defaunation in the Anthropocene." *Science* 345 (6195):401–406. doi: 10.1126/science.1251817.

Dodds, W. K. 2009. *Laws, Theories, and Patterns in Ecology*. Berkeley, CA: University of California Press.

Dodds, W. K., A. J. Lopez, W. B. Bowden, S. Gregory, N. B. Grimm, S. K. Hamilton, A. E. Hershey et al. 2002. "N uptake as a function of concentration in streams." *Journal of the North American Benthological Society* 21 (2):206–220. doi: 10.2307/1468410.

Donovan, S. K., M. D. Sutton, and J. D. Sigwart. 2011. "Crinoids for lunch? An unexpected biotic interaction from the Upper Ordovician of Scotland." *Geology* 38 (10):935–938. doi: 10.1130/g31296.1.

Drake, J. A. 1991. "Community-assembly mechanics and the structure of an experimental species ensemble." *American Naturalist* 137 (1):1–26. doi: 10.1086/285143.

Drea, C. M., and A. N. Carter. 2009. "Cooperative problem solving in a social carnivore." *Animal Behaviour* 78 (4):967–977. doi: 10.1016/j.anbehav.2009.06.030.

Droser, M. L., R. A. Fortey, and X. Li. 1996. "The Ordovician radiation." *American Scientist* 84 (2):122–131.

Dudley, T. L., and C. M. Dantonio. 1991. "The effects of substrate texture, grazing, and disturbance on macroalgal establishment in streams." *Ecology* 72 (1):297–309.

Dulvy, N. K., R. P. Freckleton, and N. V. C. Polunin. 2004. "Coral reef cascades and the indirect effects of predator removal by exploitation." *Ecology Letters* 7 (5):410–416. doi: 10.1111/j.1461-0248.2004.00593.x.

Dunne, J. A. 2006. The network structure of food webs. In *Ecological Networks: Linking Structure to Dynamics in Food Webs*, M. Pascual and J. A. Dunne, editors, pp. 27–86. Oxford: Oxford University Press.

D'Alelio, D., S. Libralato, T. Wyatt, and M. R. d'Alcala. 2016. "Ecological-network models link diversity, structure and function in the plankton food-web." *Scientific Reports* 6. doi: 10.1038/srep21806.

Ehrlich, P. R., and P. H. Raven. 1964. "Butterflies and plants: A study in coevolution." *Evolution* 18:586–608.

Elliott, J. M., and L. Persson. 1978. "The estimation of daily rate of food consumption for fish." *Journal of Animal Ecology* 47:977–991.

Ellison, A. M., and N. J. Gotelli. 2001. "Evolutionary ecology of carnivorous plants." *Trends in Ecology & Evolution* 16 (11):623–629.

Elton, C. 1927. *Animal Ecology*. London: Sidgwick and Jackson Ltd.

Emlen, J. M. 1967. "Optimal choice in animals." *American Naturalist* 102: 385–389.

Emslie, R. 2012. Diceros bicornis. The IUCN Red List of Threatened Species 2012: e.T6557 A16980917. http://dx.doi.org/10.2305/IUCN.UK.2012.RLTS.T6557A16980917.en.

Ernest, S. K. M., B. J. Enquist, J. H. Brown, E. L. Charnov, J. F. Gillooly, V. Savage, E. P. White et al. 2003. "Thermodynamic and metabolic effects on the scaling of production and population energy use." *Ecology Letters* 6 (11):990–995. doi: 10.1046/j.1461-0248 .2003.00526.x.

Estes, J. A., M. T. Tinker, T. M. Williams, and D. F. Doak. 1998. "Killer whale predation on sea otters linking oceanic and nearshore ecosystems." *Science* 282 (5388):473–476.

Evans, C. S., L. Evans, and P. Marler. 1993. "On the meaning of alarm calls—Functional reference in an avian vocal system." *Animal Behaviour* 46 (1):23–38.

Evans-White, M. A., W. K. Dodds, and M. R. Whiles. 2003. "Ecosystem significance of crayfishes and stonerollers in a prairie stream: Functional differences between co-occurring omnivores." *Journal of the North American Benthological Society* 22:423–441.

Exnerova, A., P. Stys, E. Fucikova, S. Vesela, K. Svadova, M. Prokopova, V. Jarosik, R. Fuchs, and E. Landova. 2007. "Avoidance of aposematic prey in European tits (Paridae): Learned or innate?" *Behavioral Ecology* 18 (1):148–156.

Fallows, C., A. J. Gallagher, and N. Hammerschlag. 2013. "White sharks (*Carcharodon carcharias*) scavenging on whales and its potential role in further shaping the ecology of an apex predator." *Plos One* 8 (4):e60797–e60797. doi: 10.1371/journal.pone.0060797.

FAO. 2002. Food energy–Methods of analysis and conversion factors. FAO Food and Nutrition Paper 77. United Nations, Food and Agricultural Organization.

FAO. 2016. Fisheries and Aquaculture topics. Fisheries statistics and information. Topics Fact Sheets. In: *FAO Fisheries and Aquaculture Department* [online]. Rome, Italy: Food and Agriculture Organization of the United Nations. Updated December 22, 2015. [Cited May 10, 2016].

Farine, D. R., C. J. Garroway, and B. C. Sheldon. 2012. "Social network analysis of mixed-species flocks: Exploring the structure and evolution of interspecific social behaviour." *Animal Behaviour* 84 (5):1271–1277.

Farquhar, G. D., S. V. Caemmerer, and J. A. Berry. 1980. "A biochemical-model of photosynthetic CO_2 assimilation in leaves of C-3 species." *Planta* 149 (1):78–90. doi: 10.1007 /bf00386231.

Fath, B. D. 2007. "Structural food web regimes." *Ecological Modelling* 208 (2–4):391–394.

Fath, B. D., U. M. Scharler, R. E. Ulanowicz, and B. Hannon. 2007. "Ecological network analysis: Network construction." *Ecological Modelling* 208 (1):49–55.

Fedonkin, M. A. 2003. "The origin of the Metazoa in the light of the Proterozoic fossil record." *Paleontological Research* 7 (1):9–41. doi: 10.2517/prpsj.7.9.

Fenn, M. E., M. A. Poth, J. D. Aber, J. S. Baron, B. T. Bormann, D. W. Johnson, A. D. Lemly, S. G. McNulty, D. F. Ryan, and R. Stottlemyer. 1998. "Nitrogen excess in North American ecosystems: Predisposing factors, ecosystem responses, and management strategies." *Ecological Applications* 8 (3):706–733.

Ferlian, O., S. Scheu, and M. M. Pollierer. 2012. "Trophic interactions in centipedes (Chilopoda, Myriapoda) as indicated by fatty acid patterns: Variations with life stage, forest age and season." *Soil Biology & Biochemistry* 52:33–42. doi: 10.1016/j.soilbio.2012.04.018.

Ferrari, M. C. O., B. D. Wisenden, and D. P. Chivers. 2010. "Chemical ecology of predator–prey interactions in aquatic ecosystems: A review and prospectus." *Canadian Journal of Zoology-Revue Canadienne De Zoologie* 88 (7):698–724.

Ferry, K. H., and R. A. Wright. 2002. "*Bythotrephes cederstroemi* in Ohio reservoirs: Evidence from fish diets'." *Ohio Journal of Science* 102 (5):116–118.

Field, C. B., J. T. Randerson, and C. M. Malmstrom. 1995. "Global net primary production—Combining ecology and remote-sensing." *Remote Sensing of Environment* 51 (1):74–88. doi: 10.1016/0034-4257(94)00066-v.

Field, I. C., M. G. Meekan, R. C. Buckworth, and C. J. A. Bradshaw. 2009. "Susceptibility of sharks, rays and chimaeras to global extinction." *Advances in Marine Biology* 56:275–363.

Findlay, S., J. Tank, S. Dye, H. M. Valett, P. J. Mulholland, W. H. McDowell, S. L. Johnson et al. 2002. "A cross-system comparison of bacterial and fungal biomass in detritus pools of headwater streams." *Microbial Ecology* 43 (1):55–66. doi: 10.1007/10.1007/s00248-001-1020-x.

Floreano, D., R. Pericet-Camara, S. Viollet, F. Ruffier, A. Brückner, R. Leitel, W. Buss et al. 2013. "Miniature curved artificial compound eyes." *Proceedings of the National Academy of Sciences* 110 (23):9267–9272. doi: 10.1073/pnas.1219068110.

Folch, J., M. Lees, and G. H. Sloane-Stanley. 1957. "A simple method for the isolation and purification of total lipids from animal tissues." *Journal of Biological Chemistry* 226 (1):497–509.

Fondo, E. N., M. Chaloupka, J. J. Heymans, and G. A. Skilleter. 2015. "Banning fisheries discards abruptly has a negative impact on the population dynamics of charismatic marine megafauna." *Plos One* 10 (12):11. doi: 10.1371/journal.pone.0144543.

Forbes, S. A. 1887. "The lake as a microcosm." *Bulletin of the Illinois State Natural History Survey* 15:537–550.

Fordyce, J. A., Z. Gompert, M. L. Forister, and C. C. Nice. 2011. "A hierarchical Bayesian approach to ecological count data: A flexible tool for ecologists." *Plos One* 6 (11):7. doi: 10.1371/journal.pone.0026785.

Forney, J. L. 1977. "Reconstruction of yellow perch (*Perca flavescens*) cohorts from examination of walleye (*Stizostedion vitreum vitreum*) stomachs." *Journal of the Fisheries Research Board of Canada* 34:925–932.

Fortey, R. A. 2004. "The lifestyle of the trilobites." *American Scientist* 92 (5):446–453.

Foster, J. R. 1977. "Pulsed gastric lavage: An efficient method of removing the stomach contents of live fish." *Progressive Fish-Culturist* 39:166–169.

France, R. L. 2011. "Leaves as 'crackers', biofilm as 'peanut butter': Exploratory use of stable isotopes as evidence for microbial pathways in detrital food webs." *Oceanological and Hydrobiological Studies* 40 (4):110–115. doi: 10.2478/s13545-011-0047-y.

Frank, D. A., R. D. Evans, and B. F. Tracy. 2004. "The role of ammonia volatilization in controlling the natural N-15 abundance of a grazed grassland." *Biogeochemistry* 68 (2):169–178.

Fretwell, S. D., and H. L. Lucas, Jr. 1969. "On territorial behavior and other factors influencing habitat distribution in birds. Part 1 theoretical development." *Acta Biotheoretica* 19 (1):16–36. doi: 10.1007/bf01601953.

Fretwell, S. D., and H. J. Lucas, Jr. 1970. "On territorial behaviour and other factors influencing habitat distribution in birds." *Acta Biotheory* 19:16–36.

Frost, P. C., J. P. Benstead, W. F. Cross, H. Hillebrand, J. H. Larson, M. A. Xenopoulos, and T. Yoshida. 2006. "Threshold elemental ratios of carbon and phosphorus in aquatic consumers." *Ecology Letters* 9 (7):774–779.

Fry, B. 1991. "Stable isotope diagrams of freshwater food webs." *Ecology* 72 (6):2293–2297.

Fry, B. 2006. *Stable Isotope Ecology*. Springer Press.

Fry, B. 2013. "Alternative approaches for solving underdetermined isotope mixing problems." *Marine Ecology Progress Series* 472:1–13.

Gallai, N., J.-M. Salles, J. Settele, and B. E. Vaissiere. 2009. "Economic valuation of the vulnerability of world agriculture confronted with pollinator decline." *Ecological Economics* 68 (3):810–821. doi: 10.1016/j.ecolecon.2008.06.014.

Garvey, J. E., and R. A. Stein. 1998. "Competition between larval fishes in reservoirs: The role of relative timing of appearance." *Transactions of the American Fisheries Society* 127:1023–1041.

Garvey, J. E., and S. R. Chipps. 2013. Quantifying diets and energy flow. In *Fisheries Techniques*, 3rd ed. Zale, Parrish, and Sutton, editors, pp. 733–779. American Fisheries Society.

Garvey, J. E., R. A. Stein, and H. M. Thomas. 1994. "Assessing how fish predation and interspecific prey competition influence a crayfish assemblage." *Ecology* 75:532–547.

Garvey, J. E., R. A. Wright, R. A. Stein, and K. H. Ferry. 2000. "Evaluating how local- and regional-scale processes interact to regulate growth of age-0 largemouth bass." *Transactions of the American Fisheries Society* 129:1044–1059.

Garvey, J. E., D. R. Devries, R. A. Wright, and J. G. Miner. 2003. "Energetic adaptations along a broad latitudinal gradient: Implications for widely distributed assemblages." *Bioscience* 53 (2):141–150.

Gascoigne, J., and R. N. Lipcius. 2004. "Allee effects in marine systems." *Marine Ecology— Progress Series* 269:49–59.

Gause, G. F. 1934. *The Struggle for Existence*. Baltimore: Dover Phoenix.

Gil, M. del Mar, M. Palmer, A. Grau, and S. Balle. 2015. "Many vulnerable or a few resilient specimens? Finding the optimal for reintroduction/restocking programs." *Plos One* 10 (9). doi: 10.1371/journal.pone.0138501.

Gilliam, J. F., and D. F. Fraser. 1987. "Habitat selection under predation hazard—Test of a model with foraging minnows." *Ecology* 68 (6):1856–1862. doi: 10.2307/1939877.

Gillooly, J. F., J. H. Brown, G. B. West, V. M. Savage, and E. L. Charnov. 2001. "Effects of size and temperature on metabolic rate." *Science* 293 (5538):2248–2251. doi: 10.1126 /science.1061967.

Gillooly, J. F., E. L. Charnov, J. H. Brown, V. M. Savage, and G. B. West. 2003a. "Allometry: How reliable is the biological time clock? Reply." *Nature* 424 (6946):270–270. doi: 10.1038/424270a.

Gillooly, J. F., J. H. Brown, A. P. Allen, G. B. West, E. L. Charnov, and V. M. Savage. 2003b. "The central role of metabolism from genes to ecosystems." *Integrative and Comparative Biology* 43 (6):923–923.

Gillooly, J. F., A. P. Allen, G. B. West, and J. H. Brown. 2005a. "The rate of DNA evolution: Effects of body size and temperature on the molecular clock." *Proceedings of the National Academy of Sciences of the United States of America* 102 (11):140–145. doi: 10.1073/pnas.0407735101.

Gillooly, J. F., A. P. Allen, J. H. Brown, J. J. Elser, C. M. del Rio, V. M. Savage, G. B. West, W. H. Woodruff, and H. A. Woods. 2005b. "The metabolic basis of whole-organism RNA and phosphorus content." *Proceedings of the National Academy of Sciences of the United States of America* 102 (33):11923–11927. doi: 10.1073/pnas.0504756102.

Golley, F. B. 1968. "Secondary productivity in terrestrial communities." *American Zoologist* 8:53–59.

Gomi, T. 1977. *Everyone Poops*, My Body in Science series. Trans. Amanda Mayer Stinchecum. Brooklyn, NY: Kane/Miller, and Scholastic.

Gonthier, D. J. 2012. "Do herbivores eavesdrop on ant chemical communication to avoid predation?" *Plos One* 7 (1):e28703.

Gordon, I. J., and H. T. Prins. 2008. *The Ecology of Browsing and Grazing*. New York: Cambridge University Press.

Gosselin, M., M. Levasseur, P. A. Wheeler, R. A. Horner, and B. C. Booth. 1997. "New measurements of phytoplankton and ice algal production in the Arctic Ocean." *Deep- Sea Research Part II—Topical Studies in Oceanography* 44 (8):1623. doi: 10.1016 /s0967-0645(97)00054-4.

Gossner, M. M., E. Pasalic, M. Lange, P. Lange, S. Boch, D. Hessenmoeller, J. Mueller et al. 2014. "Differential responses of herbivores and herbivory to management in temperate European beech." *Plos One* 9 (8). doi: 10.1371/journal.pone.0104876.

Grace, M. R., D. P. Giling, S. Hladyz, V. Caron, R. M. Thompson, and R. M. Nally. 2015. "Fast processing of diel oxygen curves: Estimating stream metabolism with BASE (BAyesian Single-station Estimation)." *Limnology and Oceanography—Methods* 13:103–114. doi: 10.1002/lom.10011.

Greenberg, R. 2011. "Exploration and protection of Europa's biosphere: Implications of permeable ice." *Astrobiology* 11 (2):183–191. doi: 10.1089/ast.2011.0608.

Hadj-Chikh, L. Z., M. A. Steele, and P. D. Smallwood. 1996. "Caching decisions by grey squirrels: A test of the handling time and perishability hypotheses." *Animal Behaviour* 52:941–948. doi: 10.1006/anbe.1996.0242.

Hairston, N. G., F. E. Smith, and L. B. Slobodkin. 1960. "Community structure, population control, and competition." *American Naturalist* 94:421–425.

Hall, R. O., Jr., J. L. Tank, D. J. Sobota, P. J. Mulholland, J. M. O'Brien, W. K. Dodds, J. R. Webster et al. 2009. "Nitrate removal in stream ecosystems measured by N-15 addition experiments: Total uptake." *Limnology and Oceanography* 54 (3):653–665. doi: 10.4319/lo.2009.54.3.0653.

Halpin, P. N. 1997. "Global climate change and natural-area protection: Management responses and research directions." *Ecological Applications* 7 (3):828–843.

Hambright, K. D., R. W. Drenner, S. R. McComas, and Jr. N. G. Hairston. 1991. "Gape-limited piscivores, planktivore size refuges, and the trophic cascade hypothesis." *Archive fur Hydrobiologie* 121:389–404.

Hamer, R., F. L. Lemckert, and P. B. Banks. 2011. "Adult frogs are sensitive to the predation risks of olfactory communication." *Biology Letters* 7 (3):361–363. doi: 10.1098/rsbl.2010.1127.

Hanley, T. C., and K. J. La Pierre. 2015. *Trophic Ecology: Bottom-up and Top-down Interactions across Aquatic and Terrestrial Systems*. Cambridge: Cambridge University Press.

Hansel, H. C., S. D. Duke, P. T. Lofy, and G. A. Gray. 1988. "Use of diagnostic bones to identify and estimate original lengths of ingested prey fishes." *Transactions of the American Fisheries Society* 117 (1):55–62.

Hanson, P. C., D. L. Bade, S. R. Carpenter, and T. K. Kratz. 2003. "Lake metabolism: Relationships with dissolved organic carbon and phosphorus." *Limnology and Oceanography* 48 (3):1112–1119.

Hanson, P. C., T. B. Johnson, D. E. Schindler, and J. F. Kitchell. 1997. "Fish Bioenergetics 3.0." *University of Wisconsin, Sea Grant Institute*.

Hartman, K. J., and J. F. Kitchell. 2008. "Bioenergetics modeling: Progress since the 1992 symposium." *Transactions of the American Fisheries Society* 137 (1):216–223.

Hartman, K. J., and F. J. Margraf. 2008. "Common relationships among proximate composition components in fishes." *Journal of Fish Biology* 73 (10):2352–2360. doi: 10.1111/j.1095-8649.2008.02083.x.

Harvell, C. D. 1990. "The ecology and evolution of inducible defenses." *Quarterly Review of Biology* 65 (3):323–340.

Hayward, A., and J. F. Gillooly. 2011. "The cost of sex: Quantifying energetic investment in gamete production by males and females." *Plos One* 6 (1):1–4.

He, X., J. F. Kitchell, S. R. Carpenter, J. R. Hodgson, D. E. Schindler, and K. L. Cottingham. 1993. "Food web structure and long-term phosphorus recycling: A simulation model evaluation." *Transactions of the American Fisheries Society* 122:773–783.

Hein, A. M., C. Hou, and J. F. Gillooly. 2012. "Energetic and biomechanical constraints on animal migration distance." *Ecology Letters* 15 (2):104–110. doi: 10.1111/j.146-0248.2011.01714.x.

Hein, C. L., M. J. Vander Zanden, and J. J. Magnuson. 2007. "Intensive trapping and increased fish predation cause massive population decline of an invasive crayfish." *Freshwater Biology* 52 (6):1134–1146.

Heisler, L. M., C. M. Somers, and R. G. Poulin. 2016. "Owl pellets: A more effective alternative to conventional trapping for broad-scale studies of small mammal communities." *Methods in Ecology and Evolution* 7 (1):96–103. doi: 10.1111/2041-210x.12454.

Helfield, J. M., and R. J. Naiman. 2001. "Effects of salmon-derived nitrogen on riparian forest growth and implications for stream productivity." *Ecology* 82:2403–2409.

Herman, D. P., D. G. Burrows, P. R. Wade, J. W. Durban, C. O. Matkin, R. G. LeDuc, L. G. Barrett-Lennard, and M. M. Krahn. 2005. "Feeding ecology of eastern North Pacific killer whales *Orcinus orca* from fatty acid, stable isotope, and organochlorine analyses of blubber biopsies." *Marine Ecology Progress Series* 302:275–291. doi: 10.3354/meps302275.

Herman, P. M. J., J. J. Middelburg, J. Widdows, C. H. Lucas, and C. H. R. Heip. 2000. "Stable isotopes' as trophic tracers: Combining field sampling and manipulative labelling of food resources for macrobenthos." *Marine Ecology Progress Series* 204:79–92. doi: 10.3354/meps204079.

Hesslein, R. H., K. A. Hallard, and P. Ramlal. 1993. "Replacement of sulfur, carbon, and nitrogen in tissue of growing broad whitefish (*Coregonus nasus*) in response to a change in diet traced by delta-s-34, delta-c-13 and delta-n-15." *Canadian Journal of Fisheries and Aquatic Sciences* 50 (10):2071–2076. doi: 10.1139/f93-230.

Hewett, S. W., and B. L. Johnson. 1987. A generalized bioenergetics model of fish growth for microcomputers. University of Wisconsin Sea Grant Institute, Madison, Wisconsin.

Hieber, M., and M. O. Gessner. 2002. "Contribution of stream detritivores, fungi, and bacteria to leaf breakdown based on biomass estimates." *Ecology* 83 (4):1026–1038. doi: 10.2307/3071911.

Hilborn, R., and M. Mangel. 1997. "The ecological detective. Confronting models with data." *Monographs in Population Biology* 28:i–xvii, 1–315.

Hillis, J. J., J. E. Garvey, and M. J. Lydy. 2015. "Contaminants reduce male contribution to reproduction at the population scale." *Ecosphere* 6 (4). doi: 10.1890/es14-00391.1.

Hirsch, B. T. 2012. "Within-group spatial position in ring-tailed coatis: Balancing predation, feeding competition, and social competition." *Behavioral Ecology and Sociobiology* 65 (2):391–399. doi: 10.1007/s00265-010-1056-3.

Hixon, M. A., and J. P. Beets. 1993. "Predation, prey refuges, and the structure of coral-reef fish assemblages." *Ecological Monographs* 63 (1):77–101. doi: 10.2307/2937124.

Hjort, J. 1914. "Fluctuations in the great fisheries of northern Europe viewed in light of biological research." *Rapports et Proces-Verbaux des Reunions Conseil International pour l'Exploration de la Mer* 20:1–228.

Hobson, K. A., and R. G. Clark. 1993. "Turnover of C-13 in cellular and plasma fractions of blood–Implications for nondestructive sampling in avian dietary studies." *Auk* 110 (3):638–641.

Hodges, K. E. 2008. "Defining the problem: Terminology and progress in ecology." *Frontiers in Ecology and the Environment* 6:35–42.

Hoey, A. S. 2011. "Size matters: Macroalgal height influences the feeding response of coral reef herbivores." *Marine Ecology—Progress Series* 411:299–U341. doi: 10.3354/meps 08660.

Holling, C. S. 1959 "The components of predation as revealed by a study of small mammal predation of the European Pine Sawfly." *Canadian Entomologist* 91:293–320.

Holt, D. E., and C. E. Johnston. "Can you hear the dinner bell? Response of cyprinid fishes to environmental acoustic cues." *Animal Behaviour* 82 (3):529–534.

Hone, D., T. Tsuihiji, M. Watabe, and K. Tsogtbaatr. 2012. "Pterosaurs as a food source for small dromaeosaurs." *Palaeogeography Palaeoclimatology Palaeoecology* 331:27–30. doi: 10.1016/j.palaeo.2012.02.021.

Hooker, T. D., and J. M. Stark. 2012. "Carbon flow from plant detritus and soil organic matter to microbes-linking carbon and nitrogen cycling in semiarid soils." *Soil Science Society of America Journal* 76 (3):903–914. doi: 10.2136/sssaj2011.0139.

Hooper, D. U., F. S. Chapin, J. J. Ewel, A. Hector, P. Inchausti, S. Lavorel, J. H. Lawton et al. 2005. "Effects of biodiversity on ecosystem functioning: A consensus of current knowledge." *Ecological Monographs* 75 (1):3–35. doi: 10.1890/04-0922.

Horn, H. S. (1966). "Measurement of 'overlap' in comparative ecological studies." *The American Naturalist* 100:419–424.

Houston, D. C. 1983. The adaptive radiation of the griffon vultures. In *Vulture Biology and Management*. S. R. Wilbur and J. A. Jackson, editors, pp. 135–152. Berkeley: University of California Press.

Hupfer, M., and J. Lewandowski. 2008. "Oxygen controls the phosphorus release from Lake sediments—A long-lasting paradigm in Limnology." *International Review of Hydrobiology* 93 (4–5):415–432. doi: 10.1002/iroh.200711054.

Hurlbert, S. H. 1978. "The measurement of niche overlap and some relatives." *Ecology* 59:67–77.

Huryn, A. D. 1996. "An appraisal of the Allen paradox in a New Zealand trout stream." *Limnology and Oceanography* 41:243–252.

Huryn, A. D., and J. B. Wallace. 2000. "Life history and production of stream insects." *Annual Review of Entomology* 45:83–110.

Huston, M. 1979. "General hypothesis of species-diversity." *American Naturalist* 113 (1):81–101. doi: 10.1086/283366.

Hutchinson, G. E. 1959. "Homage to Santa Rosalia or why are there so many kinds of animals?" *American Naturalist* 93:145–159.

Hynes, H. B. N. 1970. *The Ecology of Running Waters*. Toronto: University of Toronto Press.

Ims, R. A. 1990. "On the adaptive value of reproductive synchrony as a predator-swamping strategy." *American Naturalist* 136 (4):485–498.

Iverson, J. B., H. Higgins, A. Sirulnik, and C. Griffiths. 1997. "Local and geographic variation in the reproductive biology of the snapping turtle (*Chelydra serpentina*)." *Herpetologica* 53 (1):96–117.

Iverson, S. J., K. J. Frost, and L. F. Lowry. 1997. "Fatty acid signatures reveal fine scale structure of foraging distribution of harbor seals and their prey in Prince William Sound, Alaska." *Marine Ecology Progress Series* 151 (1–3):255–271.

Iverson, S. J., K. J. Frost, and S. L. C. Lang. 2002. "Fat content and fatty acid composition of forage fish and invertebrates in Prince William Sound, Alaska: Factors contributing to among and within species variability." *Marine Ecology Progress Series* 241:161–181. doi: 10.3354/meps241161.

Ivlev, V. S. 1939a. "Transformation of energy by aquatic animals." *Internationale Revue der gesamten Hydrobiologie* 38:449–458.

Ivlev, V. S. 1939b. "Balance of energy in carps." *Zoologicheskii Zhurnal* 18:449–458.

Ivlev, V. S. 1961. *Experimental Ecology of the Feeding of Fishes*. New Haven, CT: Yale University Press.

Janzen, D. H. 1977. "Why fruits rot, seeds mold, and meat spoils." *American Naturalist* 111 (980):691–713. doi: 10.1086/283200.

Jenny, H. 1980. "The soil resource. Origin and behaviour." *Ecological Studies* 37:1–377.

Jetz, W., C. Rowe, and T. Guilford. 2001. "Non-warning odors trigger innate color aversions—As long as they are novel." *Behavioral Ecology* 12 (2):134–139.

Jobling, M. 1994. *Fish Bioenergetics*. Berlin: Springer.

Johnson, B., and J. Wallace. 2005. "Bottom-up limitation of a stream salamander in a detritus-based food web." *Canadian Journal of Fisheries and Aquatic Sciences* 62:301–311.

Johnson, B., W. Cross, and J. Wallace. 2003. "Long-term resource limitation reduces insect detritivore growth in a headwater stream." *Journal of the North American Benthological Society* 22:565–574.

Johnson, Z. I., R. Shyam, A. E. Ritchie, C. Mioni, V. P. Lance, J. W. Murray, and E. R. Zinser. 2011. "The effect of iron- and light-limitation on phytoplankton communities of deep chlorophyll maxima of the western Pacific Ocean." *Journal of Marine Research* 68 (2):283–308.

Jonsson, K. I., E. Rabbow, R. O. Schill, M. Harms-Ringdahl, and P. Rettberg. 2008. "Tardigrades survive exposure to space in low Earth orbit." *Current Biology* 18 (17):R729–R731. doi: 10.1016/j.cub.2008.06.048.

Jonsson, M., S. D. Wratten, D. A. Landis, J. M. L. Tompkins, and R. Cullen. 2010. "Habitat manipulation to mitigate the impacts of invasive arthropod pests." *Biological Invasions* 12 (9):2933–2945. doi: 10.1007/s10530-010-9737-4.

Jorgensen, S. E. 2009. *Ecosystem Ecology.* Cambridge, MA: Academic Press.

Jorgensen, S. E., and H. Mejer. 1979. "Holistic approach to ecological modeling." *Ecological Modelling* 7:169–189.

Kacsoh, B. Z., Z. R. Lynch, N. T. Mortimer, and T. A. Schlenke 2013. "Fruit flies medicate offspring after seeing parasites." *Science.* New York, 339 (6122):947–950. doi: 10.1126 /science.1229625.

Karban, R. 1997. "Evolution of prolonged development: A life table analysis for periodical cicadas." *American Naturalist* 150 (4):446–461. doi: 10.1086/286075.

Kaspari, M., S. Powell, J. Lattke, and S. O'Donnell. 2011. "Predation and patchiness in the tropical litter: Do swarm-raiding army ants skim the cream or drain the bottle?" *Journal of Animal Ecology* 80 (4):818–823. doi: 10.1111/j.1365-2656.2011.01826.x.

Kasting, J. F. 1993. "Earth's early atmosphere." *Science* 259 (5097):920–926. doi: 10.1126 /science.11536547.

Kasting, J. F., and T. P. Ackerman. 1986. "Climatic consequences of very high-carbon dioxide levels in the Earth's early atmosphere." *Science* 234 (4782):1383–1385. doi: 10.1126 /science.11539665.

Kats, L. B., and L. M. Dill. 1998. "The scent of death: Chemosensory assessment of predation risk by prey animals." *Ecoscience* 5 (3):361–394.

Kearns, C. A., D. W. Inouye, and N. M. Waser. 1998. "Endangered mutualisms: The conservation of plant-pollinator interactions." *Annual Review of Ecology and Systematics* 29:83–112. doi: 10.1146/annurev.ecolsys.29.1.83.

Keefer, M. L., C. A. Peery, R. R. Ringe, and I. C. Bjornn. 2004. "Regurgitation rates of intragastric radio transmitters by adult Chinook salmon and steelhead during upstream migration in the Columbia and Snake Rivers." *North American Journal of Fisheries Management* 24 (1):47–54.

Kelley, J. L., and A. E. Magurran. 2003. "Learned predator recognition and antipredator responses in fishes." *Fish and Fisheries* 4 (3):216–226.

Kelly, J. R., and R. E. Scheibling. 2012. "Fatty acids as dietary tracers in benthic food webs." *Marine Ecology Progress Series* 446:1–22.

Kerr, R. A. 1994. Iron fertilization: A tonic, but no cure for the greenhouse. *Science* 263:1089–1090.

Kerr, S. R., and L. M. Dickie. 2001. The biomass spectrum: A predator–prey theory of aquatic production. *Complexity in Ecological Systems Series*, Columbia University Press.

Kim, G. W., and D. R. DeVries. 2001. "Adult fish predation on freshwater limnetic fish larvae: A mesocosm experiment." *Transactions of the American Fisheries Society* 130 (2):189–203.

Kirsch, P. E., S. J. Iverson, W. D. Bowen, S. R. Kerr, and R. G. Ackman. 1998. "Dietary effects on the fatty acid signature of whole Atlantic cod (*Gadus morhua*)." *Canadian Journal of Fisheries and Aquatic Sciences* 55 (6):1378–1386.

Kishida, O., G. C. Trussell, A. Ohno, S. Kuwano, T. Ikawa, and K. Nishimura. 2012. "Predation risk suppresses the positive feedback between size structure and cannibalism." *Journal of Animal Ecology* 80 (6):1278–1287.

Kitchell, J. F., D. J. Stewart, and D. Weininger. 1977. "Applications of a bioenergetics model to yellow perch (*Perca flavescens*) and walleye (*Stizostedion vitreum vitreum*)." *Journal of the Fisheries Research Board of Canada* 34:1922–1935.

Kleiber, M., and T. A. Rogers. 1961. "Energy metabolism." *Annual Review of Physiology* 23:15. doi: 10.1146/annurev.ph.23.030161.000311.

Koch, U., D. Martin-Creuzburg, H.-P. Grossart, and D. Straile. 2012. "Single dietary amino acids control resting egg production and affect population growth of a key freshwater herbivore." *Oecologia* 167 (4):981–989. doi: 10.1007/s00442-011-2047-4.

Koerselman, W., and A. F. M. Meuleman. 1996. "The vegetation N:P ratio: A new tool to detect the nature of nutrient limitation." *Journal of Applied Ecology* 33 (6):1441–1450. doi: 10.2307/2404783.

Kohler, S. L., and M. A. McPeek. 1989. "Predation risk and the foraging behavior of competing stream insects." *Ecology* 70 (6):1811–1825.

Koopman, H. N., S. M. Budge, D. R. Ketten, and S. J. Iverson. 2006. "Topographical distribution of lipids inside the mandibular fat bodies of odontocetes: Remarkable complexity and consistency." *IEEE Journal of Oceanic Engineering* 31 (1):95–106. doi: 10.1109 /joe.2006.872205.

Kotler, B. P. 1997. "Patch use by gerbils in a risky environment: Manipulating food and safety to test four models." *Oikos* 78 (2):274–282.

Kotler, B. P., and J. S. Brown. 1988. "Environmental heterogeneity and the coexistence of desert rodents." *Annual Review of Ecology and Systematics* 19:281–307.

Kotler, B. P., J. S. Brown, and O. Hasson. 1991. "Factors affecting gerbil foraging behavior and rates of owl predation." *Ecology* 72:2249–2260.

Kotler, B. P., J. Brown, S. Mukherjee, O. Berger-Tal, and A. Bouskila. 1997. "Moonlight avoidance in gerbils reveals a sophisticated interplay among time allocation, vigilance and state-dependent foraging." *Proceedings of the Royal Society B—Biological Sciences* 277 (1687):1469–1474. doi: 10.1098/rspb.2009.2036.

Krause, A. E., K. A. Frank, D. M. Mason, R. E. Ulanowicz, and W. W. Taylor. 2003. "Compartments revealed in food-web structure." *Nature* 426:282–285.

Krebs, C. J. 2001. *Ecological Methodology*, 2nd edition. San Francisco: Benjamin Cummings.

Kroon, A., R. L. Veenendaal, J. Bruin, M. Egas, and M. W. Sabelis. 2008. "Sleeping with the enemy-predator-induced diapause in a mite." *Naturwissenschaften* 95 (12):1195–1198.

Kumari, T. R. J. 2011. "Arbuscular mycorrhizal symbiosis: An overview." *Research Journal of Biotechnology* 6 (1):75–79.

Labandeira, C. 2007. "The origin of herbivory on land: Initial patterns of plant tissue consumption by arthropods." *Insect Science* 14 (4):259–275. doi: 10.1111/j.1744-7917.2007.00152.x.

Lamarra, V. A., Jr. 1975. "Digestive activities of carp as a major contributor to the nutrient loading of lakes." *Verhandlungen Internationale Vereiningung Limnologie* 19:2461–2468.

Land, M. F. 1997. "Visual acuity in insects." *Annual Review of Entomology* 42:147–177.

Larkin, P. A. 1977. "An epitaph for the concept of maximum sustained yield." *Transactions of the American Fisheries Society* 106 (1):1–11. doi: 10.1577/1548-8659(1977)106<1:AEF TCO>2.0.CO;2.

Laureillard, J., L. Mejanelle, and M. Sibuet. 2004. "Use of lipids to study the trophic ecology of deep-sea xenophyophores." *Marine Ecology—Progress Series* 270:129–140.

Lawler, S. P., and P. J. Morin. 1993. "Temporal overlap, competition, and priority effects in larval anurans." *Ecology* 74 (1):174–182.

Lawrence, M. G., S. D. Jupiter, and B. S. Kamber. 2006. "Aquatic geochemistry of the rare earth elements and yttrium in the Pioneer River catchment, Australia." *Marine and Freshwater Research* 57 (7):725–736. doi: 10.1071/mf05229.

Lawton, J. H. 1999. "Are there general laws in ecology?" *Oikos* 84 (2):177–192. doi: 10 .2307/3546712.

Lee, Y. F., and L. L. Severinghaus. 2004. "Sexual and seasonal differences in the diet of Lanyu scops owls based on fecal analysis." *Journal of Wildlife Management* 68 (2):299–306. doi: 10.2193/0022-541x(2004)068[0299:sasdit]2.0.co;2.

Legezynska, J., M. Kedra, and W. Walkusz. 2012. "When season does not matter: Summer and winter trophic ecology of Arctic amphipods." *Hydrobiologia* 684 (1):189–214. doi: 10.1007/s10750-011-0982-z.

Leggett, W. C., and E. DeBlois. 1994. "Recruitment in marine fishes: Is it regulated by starvation and predation in the egg and larval stages?" *Netherlands Journal of Sea Research* 32:119–134.

Lehninger, A. L. 1960. *Bioenergetics: The Molecular Basis of Biological Energy Transformations*. San Francisco: W.A. Benjamin.

Leibold, M. A., M. Holyoak, N. Mouquet, P. Amarasekare, J. M. Chase, M. F. Hoopes, R. D. Holt, J. B. Shurin, R. Law, D. Tilman, M. Loreau, and A. Gonzalez. 2004. "The metacommunity concept: A framework for multi-scale community ecology." *Ecology Letters* 7:601–613.

Leong, W., and J. R. Pawlik. 2010. "Evidence of a resource trade-off between growth and chemical defenses among Caribbean coral reef sponges." *Marine Ecology—Progress Series* 406:71–78.

Leroux, S. J., D. Hawlena, and O. J. Schmitz. 2012. "Predation risk, stoichiometric plasticity and ecosystem elemental cycling." *Proceedings of the Royal Society B: Biological Sciences* 279 (1745):4183–4191. doi: 10.1098/rspb.2012.1315.

Leveau, L. M., C. M. Leveau, and U. F. J. Pardinas. 2004. "Trophic relationships between white-tailed kites (*Elanus leucurus*) and barn owls (*Tyto alba*) in southern Buenos Aires Province, Argentina." *Journal of Raptor Research* 38 (2):178–181.

Levick, S. R., G. P. Asner, T. Kennedy-Bowdoin, and D. E. Knapp. 2010. "The spatial extent of termite influences on herbivore browsing in an African savanna." *Biological Conservation* 143 (11):2462–2467. doi: 10.1016/j.biocon.2010.06.012.

Levins, R. 1968. *Evolution in Changing Environments.* Princeton: Princeton University Press.

Ley, R. E., M. Hamady, C. Lozupone, P. J. Turnbaugh, R. R. Ramey, J. S. Bircher, M. L. Schlegel et al. 2008. "Evolution of mammals and their gut microbes." *Science* 320 (5883):1647–1651. doi: 10.1126/science.1155725.

Liere, H., and A. Larsen. 2012. "Cascading trait-mediation: Disruption of a trait-mediated mutualism by parasite-induced behavioral modification." *Oikos* 119 (9):1394–1400.

Lijklema, L. 1994. "Nutrient dynamics in shallow lakes—Effects of changes in loading and role of sediment–water interactions." *Hydrobiologia* 275:335–348.

Likens, G. E., F. H. Bormann, N. M. Johnson, D. W. Fisher, and R. S. Pierce. 1970. "Effects of forest cutting and herbicide treatment on nutrient budgets in the Hubbard Brook watershed-ecosystem." *Ecological Monographs* 40:23–47.

Lima, S. L. 1995. "Back to the basics of antipredatory vigilance—The group-size effect." *Animal Behaviour* 49 (1):11–20.

Lindeman, R. A. 1942. "The trophic-dynamic aspect of ecology." *Ecology* 23:399–418.

Lindenmayer, D. B., A. D. Manning, P. L. Smith, H. P. Possingham, J. Fischer, I. Oliver, and M. A. McCarthy. 2002. "The focal-species approach and landscape restoration: A critique." *Conservation Biology* 16 (2):338–345.

Lindroth, R. L. 1989. Mammalian herbivore–plant interactions. In *Plant–Animal Interactions.* W. G. Abrahamson, editor, pp. 163–206, New York: McGraw-Hill Publishing Company.

Lindsey, C. C. 1966. "Body sizes of poikilotherm vertebrates at different latitudes." *Evolution* 20:456–465.

Litvak, M. K., and R. I. C. Hansell. 1990. "Investigation of food habit and niche relationships in a cyprinid community." *Canadian Journal of Zoology* 68:1873–1879.

Litvak, M. K., and W. C. Leggett. 1992. "Age and size-selective predation on larval fishes: The bigger-is-better hypothesis revisited." *Marine Ecology Progress Series* 81:13–24.

Lodge, D. M. 1993. "Biological invasions: Lessons for ecology." *Trends in Ecology and Evolution* 8:133–136.

Lodge, D. M., J. W. Barko, D. Strayer, J. M. Melack, G. G. Mittelbach, R. W. Howarth, B. Menge, and J. E. Titus. 1988. Spatial heterogeneity and habitat interactions in lake communities. In *Complex Interactions in Lake Communities.* S. R. Carpenter, editor. New York: Springer-Verlag.

Lodge, D. M., M. W. Kershner, J. E. Aloi, and A. P. Covich. 1994. "Effects of an omnivorous crayfish (*Orconectes rusticus*) on a freshwater littoral food web." *Ecology* 75:1265–1281.

Lodge, D. M., R. A. Stein, K. M. Brown, A. P. Covich, C. Bronmark, J. E. Garvey, and S. P. Klosiewski. 1998. "Predicting impact of freshwater exotic species on native biodiversity: Challenges in spatial scaling." *Australian Journal of Ecology* 23 (1):53–67.

LoDuca, S. T., and E. R. Behringer. 2009. "Functional morphology and evolution of early Paleozoic dasycladalean algae (Chlorophyta)." *Paleobiology* 35 (1):63–76.

Loladze, I., and J. J. Elser. 2011. "The origins of the Redfield nitrogen-to-phosphorus ratio are in a homoeostatic protein-to-rRNA ratio." *Ecology Letters* 14 (3):244–250. doi: 10 .1111/j.1461-0248.2010.01577.x.

Lonnstedt, O. M., M. I. McCormick, and D. P. Chivers. 2012. "Well-informed foraging: Damage-released chemical cues of injured prey signal quality and size to predators." *Oecologia* 168 (3):651–658.

Loo, J. 2009. "Ecological impacts of non-indigenous invasive fungi as forest pathogens." *Biological Invasions* 11 (1):81–96. doi: 10.1007/s10530-008-9321-3.

Lotka, A. J. 1925. *Elements of Physical Biology*. Baltimore: Williams and Wilkins.

Lourenco, P. M., J. P. Granadeiro, J. L. Guilherme, and T. Catry. 2015. "Turnover rates of stable isotopes in avian blood and toenails: Implications for dietary and migration studies." *Journal of Experimental Marine Biology and Ecology* 472:89–96. doi: 10.1016/j .jembe.2015.07.006.

Lovenberg, W. 1973. Some vaso- and psychoactive substances in food: Amines, stimulants, depressants, and hallucinogens. In: *Toxicants Naturally Occurring in Foods*, Washington DC: National Academy of Sciences Press, 2nd Edition, pp. 170–188.

Lovern, J. A. 1934. "Fat metabolism in fishes: Selective formation of fat deposits." *Biochemical Journal* 28 (2):394–402.

Lowe, C. G. 2002. "Bioenergetics of free-ranging juvenile scalloped hammerhead sharks (Sphyrna lewini) in Kane'ohe Bay, O'ahu, HI." *Journal of Experimental Marine Biology and Ecology* 278 (2):141–156. doi: 10.1016/s0022-0981(02)00331-3.

Loxdale, H. D., G. Lushai, and J. A. Harvey. 2011. "The evolutionary improbability of 'generalism' in nature, with special reference to insects." *Biological Journal of the Linnean Society* 103 (1):1–18. doi: 10.1111/j.1095-8312.2011.01627.x.

Lozano, G. A. 1991. "Optimal foraging theory—A possible role for parasites." *Oikos* 60 (3):391–395. doi: 10.2307/3545084.

Ludsin, S. A., M. W. Kershner, K. A. Blocksom, R. L. Knight, and R. A. Stein. 2001. "Life after death in Lake Erie: Nutrient controls drive fish species richness, rehabilitation." *Ecological Applications* 11 (3):731–746.

MacArthur, R. H. 1958. "Population ecology of some warblers of northeastern coniferous forests." *Ecology* 39 (4):599–619. doi: 10.2307/1931600.

MacArthur, R. H., and E. R. Pianka. 1966. "On optimal use of a patchy environment." *American Naturalist* 100:603–609.

MacArthur, R. H., and E. O. Wilson. 1967. *The Theory of Island Biogeography*. Princeton, New Jersey: Princeton University Press.

Madenjian, C. P., J. T. Tyson, R. L. Knight, M. W. Kershner, and M. J. Hansen. 1996. "First-year growth, recruitment, and maturity of walleyes in Western Lake Erie." *Transactions of the American Fisheries Society* 125 (6):821–830.

Majdi, N., A. Boiche, W. Traunspurger, and A. Lecerf. 2014. "Predator effects on a detritus-based food web are primarily mediated by non-trophic interactions." *Journal of Animal Ecology* 83:953–962.

Makkonen, M., M. P. Berg, I. T. Handa, S. Haettenschwiler, J. van Ruijven, P. M. van Bodegom, and R. Aerts. 2012. "Highly consistent effects of plant litter identity and functional traits on decomposition across a latitudinal gradient." *Ecology Letters* 15 (9):1033–1041. doi: 10.1111/j.1461-0248.2012.01826.x.

Mangel, M., and C. W. Clark. 1989. *Dynamic Modeling in Behavioral Ecology*. Princeton, NJ: Princeton University Press.

Marchetti, C., and P. J. Drent. 2000. "Individual differences in the use of social information in foraging by captive great tits." *Animal Behaviour* 60:131–140. doi: 10.1006/anbe .2000.1443.

Marek, P. E., and J. E. Bond. 2009. "A Mullerian mimicry ring in Appalachian millipedes." *Proceedings of the National Academy of Sciences of the United States of America* 106 (24):9755–9760.

Mariette, M. M., E. C. Pariser, A. J. Gilby, M. J. L. Magrath, S. R. Pryke, and S. C. Griffith. 2011. "Using an electronic monitoring system to link offspring provisioning and foraging behavior of a wild passerine." *Auk* 128 (1):26–35. doi: 10.1525/auk.2011.10117.

Marrow, P., R. Law, and C. Cannings. 1992. "The coevolution of predator prey interactions— ESSS and red queen dynamics." *Proceedings of the Royal Society of London Series B— Biological Sciences* 250 (1328):133–141. doi: 10.1098/rspb.1992.0141.

Martin, J. L., S. A. Stockton, S. Allombert, and A. J. Gaston. 2011. "Top-down and bottom-up consequences of unchecked ungulate browsing on plant and animal diversity in temperate forests: Lessons from a deer introduction." *Biological Invasions* 12 (2):353–371. doi: 10.1007/s10530-009-9628-8.

Martin, L. D., B. M. Rothschild, and D. A. Burnham. 2016. "Hesperornis escapes plesiosaur attack." *Cretaceous Research* 63:23–27. doi: http://dx.doi.org/10.1016/j.cretres.2016.02.005.

Martin, T. G., B. A. Wintle, J. R. Rhodes, P. M. Kuhnert, S. A. Field, S. J. Low-Choy, A. J. Tyre, and H. P. Possingham. 2005. "Zero tolerance ecology: Improving ecological inference by modelling the source of zero observations." *Ecology Letters* 8 (11):1235–1246. doi: 10.1111/j.1461-0248.2005.00826.x.

Martinez, N. D. 1991. "Artifacts or attributes—Effects of resolution on the Little Rock lake food web." *Ecological Monographs* 61:367–392.

Martinoia, E., M. Maeshima, and H. E. Neuhaus. 2007. "Vacuolar transporters and their essential role in plant metabolism." *Journal of Experimental Botany* 58 (1):83–102. doi: 10.1093/jxb/erl183.

Matassa, C. M., and G. C. Trussell. 2011. "Landscape of fear influences the relative importance of consumptive and nonconsumptive predator effects." *Ecology* 92 (12):2258–2266.

Mathis, A., and R. J. F. Smith. 1992. "Avoidance of areas marked with a chemical alarm substance by fathead minnows (*Pimephales promelas*) in a natural habitat." *Canadian Journal of Zoology-Revue Canadienne De Zoologie* 70 (8):1473–1476.

May, R. M. 1972. "Will a large complex system be stable." *Nature* 238 (5364):413–414. doi: 10.1038/238413a0.

Maynard Smith, J. 1982. *Evolution and Theory of Games*. Cambridge, MA: Cambridge University Press.

McCarthy, J. M., C. L. Hein, J. D. Olden, and M. J. Vander Zanden. 2006. "Coupling long-term studies with meta-analysis to investigate impacts of non-native crayfish on zoobenthic communities." *Freshwater Biology* 51 (2):224–235.

McGlynn, T. P., and E. K. Poirson. 2012. "Ants accelerate litter decomposition in a Costa Rican lowland tropical rain forest." *Journal of Tropical Ecology* 28:437–443. doi: 10.1017/s0266467412000375.

McGuire, K. L., and K. K. Treseder. 2010. "Microbial communities and their relevance for ecosystem models: Decomposition as a case study." *Soil Biology & Biochemistry* 42 (4):529–535. doi: 10.1016/j.soilbio.2009.11.016.

McNaughton, S. J. 1985. "Ecology of a grazing ecosystem—The Serengeti." *Ecological Monographs* 55 (3):259–294. doi: 10.2307/1942578.

Menge, B. A. 1995. "Joint bottom-up and top-down regulation of rocky intertidal algal beds in South Africa." *Trends in Ecology & Evolution* 10 (11):431–432.

Menge, B. A., and J. P. Sutherland. 1987. "Community regulation: Variation in disturbance, competition, and predation in relation to environmental stress and recruitment." *American Naturalist* 130:730–757.

Mills, E. L., J. M. Casselman, R. Dermott, J. D. Fitzsimons, G. Gal, K. T. Holeck, J. A. Hoyle et al. 2003. "Lake Ontario: Food web dynamics in a changing ecosystem (1970–2000)." *Canadian Journal of Fisheries and Aquatic Sciences* 60 (4):471–490.

Mitani, J. C., D. A. Merriwether, and C. B. Zhang. 2000. "Male affiliation, cooperation and kinship in wild chimpanzees." *Animal Behaviour* 59:885–893. doi: 10.1006/anbe .1999.1389.

Mittelbach, G. G. 1981. "Foraging efficiency and body size: A study of optimal diet and habitat use by bluegill." *Ecology* 62:1370–1386.

Mittelbach, G. G. 2002. Fish foraging and habitat choice: A theoretical perspective. In: *Handbook of Fish Biology and Fisheries: Volume 1*, edited by P. J. B. Hart and J. D. Reynolds, pp. 251–266. Malden, MA: Blackwell Science.

Mittelbach, G. G. 2012. *Community Ecology.* Berlin: Sinuaer.

Mittelbach, G. G., and L. Persson. 1998. "The ontogeny of piscivory and its ecological consequences." *Canadian Journal of Fisheries and Aquatic Sciences* 55 (6):1454–1465.

Mittelbach, G. G., A. M. Turner, D. J. Hall, J. E. Rettig, and C. W. Osenberg. 1995. "Perturbation and resilience: A long-term, whole-lake study of predator extinction and reintroduction." *Ecology* 76:2347–2360.

Mittelbach, G. G., C. F. Steiner, S. M. Scheiner, K. L. Gross, H. L. Reynolds, R. B. Waide, M. R. Willig, S. I. Dodson, and L. Gough. 2001. "What is the observed relationship between species richness and productivity?" *Ecology* 82 (9):2381–2396.

Moore, J. C., and P. C. de Ruiter. 2012. *Energetic Food Webs: An Analysis of Real and Model Ecosystems.* Oxford: Oxford University Press.

Mora, C., D. P. Tittensor, S. Adl, A. G. B. Simpson, and B. Worm. 2011. "How many species are there on earth and in the ocean?" *Plos Biology* 9 (8):e1001127. doi: 10.1371/journal .pbio.1001127.

Morisita, M. 1959. "Measuring of the dispersion and analysis of distribution patterns." *Memoires of the Faculty of Science, Kyushu University, Series E. Biology* 2:215–235.

Muller-Navarra, D. C., M. T. Brett, A. M. Liston, and C. R. Goldman. 2000. "A highly unsaturated fatty acid predicts carbon transfer between primary producers and consumers." *Nature* 403 (6765):74–77. doi: 10.1038/47469.

Muller-Navarra, D. C., M. T. Brett, S. Park, S. Chandra, A. P. Ballantyne, E. Zorita, and C. R. Goldman. 2004. "Unsaturated fatty acid content in seston and tropho-dynamic coupling in lakes." *Nature* 427 (6969):69–72. doi: 10.1038/nature02210.

Munch, S. B., and D. O. Conover. 2002. "Accounting for local physiological adaptation in bioenergetic models: Testing hypotheses for growth rate evolution by virtual transplant experiments." *Canadian Journal of Fisheries and Aquatic Sciences* 59 (2):393–403.

Nager, R. G., P. Monaghan, R. Griffiths, D. C. Houston, and R. Dawson. 1999. "Experimental demonstration that offspring sex ratio varies with maternal condition." *Proceedings of the National Academy of Sciences of the United States of America* 96 (2):570–573. doi: 10 .1073/pnas.96.2.570.

Naguib, M., A. Kazek, S. V. Schaper, K. van Oers, and M. E. Visser. 2010. "Singing activity reveals personality traits in great tits." *Ethology* 116 (8):763–769. doi: 10.1111/j.1439-0310.2010.01791.x.

Naguib, M., C. Florcke, and K. van Oers. 2011. "Effects of social conditions during early development on stress response and personality traits in great tits (*Parus major*)." *Developmental Psychobiology* 53 (6):592–600. doi: 10.1002/dev.20533.

Naiman, R. J., S. R. Elliott, J. M. Helfield, and T. C. O'Keefe. 1999. "Biophysical interactions and the structure and dynamics of riverine ecosystems: The importance of biotic feedbacks." *Hydrobiologia* 410:79–86.

Naiman, R., R. Bilby, D. Schindler, and J. Helfield. 2002. "Pacific salmon, nutrients, and the dynamics of freshwater and riparian ecosystems." *Ecosystems* 5:399–417.

Naor, M., E. S. Bernardes, and A. Coman. 2012. "Theory of constraints: Is it a theory and a good one?" *International Journal of Production Research* 51 (2):542–554. doi: 10.1080/00207543.2011.654137.

Nelson, M. P., J. A. Vucetich, R. O. Peterson, and L. M. Vucetich. 2011. "The Isle Royale Wolf-Moose Project (1958–present) and the wonder of long-term ecological research." *Endeavour* 35 (1):30–38. doi: 10.1016/j.endeavour.2010.09.002.

Nemeth, A., and K. Takacs. 2010. "The paradox of cooperation benefits." *Journal of Theoretical Biology* 264 (2):301–311. doi: 10.1016/j.jtbi.2010.02.005.

Neubauer, P., and Jensen, O. P. 2015. "Bayesian estimation of predator diet composition from fatty acids and stable isotopes." *PeerJ* 3:e920; doi 10.7717/peerj.920

Newbold, J. D., J. W. Elwood, R. V. Oneill, and A. L. Sheldon. 1983. "Phosphorus dynamics in a woodland stream ecosystem—A study of nutrient spiralling." *Ecology* 64 (5):1249–1265. doi: 10.2307/1937833.

Ney, J. J. 1990. "Trophic economics in fisheries—Assessment of demand–supply relationships between predators and prey." *Reviews in Aquatic Sciences* 2 (1):55–81.

Ney, J. J. 1993. "Bioenergetics modeling today: Growing pains on the cutting edge." *Transactions of the American Fisheries Society* 122:736–748.

Nichol, A. A., V. Douglas, and L. Peck. 1933. "On the immunity of rattlesnakes to their venom." *Copeia* 1933 (4):211–213.

Nicolaus, M., J. M. Tinbergen, K. M. Bouwman, S. P. M. Michler, C. Ubels, C. Both, B. Kempenaers, and N. J. Dingemanse. 2012. "Experimental evidence for adaptive personalities in a wild passerine bird." *Proceedings of the Royal Society B—Biological Sciences* 279 (1749):4885–4892. doi: 10.1098/rspb.2012.1936.

Noonburg, E. G., L. A. Newman, M. Lewis, R. L. Crabtree, and A. B. Potapov. 2007. "Sequential decision-making in a variable environment: Modeling elk movement in Yellowstone National Park as a dynamic game." *Theoretical Population Biology* 71 (2):182–195. doi: 10.1016/j.tpb.2006.09.004.

Noren, S. R., M. S. Udevitz, and C. V. Jay. 2012. "Bioenergetics model for estimating food requirements of female Pacific walruses Odobenus rosmarus divergens." *Marine Ecology-Progress Series* 460:261–275. doi: 10.3354/meps09706.

Odum, E. P. 1969. "The strategy of ecosystem development." *Science* 164:262–270.

Odum, H. T. 1957. "Trophic structure and productivity of Silver Springs, Florida." *Ecological Monographs* 27:55–112.

Ogutu, J. O., H. P. Piepho, R. S. Reid, M. E. Rainy, R. L. Kruska, J. S. Worden, M. Nyabenge, and N. T. Hobbs. 2011. "Large herbivore responses to water and settlements in savannas." *Ecological Monographs* 80 (2):241–266.

Paerl, H. W., and J. L. Pinckney. 1996. "A mini-review of microbial consortia: Their roles in aquatic production and biogeochemical cycling." *Microbial Ecology* 31 (3):225–247.

Paine, R. T. 1966. "Food web complexity and species diversity." *American Naturalist* 100:65–75.

Paine, R. T. 1969. "A note on trophic complexity and community stability." *The American Naturalist* 103 (929):91–93. doi: 10.1086/282586.

Paine, R. T. 1980. "Food webs: Linkage, interaction strength, and community infrastructure." *Journal of Animal Ecology* 49:667–685.

Paine, R. T. 2010. "Macroecology: Does it ignore or can it encourage further ecological syntheses based on spatially local experimental manipulations?" *American Naturalist* 176:385–393.

Parnell, A. C., R. Inger, S. Bearhop, and A. L. Jackson. 2010. "Source partitioning using stable isotopes: Coping with too much variation." *Plos One* 5 (3):e9672. doi: 10.1371/journal.pone.0009672.

Parnell, A. C., D. L. Phillips, S. Bearhop, B. X. Semmens, E. J. Ward, J. W. Moore, A. L. Jackson, J. Grey, D. J. Kelly, and R. Inger. 2013. "Bayesian stable isotope mixing models." *Environmetrics* 24 (6):387–399. doi: 10.1002/env.2221.

Pascual, M., and J. A. Dunne. 2006. *Ecological Networks: Linking Structure to Dynamics in Food Webs.* Oxford: Oxford University Press.

Pastor, J., B. Dewey, R. J. Naiman, P. F. McInnes, and Y. Cohen. 1993. "Moose browsing and soil fertility in the boreal forests of Isle-Royale-National-Park." *Ecology* 74 (2):467–480. doi: 10.2307/1939308.

Paukert, C. P., and J. H. Petersen. 2007. "Comparative growth and consumption potential of rainbow trout and humpback chub in the Colorado River, Grand Canyon, Arizona, under different temperature scenarios." *Southwestern Naturalist* 52 (2):234–242. doi: 10.1894/0038-4909(2007)52[234:cgacpo]2.0.co;2.

Pauly, D., and V. Christensen. 1995. "Primary production required to sustain global fisheries." *Nature* 374 (6519):255–257. doi: 10.1038/374255a0.

Pauly, D., V. Christensen, and C. Walters. 2000. "Ecopath, Ecosim, and Ecospace as tools for evaluating ecosystem impact of fisheries." *ICES Journal of Marine Science* 57 (3):697–706. doi: 10.1006/jmsc.2000.0726.

Pauly, D., V. Christensen, J. Dalsgaard, R. Froese, and F. Torres. 1998. "Fishing down marine food webs." *Science* 279 (5352):860–863. doi: 10.1126/science.279.5352.860.

Pauly, D., V. Christensen, S. Guenette, T. J. Pitcher, U. R. Sumaila, C. J. Walters, R. Watson, and D. Zeller. 2002. "Towards sustainability in world fisheries." *Nature* 418 (6898):689–695.

Pearl, R. 1927. "The biology of superiority." *American Mercury* 12:257–266.

Peckarsky, B. L. 1980. "Predator–prey interactions between stoneflies and mayflies—Behavioral observations." *Ecology* 61 (4):932–943.

Peden, D. G., G. M. Van Dyne, R. W. Rice, and R. M. Hansen. 1974. "The trophic ecology of Bison bison L. on shortgrass plains." *Journal of Applied Ecology* 11 (2):489–497. doi: 10.2307/2402203.

Pekar, S., D. Mayntz, T. Ribeiro, and M. E. Herberstein. 2012. "Specialist ant-eating spiders selectively feed on different body parts to balance nutrient intake." *Animal Behaviour* 79 (6):1301–1306. doi: 10.1016/j.anbehav.2010.03.002.

Pellmyr, O. 2002. "Pollination by animals." In *Plant–Animal Interactions: An Evolutionary Approach*, C. M. Herrera and O. Pellmyr, editors, pp. 157–184. Oxford, UK: Blackwell Science.

Perga, M.-E., I. Domaizon, J. Guillard, V. Hamelet, and O. Anneville. 2013. "Are cyanobacterial blooms trophic dead ends?" *Oecologia* 172 (2):551–562. doi: 10.1007/s00442-012-2519-1.

Peron, F., L. Rat-Fischer, M. Lalot, L. Nagle, and D. Bovet. 2010. "Cooperative problem solving in African grey parrots (Psittacus erithacus)." *Animal Cognition* 14 (4):545–553. doi: 10.1007/s10071-011-0389-2.

Perry, A. L., P. J. Low, J. R. Ellis, and J. D. Reynolds. 2005. "Climate change and distribution shifts in marine fishes." *Science* 308 (5730):1912–1915. doi: 10.1126/science.1111322.

Persson, L., J. Andersson, E. Wahlstrom, and P. Eklov. 1996. "Size-specific interactions in lake systems: Predator gape limitation and prey growth rate and mortality." *Ecology* 77 (3):900–911.

Peters, F. E. 1968. *Aristotle and the Arabs: The Aristotelian Tradition in Islam*. New York: New York University Press.

Peters, R. H. 1980. *A Critique for Ecology*. New York: Cambridge University Press.

Pfennig, D. W., and S. P. Mullen. 2010. "Mimics without models: Causes and consequences of allopatry in Batesian mimicry complexes." *Proceedings of the Royal Society B—Biological Sciences* 277 (1694):2577–2585.

Pfennig, D. W., and S. P. Mullen. 2010. "Mimics without models: Causes and consequences of allopatry in Batesian mimicry complexes." *Proceedings of the Royal Society B-Biological Sciences* 277 (1694):2577–2585. doi: 10.1098/rspb.2010.0586.

Phelps, Q. E., G. W. Whitledge, S. J. Tripp, K. T. Smith, J. E. Garvey, D. P. Herzog, D. E. Ostendorf et al. 2012. "Identifying river of origin for age-0 Scaphirhynchus sturgeons in the Missouri and Mississippi rivers using fin ray microchemistry." *Canadian Journal of Fisheries and Aquatic Sciences* 69 (5):930–941. doi: 10.1139/f2012-038.

Phillips, D. L. 2012. "Converting isotope values to diet composition: The use of mixing models." *Journal of Mammalogy* 93 (2):342–352. doi: 10.1644/11-mamm-s-158.1.

Phillips, D. L., R. Inger, S. Bearhop, A. L. Jackson, J. W. Moore, A. C. Parnell, B. X. Semmens, and E. J. Ward. 2014. "Best practices for use of stable isotope mixing models in food-web studies." *Canadian Journal of Zoology* 92 (10):823–835. doi: 10.1139/cjz-2014-0127.

Pierce, G. J., and J. G. Ollason. 1987. "8 reasons why optimal foraging theory is a complete waste of time." *Oikos* 49 (1):111–118. doi: 10.2307/3565560.

Pierce, W. D., R. A. Cushman, and C. E. Hood. 1912. U.S. Department of Agriculture Bulletin, 100.

Pimm, S. L., and J. H. Lawton. 1977. "Number of trophic levels in ecological communities." *Nature* 268:329–331.

Pimm, S. L., and J. H. Lawton. 1980. "Are food webs divided into compartments?" *Journal of Animal Ecology* 49 (3):879–898.

Pimm, S. L. 1982. *Food Webs.* London: Chapman and Hall.

Pimm, S. L. 2001. *Food Webs.* Chicago: University of Chicago Press.

Polis, G. A. 1991. "Complex trophic interactions in deserts—An empirical critique of food-web theory." *American Naturalist* 138 (1):123–155. doi: 10.1086/285208.

Polis, G. A., and R. D. Holt. 1992. "Intraguild predation—The dynamics of complex trophic interactions." *Trends in Ecology & Evolution* 7 (5):151–154.

Polis, G. A., and S. D. Hurd. 1996. "Linking marine and terrestrial food webs: Allochthonous input from the ocean supports high secondary productivity on small islands and coastal land communities." *American Naturalist* 147 (3):396–423.

Polis, G. A., C. A. Myers, and R. D. Holt. 1989. "The ecology and evolution of intraguild predation: Potential competitors that eat each other." *Annual Review of Ecology and Systematics* 20:297–330.

Polis, G. A., R. D. Holt, B. A. Menge, and K. O. Winemiller. 1996. Time, space, and life history: Influences on food webs. In *Food Webs: Integration of Patterns and Dynamics*, G. A. Polis and K. O. Winemiller, editors. New York: Chapman and Hall.

Polis, G. A., W. B. Anderson, and R. D. Holt. 1997. "Toward an integration of landscape and food web ecology: The dynamics of spatially subsidized food webs." *Annual Review of Ecology And Systematics* 28:289–316.

Pond, W. G., D. B. Chruch, K. R. Pond, and P. A. Schoknecht. 1996. *Basic Animal Nutrition.* New York: Wiley.

Porazinska, D. L., R. D. Bardgett, M. B. Blaauw, H. W. Hunt, A. N. Parsons, T. R. Seastedt, and D. H. Wall. 2003. "Relationships at the aboveground-belowground interface: Plants, soil biota, and soil processes." *Ecological Monographs* 73 (3):377–395. doi: 10.1890/0012-9615(2003)073[0377:rataip]2.0.co;2.

Post, D. M. 2002. "Using stable isotopes to estimate trophic position: Models, methods, and assumptions." *Ecology* 83 (3):703–718. doi: 10.2307/3071875.

Post, D. M., E. Conners, and D. S. Goldberg. 2000. "Prey preference by a top predator and the stability of linked food chains." *Ecology* 81 (1):8–14.

Post, J. R., M. Sullivan, S. Cox, N. P. Lester, C. J. Walters, E. A. Parkinson, A. J. Paul, L. Jackson, and B. J. Shuter. 2002. "Canada's recreational fisheries: The invisible collapse?" *Fisheries* 27 (1):6–17.

Pough, F. H. 1980. "Advantages of ectothermy for tetrapods." *American Naturalist* 115:92–112.

Poulos, D. E., and M. I. McCormick. 2015. "Asymmetries in body condition and order of arrival influence competitive ability and survival in a coral reef fish." *Oecologia* 179 (3):719–728. doi: 10.1007/s00442-015-3401-8.

Power, M. E. 1992. "Top-down and bottom-up forces in food webs—Do plants have primacy." *Ecology* 73 (3):733–746.

Power, M. E., D. Tilman, J. A. Estes, B. A. Menge, W. J. Bond, L. S. Mills, G. Daily, J. C. Castilla, J. Lubchenco, and R. T. Paine. 1996. "Challenges in the quest for keystones." *Bioscience* 46 (8):609–620.

Prado, S. S., K. Y. Hung, M. P. Daugherty, and R. P. P. Almeida. 2012. "Indirect effects of temperature on stink bug fitness, via maintenance of gut-associated symbionts." *Applied and Environmental Microbiology* 76 (4):1261–1266. doi: 10.1128/aem.02034-09.

Pratt, D. M., and V. H. Anderson. 1985. "Giraffe social-behavior." *Journal of Natural History* 19 (4):771–781. doi: 10.1080/00222938500770471.

Preisser, E. L., D. I. Bolnick, and M. F. Benard. 2005. "Scared to death? The effects of intimidation and consumption in predator-prey interactions." *Ecology* 86 (2):501–509.

Pretorius, Y., W. F. de Boer, C. van der Waal, H. J. de Knegt, R. C. Grant, N. M. Knox, E. M. Kohi et al. 2011. "Soil nutrient status determines how elephant utilize trees and shape environments." *Journal of Animal Ecology* 80 (4):875–883. doi: 10.1111/j.1365-2656.2011.01819.x.

Probst, W. E., C. F. Rabeni, W. G. Covington, and R. E. Marteney. 1984. "Resource use by stream-dwelling rock bass and smallmouth bass." *Transactions of the American Fisheries Society* 113 (3):283–294. doi: 10.1577/1548-8659(1984)113<283:rubsrb>2.0.co;2.

Putman, R. J. 1977. "Dynamics of blowfly, *Calliphora erythrocephala*, within carrion." *Journal of Animal Ecology* 46 (3):853–866. doi: 10.2307/3645.

Putman, R. J. 1978. "Flow of energy and organic-matter from a carcass during decomposition. 2. Decomposition of small mammal carrion in temperate systems." *Oikos* 31 (1):58–68. doi: 10.2307/3543384.

Pyke, G. H. 1978. "Optimal foraging in hummingbirds—Testing the marginal value theorem." *American Zoologist* 18 (4):739–752.

Pyke, G. H. 1984. "Optimal foraging theory—A critical review." *Annual Review of Ecology and Systematics* 15:523–575. doi: 10.1146/annurev.es.15.110184.002515.

Rabalais, N. N. 2002. "Nitrogen in aquatic ecosystems." *Ambio* 31 (2):102–112.

Rabalais, N. N., R. E. Turner, and W. J. Wiseman. 2002. "Gulf of Mexico hypoxia, aka 'The dead zone'." *Annual Review of Ecology and Systematics* 33:235–263. doi: 10.1146/annurev.ecolysis.33.010802.150513.

Real, L. A. 1981. "Uncertainty and pollinator-plant interactions: The foraging behavior of bees and wasps on artificial flowers." *Ecology* 62:20–26.

Real, L. A., and J. A. Brown. 1991. *Foundations of Ecology*. Chicago: University of Chicago Press.

Redfield, A. C. 1934. On the proportions of organic derivations in sea water and their relation to the composition of plankton. In *James Johnstone Memorial Volume*. R. J. Daniel, editor, pp. 176–192. London: University Press of Liverpool.

Relyea, R. A. 2001. "Morphological and behavioral plasticity of larval anurans in response to different predators." *Ecology* 82 (2):523–540. doi: 10.1890/0012-9658(2001)082[0523:mabpol]2.0.co;2.

Relyea, R. A., and E. E. Werner. 1999. "Quantifying the relation between predator-induced behavior and growth performance in larval anurans." *Ecology* 80 (6):2117–2124.

Rengefors, K., I. Karlsson, and L. A. Hansson. 1998. "Algal cyst dormancy: A temporal escape from herbivory." *Proceedings of the Royal Society of London Series B—Biological Sciences* 265 (1403):1353–1358.

Renkonen, O. 1938. "Statisch-ökologische Untersuchungen über die terrestrische Käferwelt der finnischen Bruchmoore." *Annales Zoologici Societatis Zoologicae-Botanicae Fennicae 'Vanamo'* 6:1–231.

Rhee, G. Y. 1978. "Effects of N-P atomic ratios and nitrate limitation on algal growth, cell composition, and nitrate uptake." *Limnology and Oceanography* 23 (1):10–25.

Rice, J. A., and P. A. Cochran. 1984. "Independent evaluation of a bioenergetics model for largemouth bass." *Ecology* 65:732–739.

Ricker, W. E. 1946. "Production and utilization of fish populations." *Ecological Monographs* 16:373–391.

Ricker, W. E. 1975. "Computation and intepretation of biological statistics of fish populations." *Bulletin of the Fisheries Research Board of Canada* 191.

Riley, A. J., and W. K. Dodds. 2013. "Whole-stream metabolism: Strategies for measuring and modeling diel trends of dissolved oxygen." *Freshwater Science* 32 (1):56–69. doi: 10.1899/12-058.1.

Ripple, W. J., and B. Van Valkenburgh. 2010. "Linking top-down forces to the Pleistocene megafaunal extinctions." *BioScience* 60 (7):516–526. doi: 10.1525/bio.2010.60.7.7.

Ritland, D. B., and L. P. Brower. 1991. "The viceroy butterfly is not a Batesian mimic." *Nature* 350 (6318):497–498.

Rivers, D. B., C. Thompson, and R. Brogan. 2011. "Physiological trade-offs of forming maggot masses by necrophagous flies on vertebrate carrion." *Bulletin of Entomological Research* 101 (5):599–611. doi: 10.1017/s0007485311000241.

Roberts, C. M., J. A. Bohnsack, F. Gell, J. P. Hawkins, and R. Goodridge. 2001. "Effects of marine reserves on adjacent fisheries." *Science* 294 (5548):1920–1923. doi: 10.1126/science.294.5548.1920.

Roberts, G. 1996. "Why individual vigilance declines as group size increases." *Animal Behaviour* 51:1077–1086.

Rockwell, N. C., D. Duanmu, S. S. Martin, C. Bachy, D. C. Price, D. Bhattacharya, A. Z. Worden, and J. C. Lagarias. 2014. "Eukaryotic algal phytochromes span the visible spectrum." *Proceedings of the National Academy of Sciences of the United States of America* 111 (10):3871–3876. doi: 10.1073/pnas.1401871111.

Roell, M. J., and D. J. Orth. 1993. "Trophic basis of production of stream-dwelling smallmouth bass, rock bass, and flathead catfish in relation to invertebrate bait harvest." *Transactions of the American Fisheries Society* 122:46–62.

Rosas, C. A., D. M. Engle, J. H. Shaw, and M. W. Palmer. 2008. "Seed dispersal by *Bison bison* in a tallgrass prairie." *Journal of Vegetation Science* 19 (6):769–778. doi: 10.3170/2008-8-18447.

Rosenzweig, M. L. 1971. "Paradox of enrichment—Destabilization of exploitation ecosystems in ecological time." *Science* 171 (3969):385–387.

Ross, D., and J. B. Wallace. 1981. "Production of *Brachycentrus spinae* ross (Trichoptera, Brachycentridae) and its role in seston dynamics of a southern Appalachian stream (USA)." *Environmental Entomology* 10:240–246.

Rozen, D. E., D. J. P. Engelmoer, and P. T. Smiseth. 2008. "Antimicrobial strategies in burying beetles breeding on carrion." *Proceedings of the National Academy of Sciences of the United States of America* 105 (46):17890–17895. doi: 10.1073/pnas.0805403105.

Ruben, J. 1995. "The evolution of endothermy in mammals and birds—From physiology to fossils." *Annual Review of Physiology* 57:69–95. doi: 10.1146/annurev.ph.57.030195.000441.

Rude, N. P., J. T. Trushenski, and G. W. Whitledge. 2016. "Fatty acid profiles are biomarkers of fish habitat use in a river-floodplain ecosystem." *Hydrobiologia* 773 (1):63–75. doi: 10.1007/s10750-016-2679-9.

Ruess, L., and P. M. Chamberlain. 2010. "The fat that matters: Soil food web analysis using fatty acids and their carbon stable isotope signature." *Soil Biology & Biochemistry* 42 (11):1898–1910. doi: 10.1016/j.soilbio.2010.07.020.

Ruess, L., M. M. Haggblom, E. J. G. Zapata, and J. Dighton. 2002. "Fatty acids of fungi and nematodes—Possible biomarkers in the soil food chain?" *Soil Biology & Biochemistry* 34 (6):745–756. doi: 10.1016/s0038-0717(01)00231-0.

Ruess, L., M. M. Haggblom, R. Langel, and S. Scheu. 2004. "Nitrogen isotope ratios and fatty acid composition as indicators of animal diets in belowground systems." *Oecologia* 139 (3):336–346.

Ruxton, G. D., and D. M. Wilkinson. 2013. "Endurance running and its relevance to scavenging by early hominids." *Evolution* 67 (3):861–867. doi: 10.1111/j.1558-5646.2012.01815.x.

Ruxton, G. D., M. P. Speed, and D. J. Kelly. 2004. "What, if anything, is the adaptive function of countershading?" *Animal Behaviour* 68:445–451.

Ruzicka, K. J., J. W. Groninger, and J. J. Zaczek. 2010. "Deer browsing, forest edge effects, and vegetation dynamics following bottomland forest restoration." *Restoration Ecology* 18 (5):702–710. doi: 10.1111/j.1526-100X.2008.00503.x.

Sachs, J. L., C. J. Essenberg, and M. M. Turcotte. 2011. "New paradigms for the evolution of beneficial infections." *Trends in Ecology & Evolution* 26 (4):202–209. doi: 10.1016/j.tree.2011.01.010.

Sahney, S., M. J. Benton, and H. J. Falcon-Lang. 2011. "Rainforest collapse triggered Carboniferous tetrapod diversification in Euramerica." *Geology* 38 (12):1079–1082. doi: 10.1130/g31182.1.

Sarnelle, O. 1993. "Herbivore effects on phytoplankton succession in a eutrophic lake." *Ecological Monographs* 63 (2):129–149.

Savage, V. M., J. F. Gillooly, J. H. Brown, G. B. West, and E. L. Charnov. 2004a. "Effects of body size and temperature on population growth." *American Naturalist* 163 (3):429–441. doi: 10.1086/381872.

Savage, V. M., J. F. Gillooly, W. H. Woodruff, G. B. West, A. P. Allen, B. J. Enquist, and J. H. Brown. 2004b. "The predominance of quarter-power scaling in biology." *Functional Ecology* 18 (2):257–282. doi: 10.1111/j.0269-8463.2004.00856.x.

Savage, V. M., A. P. Allen, J. H. Brown, J. F. Gillooly, A. B. Herman, W. H. Woodruff, and G. B. West. 2007. "Scaling of number, size, and metabolic rate of cells with body size in mammals." *Proceedings of the National Academy of Sciences of the United States of America* 104 (11):4718–4723. doi: 10.1073/pnas.0611235104.

Scharf, F. S., J. A. Buckel, F. Juanes, and D. O. Conover. 1997. "Estimating piscine prey size from partial remains: Testing for shifts in foraging mode by juvenile bluefish." *Environmental Biology of Fishes* 49 (3):377–388.

Scharf, F. S., F. Juanes, and M. Sutherland. 1998a. "Inferring ecological relationships from the edges of scatter diagrams: Comparison of regression techniques." *Ecology* 79:448–460.

Scharf, F. S., R. M. Yetter, A. P. Summers, and F. Juanes. 1998b. "Enhancing diet analyses of piscivorous fishes in the Northwest Atlantic through identification and reconstruction of original prey sizes from ingested remains." *Fishery Bulletin* 96 (3):575–588.

Scheffer, M., and S. R. Carpenter. 2003. "Catastrophic regime shifts in ecosystems: Linking theory to observation." *Trends in Ecology & Evolution* 18 (12):648–656.

Scheuring, I., and D. W. Yu. 2002. "How to assemble a beneficial microbiome in three easy steps." *Ecology Letters* 15 (11):1300–1307. doi: 10.1111/j.1461-0248.2012.01853.x.

Schindler, D. E., and L. A. Eby. 1997. "Stoichiometry of fishes and their prey: Implications for nutrient recycling." *Ecology* 78 (6):1816–1831. doi: 10.1890/0012-9658(1997)078[1816:sofatp]2.0.co;2.

Schindler, D. E., and M. D. Scheuerell. 2002. "Habitat coupling in lake ecosystems." *Oikos* 98 (2):177–189.

Schindler, D. E., J. F. Kitchell, X. He, S. R. Carpenter, J. R. Hodgson, and K. L. Cottingham. 1993. "Food web structure and phosphorus cycling in lakes." *Transactions of the American Fisheries Society* 122 (5):756–772.

Schlacher, T. A., and T. H. Wooldridge. 1996. "Origin and trophic importance of detritus— Evidence from stable isotopes in the benthos of a small, temperate estuary." *Oecologia* 106 (3):382–388.

Schmidt-Nielsen, K. 1997. *Animal Physiology: Adaptation and Environment*. Cambridge, MA: Cambridge University Press.

Schmitz, O. J., A. P. Beckerman, and K. M. Obrien. 1997. "Behaviorally mediated trophic cascades: Effects of predation risk on food web interactions." *Ecology* 78 (5):1388–1399.

Schmitz, O. J., J. H. Grabowski, B. L. Peckarsky, E. L. Preisser, G. C. Trussell, and J. R. Vonesh. 2008. "From individuals to ecosystem function: Toward an integration of evolutionary and ecosystem ecology." *Ecology* 89 (9):2436–2445. doi: 10.1890/07-1030.1.

Schneider, D. W., S. P. Madon, J. A. Stoeckel, and R. E. Sparks. 1998. "Seston quality controls zebra mussel (*Dreissena polymorpha*) energetics in turbid rivers." *Oecologia* 117 (3):331–341.

Schoeller, D. A. 1999. "Recent advances from application of doubly labeled water to measurement of human energy expenditure." *Journal of Nutrition* 129 (10):1765–1768.

Schoener, T. W. 1971. "Theory of feeding strategies." *Annual Review of Ecology and Systematics* 2:369–404.

Schoener, T. W. 1974. "Resource partitioning in ecological communities." *Science* 185:27–39.

Schrama, M., J. Jouta, M. P. Berg, and H. Olff. 2013. "Food web assembly at the landscape scale: Using stable isotopes to reveal changes in trophic structure during succession." *Ecosystems* 16 (4):627–638. doi: 10.1007/s10021-013-9636-5.

Searle, K. R., N. T. Hobbs, and S. T. Jaronski. 2011. "Asynchrony, fragmentation, and scale determine benefits of landscape heterogeneity to mobile herbivores." *Oecologia* 163 (3):815–824. doi: 10.1007/s00442-010-1610-8.

Sekercioglu, C. H., G. C. Daily, and P. R. Ehrlich. 2004. "Ecosystem consequences of bird declines." *Proceedings of the National Academy of Sciences of the United States of America* 101 (52):18042–18047. doi: 10.1073/pnas.0408049101.

Selch, T. M., and S. R. Chipps. 2007. "The cost of capturing prey: Measuring largemouth bass (*Micropterus salmoides*) foraging activity using glycolytic enzymes (lactate dehydrogenase)." *Canadian Journal of Fisheries and Aquatic Sciences* 64 (12):1761–1769. doi: 10.1139/f07-133.

Selva, N., B. Jedrzejewska, W. Jedrzejewski, and A. Wajrak. 2005. "Factors affecting carcass use by a guild of scavengers in European temperate woodland." *Canadian Journal of Zoology—Revue Canadienne De Zoologie* 83 (12):1590–1601. doi: 10.1139/z05-158.

Sepple, C. P., and N. W. Read. 1989. "Gastrointestinal correlates of the development of hunger in man." *Appetite* 13 (3):183–191. doi: 10.1016/0195-6663(89)90011-1.

Shapiro, J. V. Lamarra, and M. Lynch. 1975. Biomanipulation: An ecosystem approach to lake restoration. In *Water Quality Management through Biological Control*. P. L. Brezonik and J. L. Fox, editors. University of Florida.

Shellmanreeve, J. S. 1994. "Limited nutrients in a dampwood termite—Nest preference, competition and cooperative nest defense." *Journal of Animal Ecology* 63 (4):921–932. doi: 10.2307/5269.

Sherry, D. F. 2005. Brain and behavior. In *The Behavior of Animals: Mechanisms, Function, and Evolution*. J. Bolhuis and L.-A. Giraldeau, editors. New York: Wiley-Blackwell.

Sherry, D. F., and J. S. Hoshooley. 2009. "The seasonal hippocampus of food-storing birds." *Behavioural Processes* 80 (3):334–338. doi: 10.1016/j.beproc.2008.12.012.

Shiels, H. A., A. Di Maio, S. Thompson, and B. A. Block. 2011. "Warm fish with cold hearts: Thermal plasticity of excitation–contraction coupling in bluefin tuna." *Proceedings of the Royal Society B—Biological Sciences* 278 (1702):18–27. doi: 10.1098/rspb.2010.1274.

Showalter, A. M., M. J. Vanni, and M. J. González. 2016. "Ontogenetic diet shifts produce trade-offs in elemental imbalance in bluegill sunfish." *Freshwater Biology* Vol. 61 (5):800–813. doi: 10.1111/fwb.12751.

Sih, A., P. Crowley, M. McPeek, J. Petranka, and K. Strohmeier. 1985. "Predation, competition, and prey communities: A review of field experiments." *Annual Review of Ecology and Systematics* 16:269–311.

Simberloff, D. 2004. "Community ecology: Is it time to move on?" *American Naturalist* 163 (6):787–799. doi: 10.1086/420777.

Skelhorn, J., and C. Rowe. 2009. "Distastefulness as an antipredator defence strategy." *Animal Behaviour* 78 (3):761–766. doi: 10.1016/j.anbehav.2009.07.006.

Skelly, D. K. 1996. "Pond drying, predators, and the distribution of *Pseudacris* tadpoles." *Copeia* 1996 (3):599–605.

Skelly, D. K., and E. E. Werner. 1990. "Behavioral and life-historical responses of larval American toads to an odonate predator." *Ecology* 71 (6):2313–2322.

Skov, C., B. B. Chapman, H. Baktoft, J. Brodersen, C. Brönmark, L.-A. Hansson, K. Hulthén, and P. A. Nilsson. 2013. "Migration confers survival benefits against avian predators for partially migratory freshwater fish." *Biology Letters* 9 (2):20121178. http://dx.doi.org/10.1098/rsbl.2012.1178.

Smith, C. R., and A. R. Baco. 2003. "Ecology of whale falls at the deep-sea floor." *Oceanography and Marine Biology* 41:311–354.

Smith, C. R., F. C. De Leo, A. F. Bernardino, A. K. Sweetman, and P. Martinez Arbizu. 2008. "Abyssal food limitation, ecosystem structure and climate change." *Trends in Ecology & Evolution* 23 (9):518–528. doi: 10.1016/j.tree.2008.05.002.

Smith, E. P., and T. M. Zaret. 1982. "Bias in estimating niche overlap." *Ecology* 63 (5):1248–1253. doi: 10.2307/1938851.

Smith, G. R., J. E. Rettig, G. G. Mittelbach, J. L. Valiulis, and S. R. Schaack. 1999. "The effects of fish on assemblages of amphibians in ponds: A field experiment." *Freshwater Biology* 41 (4):829–837.

Smith, V. H., G. D. Tilman, and J. C. Nekola. 1999. "Eutrophication: Impacts of excess nutrient inputs on freshwater, marine, and terrestrial ecosystems." *Environmental Pollution* 100:179–196.

Sogard, S. M. 1997. "Size-selective mortality in the juvenile stage of teleost fishes: A review." *Bulletin of Marine Science* 60 (3):1129–1157.

Solomon, M. E. 1949. "The natural control of animal populations." *Journal of Animal Ecology* 18 (1):1–35.

Speakman, J. R. 2000. "The cost of living: Field metabolic rates of small mammals." In *Advances in Ecological Research, Vol 30*, A. H. Fitter and D. G. Raffaelli, editors, pp. 177–297. Cambridge, MA: Academic Press.

Stainier, D. Y. R. 2005. "No organ left behind: Tales of gut development and evolution." *Science* 307 (5717):1902–1904. doi: 10.1126/science.1108709.

Stankowich, T. 2012. "Armed and dangerous: Predicting the presence and function of defensive weaponry in mammals." *Adaptive Behavior* 20 (1):32–43.

Steele, J. H. 1998. "Regime shifts in marine ecosystems." *Ecological Applications* 8 (Suppl.):S33–S36.

Stein, R. A. 1977. "Selective predation, optimal foraging, and the predator-prey interaction between fish and crayfish." *Ecology* 58:1237–1253.

Stein, R. A., and J. J. Magnuson. 1976. "Behavioral response of crayfish to a fish predator." *Ecology* 57 (4):751–761.

Stein, R. A., D. R. DeVries, and J. M. Dettmers. 1995. "Food web regulation by a planktivore: Exploring the generality of the trophic cascade hypothesis." *Canadian Journal of Fisheries and Aquatic Sciences* 52:2518–2526.

Steneck, R. S., S. D. Hacker, and M. N. Dethier. 1991. "Mechanisms of competitive dominance between crustose coralline algae—An herbivore-mediated competitive reversal." *Ecology* 72 (3):938–950.

Steneck, R. S., M. H. Graham, B. J. Bourque, D. Corbett, J. M. Erlandson, J. A. Estes, and M. J. Tegner. 2002. "Kelp forest ecosystems: Biodiversity, stability, resilience and future." *Environmental Conservation* 29 (4):436–459. doi: 10.1017/s037689 2902000322.

Stephens, D. W., and J. R. Krebs. 1986. *Risk-Sensitive Foraging*. Princeton, NJ: Princeton University Press.

Sterner, R. W., and J. J. Elser. 2002. *Ecological Stoichiometry: The Biology of Elements from Molecules to the Biosphere*. New Jersey: Princeton University Press.

Sterner, R. W., T. Andersen, J. J. Elser, D. O. Hessen, J. M. Hood, E. McCauley, and J. Urabe. 2008. "Scale-dependent carbon:nitrogen:phosphorus seston stoichiometry in marine and freshwaters." *Limnology and Oceanography* 53 (3):1169–1180. doi: 10.4319/lo .2008.53.3.1169.

Stevens, J. R., F. A. Cushman, and M. D. Hauser. 2005. "Evolving the psychological mechanisms for cooperation." *Annual Review of Ecology Evolution and Systematics* 36:499–518. doi: 10.1146/annurev.ecolsys.36.113004.083814.

Stevens, M. T., D. M. Waller, and R. L. Lindroth. 2007. "Resistance and tolerance in *Populus tremuloides*: Genetic variation, costs, and environmental dependency." *Evolutionary Ecology* 21 (6):829–847. doi: 10.1007/s10682-006-9154-4.

Stiner, M. C. 2002. "Carnivory, coevolution, and the geographic spread of the genus *Homo*." *Journal of Archaeological Research* 10 (1):1–63.

Stomp, M., J. Huisman, F. de Jongh, A. J. Veraart, D. Gerla, M. Rijkeboer, B. W. Ibelings, U. I. A. Wollenzien, and L. J. Stal. 2004. "Adaptive divergence in pigment composition promotes phytoplankton biodiversity." *Nature* 432 (7013):104–107. doi: 10.1038 /nature03044.

Stone, G. N., Rwjm van der Ham, and J. G. Brewer. 2008. "Fossil oak galls preserve ancient multitrophic interactions." *Proceedings of the Royal Society B-Biological Sciences* 275 (1648):2213–2219. doi: 10.1098/rspb.2008.0494.

Strauss, R. E. 1979. "Reliability estimates for Ivlev electivity index, the forage ratio, and a proposed linear index of food selection." *Transactions of the American Fisheries Society* 108 (4):344–352. doi: 10.1577/1548-8659(1979)108<344:refiei>2.0.co;2.

Strong, D. R. 1992. "Are trophic casades all wet? Differentiation and donor-control in speciose ecosystems." *Ecology* 73:747–754.

Suring, E., and S. R. Wing. 2009. "Isotopic turnover rate and fractionation in multiple tissues of red rock lobster (*Jasus edwardsii*) and blue cod (*Parapercis colias*): Consequences for ecological studies." *Journal of Experimental Marine Biology and Ecology* 370 (1–2):56–63. doi: 10.1016/j.jembe.2008.11.014.

Sutherland, W. J. 1983. "Aggregation and the 'ideal free' distribution." *Journal of Animal Ecology* 52:821–828.

Swanson, H. K., M. Lysy, M. Power, A. D. Stasko, J. D. Johnson, and J. D. Reist. 2015. "A new probabilistic method for quantifying n-dimensional ecological niches and niche overlap." *Ecology* 96 (2):318–324. doi: 10.1890/14-0235.1.

Symmonds, M., J. J. Emmanuel, M. E. Drew, R. L. Batterham, and R. J. Dolan. 2011. "Metabolic state alters economic decision making under risk in humans." *Plos One* 5 (6):e11090. doi: e1109010.1371/journal.pone.0011090.

Tamburri, M. N., and J. P. Barry. 1999. "Adaptations for scavenging by three diverse bathyla species, *Eptatretus stouti*, *Neptunea amianta* and *Orchomene obtusus*." *Deep-Sea Research Part I—Oceanographic Research Papers* 46 (12):2079–2093. doi: 10.1016 /s0967-0637(99)00044-8.

Tansley, A. G. 1935. "The use and abuse of vegetational terms and concepts." *Ecology* 16 (3): 284–307.

Terborgh, J., and J. A. Estes. 2010. *Trophic Cascades: Predators, Prey, and the Changing Dynamics of Nature*. Washington: Island Press.

Thiebaux, M. L., and L. M. Dickie. 1993. "Structure of the body-size spectrum of the biomass in aquatic ecosystems—A consequence of allometry in predator-prey interactions." *Canadian Journal of Fisheries and Aquatic Sciences* 50 (6):1308–1317. doi: 10.1139/f93-148.

Thomson, J. D., G. Weiblen, B. A. Thomson, S. Alfaro, and P. Legendre. 1996. "Untangling multiple factors in spatial distributions: Lilies, gophers, and rocks." *Ecology* 77 (6):1698–1715. doi: 10.2307/2265776.

Tiede, J., B. Wemheuer, M. Traugott, R. Daniel, T. Tscharntke, A. Ebeling, and C. Scherber. 2016. "Trophic and non-trophic interactions in a biodiversity experiment assessed by next-generation sequencing." *Plos One* 11 (2). doi: 10.1371/journal.pone.0148781.

Tilman, D., S. S. Kilham, and P. Kilham. 1982. "Phytoplankton community ecology: The role of limiting nutrients." *Annual Review of Ecology and Systematics* 13:349–372.

Tome, M. W. 1988. "Optimal foraging—Food patch depletion by ruddy ducks." *Oecologia* 76 (1):27–36.

Tragust, S., B. Mitteregger, V. Barone, M. Konrad, L. V. Ugelvig, and S. Cremer. 2013. "Ants disinfect fungus-exposed brood by oral uptake and spread of their poison." *Current Biology* 23 (1):76–82. doi: 10.1016/j.cub.2012.11.034.

Tregenza, T. 1994. "Common misconceptions in applying the ideal free distribution." *Animal Behaviour* 47 (2):485–487.

Tregenza, T., J. J. Shaw, and D. J. Thompson. 1996a. "An experimental investigation of a new ideal free distribution model." *Evolutionary Ecology* 10 (1):45–49.

Tregenza, T., D. J. Thompson, and G. A. Parker. 1996b. "Interference and the ideal free distribution: Oviposition in a parasitoid wasp." *Behavioral Ecology* 7 (4):387–394.

Trudgill, D. L. 1991. "Resistance to and tolerance of plant parasitic nematodes in plants." *Annual Review of Phytopathology* 29:167–192.

Tscharntke, T., A. M. Klein, A. Kruess, I. Steffan-Dewenter, and C. Thies. 2005. "Landscape perspectives on agricultural intensification and biodiversity—Ecosystem service management." *Ecology Letters* 8 (8):857–874. doi: 10.1111/j.1461-0248.2005.00782.x.

Turner, J. T. 2002. "Zooplankton fecal pellets, marine snow and sinking phytoplankton blooms." *Aquatic Microbial Ecology* 27 (1):57–102. doi: 10.3354/ame027057.

Ulanowicz, R. E. 2004. "Quantitative methods for ecological network analysis." *Computational Biology and Chemistry* 28 (5–6):321–339. doi: 10.1016/j.compbiolchem.2004.09.001.

Ulanowicz, R. E., R. D. Holt, and M. Barfield. 2014. "Limits on ecosystem trophic complexity: Insights from ecological network analysis." *Ecology Letters* 17 (2):127–136. doi: 10.1111/ele.12216.

Urabe, J., and Y. Watanabe. 1992. "Possibility of N-limitation or P-limitation for planktonic cladocerans—An experimental test." *Limnology and Oceanography* 37 (2):244–251.

van Alpen, J. J. M., C. Bernstein, and G. Driessen. 2003. "Information acquisition and time allocation in insect parasitoids." *Trends in Ecology & Evolution* 18 (2):81–87.

Van Dooren, T. J. M., H. A. van Goor, and M. van Putten. 2010. "Handedness and asymmetry in scale-eating cichlids: Antisymmetries of different strength." *Evolution* 64 (7):2159–2165. doi: 10.1111/j.1558-5646.2010.00977.x.

Van Dover, C. L., T. L. Harmer, Z. P. McKiness, and C. Meredith. 2002. "Biogeography and ecological setting of Indian Ocean hydrothermal vents (vol 294, pg 818, 2001)." *Science* 295 (5554):442–442.

Van Valen, L. 1977. "Red queen." *American Naturalist* 111 (980):809–810.

Van Valkenburgh, B. 2007. "Deja vu: The evolution of feeding morphologies in the Carnivora." *Integrative and Comparative Biology* 47 (1):147–163. doi: 10.1093/icb/icm016.

Van Valkenburgh, B. 2009. "Costs of carnivory: Tooth fracture in Pleistocene and Recent carnivorans." *Biological Journal of the Linnean Society* 96 (1):68–81. doi: 10.1111/j.1095-8312.2008.01108.x.

Van Valkenburgh, B., and R. E. Molnar. 2002. "Dinosaurian and mammalian predators compared." *Paleobiology* 28 (4):527–543.

Vander Zanden, M. J., and J. B. Rasmussen. 2001. "Variation in delta N-15 and delta C-13 trophic fractionation: Implications for aquatic food web studies." *Limnology and Oceanography* 46 (8):2061–2066.

Vander Zanden, M. J., J. D. Olden, J. H. Thorne, and N. E. Mandrak. 2004. "Predicting occurrences and impacts of smallmouth bass introductions in north temperate lakes." *Ecological Applications* 14 (1):132–148.

Vandover, C. L., and B. Fry. 1994. "Microorganisms as food resources at deep-sea hydrothermal vents." *Limnology and Oceanography* 39 (1):51–57.

Vanni, M. J. 2002. "Nutrient cycling by animals in freshwater ecosystems." *Annual Review of Ecology and Systematics* 33:341–370.

Vanni, M. J., and C. D. Layne. 1997. "Nutrient recycling and herbivory as mechanisms in the "top-down" effect of fish on algae in lakes." *Ecology* 78 (1):21–40.

Vanni, M. J., C. D. Layne, and S. E. Arnott. 1997. "'Top-down' trophic interactions in lakes: Effects of fish on nutrient dynamics." *Ecology* 78:1–20.

Vanni, M. J., K. K. Arend, M. T. Bremigan, D. B. Bunnell, J. E. Garvey, M. J. Gonzalez, W. H. Renwick, P. A. Soranno, and R. A. Stein. 2005. "Linking landscapes and food webs: Effects of omnivorous fish and watersheds on reservoir ecosystems." *Bioscience* 55 (2):155–167.

Vannote, R. L., G. W. Minshall, K. W. Cummins, J. R. Sedell, and C. E. Cushing. 1980. "The river continuum concept." *Canadian Journal of Fisheries and Aquatic Sciences* 37:130–137.

Vatland, S., P. Budy, and G. P. Thiede. 2008. "A bioenergetics approach to modeling striped bass and threadfin shad predator-prey dynamics in Lake Powell, Utah–Arizona." *Transactions of the American Fisheries Society* 137 (1):262–277. doi: 10.1577/t05-146.1.

Vitousek, P. M. 1990. "Biological invasions and ecosystem processes: Towards an integration of population biology and ecosystem studies." *Oikos* 57.

Vitousek, P. M., H. A. Mooney, J. Lubchenco, and J. M. Mellilo. 1997a. "Human domination of earth's ecosystems." *Science* 277.

Vitousek, P. M., J. D. Aber, R. W. Howarth, G. E. Likens, P. A. Matson, D. W. Schindler, W. H. Schlesinger, and G. D. Tilman. 1997b. "Human alteration of the global nitrogen cycle: Sources and consequences." *Ecological Applications* 7:737–750.

Volman, S. F., T. C. Grubb, and K. C. Schuett. 1997. "Relative hippocampal volume in relation to food-storing behavior in four species of woodpeckers." *Brain Behavior and Evolution* 49 (2):110–120. doi: 10.1159/000112985.

Vos, J. G., E. Dybing, H. A. Greim, O. Ladefoged, C. Lambre, J. V. Tarazona, I. Brandt, and A. D. Vethaak. 2000. "Health effects of endocrine-disrupting chemicals on wildlife, with special reference to the European situation." *Critical Reviews in Toxicology* 30 (1):71–133.

Wahle, K. W. J., S. D. Heys, and D. Rotondo. 2004. "Conjugated linoleic acids: Are they beneficial or detrimental to health?" *Progress in Lipid Research* 43 (6):553–587. doi: 10.1016/j.plipres.2004.08.002.

Wahlstrom, E., L. Persson, S. Diehl, and P. Bystrom. 2000. "Size-dependent foraging efficiency, cannibalism and zooplankton community structure." *Oecologia* 123 (1):138–148.

Wajnberg, E., P. Coquillard, L. E. M. Vet, and T. Hoffmeister. 2012. "Optimal resource allocation to survival and reproduction in parasitic wasps foraging in fragmented habitats." *Plos One* 7 (6):9. doi: 10.1371/journal.pone.0038227.

Walker, J. C. G. 1985. "Carbon-dioxide on the early earth." *Origins of Life and Evolution of the Biosphere* 16 (2):117–127. doi: 10.1007/bf01809466.

Wallace, J. B., and J. R. Webster. 1996. "The role of macroinvertebrates in stream ecosystem function." *Annual Review of Entomology* 41:115–139.

Wallace, J. B., S. L. Eggert, J. L. Meyer, and J. R. Webster. 1997. "Multiple trophic levels of a forest stream linked to terrestrial litter inputs." *Science* 277:102–104.

Wallace, R. K. 1981. "An assessment of diet-overlap indexes." *Transactions of the American Fisheries Society* 110 (1):72–76. doi: 10.1577/1548-8659(1981)110<72:aaodi>2.0 .co;2.

Walther, D. A., M. R. Whiles, M. B. Flinn, and D. W. Butler. 2006. "Assemblage-level estimation of nontanypodine chironomid growth and production in a southern Illinois stream." *Journal of the North American Benthological Society* 25:444–452.

Wang, I. J. 2012. "Inversely related aposematic traits: Reduced conspicuousness evolves with increased toxicity in a polymorphic poison-dart frog." *Evolution* 65 (6):1637–1649.

Ward, D. 2010. "The effects of apical meristem damage on growth and defences of two Acacia species in the Negev Desert." *Evolutionary Ecology Research* 12 (5):589–602.

Ward, E. J., B. X. Semmens, E. E. Holmes, and K. C. Balcomb. 2011. "Effects of multiple levels of social organization on survival and abundance." *Conservation Biology* 25 (2):350–355. doi: 10.1111/j.1523-1739.2010.01600.x.

Waters, T. F. 1977. "Secondary production in inland waters." *Advances in Ecological Research* 10:91–164.

Waters, T. F. 1988. "Fish production–benthos production relationships in trout streams." *Polish Archives of Hydrobiology* 35:545–561.

Waterson, E. J., and E. A. Canuel. 2008. "Sources of sedimentary organic matter in the Mississippi River and adjacent Gulf of Mexico as revealed by lipid biomarker and delta(13) C-TOC analyses." *Organic Geochemistry* 39 (4):422–439. doi: 10.1016/j .orggeochem.2008.01.011.

Webster, J. R., P. J. Mulholland, J. L. Tank, H. M. Valett, W. K. Dodds, B. J. Peterson, W. B. Bowden et al. 2003. "Factors affecting ammonium uptake in streams—An inter-biome perspective." *Freshwater Biology* 48 (8):1329–1352. doi: 10.1046/j.1365-2427.2003.01094.x.

Wellborn, G. A., D. K. Skelly, and E. E. Werner. 1996. "Mechanisms creating community structure across a freshwater habitat gradient." *Annual Review of Ecology and Systematics* 27:337–363.

Werner, E. E. 1986. "Amphibian metamorphosis—Growth-rate, predation risk, and the optimal size at transformation." *American Naturalist* 128 (3):319–341.

Werner, E. E., and J. F. Gilliam. 1984. "The ontogenetic niche and species interactions in size-structured populations." *Annual Review of Ecology and Systematics* 15:393–425.

Werner, E. E., and S. D. Peacor. 2003. "A review of trait-mediated indirect interactions in ecological communities." *Ecology* 84 (5):1083–1100. doi: 10.1890/0012-9658 (2003)084[1083:arotii]2.0.co;2.

Whiles, M. R., M. A. Callaham, C. K. Meyer, B. L. Brock, and R. E. Charlton. 2001. "Emergence of periodical cicadas (Magicicada cassini) from a Kansas riparian forest: Densities, biomass and nitrogen flux." *American Midland Naturalist* 145 (1):176–187.

White, K. S., J. W. Testa, and J. Berger. 2001. "Behavioral and ecologic effects of differential predation pressure on moose in Alaska." *Journal of Mammalogy* 82 (2):422–429.

Whitledge, G. W., P. G. Bajer, and R. Hayward. 2006. "Improvement of bioenergetics model predictions for fish undergoing compensatory growth." *Transactions of the American Fisheries Society* 135 (1):49–54. doi: 10.1577/t05-003.1.

Whittaker, R. H., and G. E. Likens. 1973. "Primary production–biosphere and man." *Human Ecology* 1 (4):357–369. doi: 10.1007/bf01536732.

Whittaker, R. J., K. J. Willis, and R. Field. 2001. "Scale and species richness: Towards a general, hierarchical theory of species diversity." *Journal of Biogeography* 28:453–470.

Whittingham, L. A., and R. J. Robertson. 1993. "Nestling hunger and parental care in red-winged blackbirds." *Auk* 110 (2):240–246.

Whittington, H. B., and D. E. G. Briggs. 1985. "The largest Cambrian animal, Anomalocaris, burgess shale, British Columbia." *Philosophical Transactions of the Royal Society of London Series B—Biological Sciences* 309 (1141):569.

WHO. 2004. Iodine status worldwide: WHO global database on iodine deficiency. Editors: B. de Benoist, M. Andersson, I. Egli, B. Takkouche, and H. Allen. Department of Nutrition for Health and Development, Geneva: World Health Organization.

Wikelski, M., V. Carrillo, and F. Trillmich. 1997. "Energy limits to body size in a grazing reptile, the Galapagos marine iguana." *Ecology* 78 (7):2204–2217.

Wikelski, M., D. Moskowitz, J. S. Adelman, J. Cochran, D. S. Wilcove, and M. L. May. 2006. "Simple rules guide dragonfly migration." *Biology Letters* 2 (3):325–329. doi: 10.1098 /rsbl.2006.0487.

Wilbur, H. M. 1988. Interactions between growing predators and growing prey. In *Size-Structured Populations: Ecology and Evolution*. B. Ebenman and L. Persson, editors. Berlin: Springer-Verlag.

Wilbur, H. M. 1997. "Experimental ecology of food webs: Complex systems in temporary ponds—The Robert H. MacArthur Award Lecture—Presented July 31, 1995 Snowbird, Utah." *Ecology* 78 (8):2279–2302.

Wilder, S. M., M. Norris, R. W. Lee, D. Raubenheimer, and S. J. Simpson. 2013. "Arthropod food webs become increasingly lipid-limited at higher trophic levels." *Ecology Letters* 16 (7):895–902. doi: 10.1111/ele.12116.

Wilken, S., J. Huisman, S. Naus-Wiezer, and E. Van Donk. 2013. "Mixotrophic organisms become more heterotrophic with rising temperature." *Ecology Letters* 16 (2):225–233. doi: 10.1111/ele.12033.

Williams, R. J., and N. D. Martinez. 2000. "Simple rules yield complex food webs." *Nature* 404:180–183.

Wilson, D. S., and E. O. Wilson. 2008. "Evolution for the good of the group." *American Scientist* 96 (5):380–389.

Wilson, E. E., and E. M. Wolkovich. 2011. "Scavenging: How carnivores and carrion structure communities." *Trends in Ecology & Evolution* 26 (3):129–35. doi: 10.1016/j.tree .2010.12.011.

Winberg, G. G. 1960. "Rate of metabolism and food requirements of fishes." *Translation Series Fisheries Research Board of Canada* 194:1–234.

Winemiller, K. O. 1990. "Spatial and temporal variation in tropical fish trophic networks." *Ecological Monographs* 60 (3):331–367. doi: 10.2307/1943061.

Winemiller, K. O., and K. A. Rose. 1992. "Patterns of life-history diversification in North American fishes: Implications for population regulation." *Canadian Journal of Fisheries and Aquatic Sciences* 49:2196–2218.

Winfree, R., B. J. Gross, and C. Kremen. 2011. "Valuing pollination services to agriculture." *Ecological Economics* 71:80–88. doi: 10.1016/j.ecolecon.2011.08.001.

Winkler, J. D., and J. Van Buskirk. 2012. "Influence of experimental venue on phenotype: Multiple traits reveal multiple answers." *Functional Ecology* 26 (2):513–521. doi: 10 .1111/j.1365-2435.2012.01965.x.

Wittmer, H. U., R. Serrouya, L. M. Elbroch, and A. J. Marshall. 2013. "Conservation strategies for species affected by apparent competition." *Conservation Biology: The Journal of the Society for Conservation Biology* 27 (2):254–60.

Wong, E. S. W., and K. Belov. 2012. "Venom evolution through gene duplications." *Gene* 496 (1):1–7.

Woodwell, G. M., R. H. Whittaker, W. A. Reiners, G. E. Likens, C. C. Delwiche, and D. B. Botkin. 1978. "Biota and world carbon budget." *Science* 199 (4325):141–146. doi: 10.1126/science.199.4325.141.

Wooldridge, S. A. 2010. "Is the coral-algae symbiosis really 'mutually beneficial' for the partners?" *BioEssays* 32:615–625.

Wootton, J. T., and M. Emmerson. 2005. "Measurement of interaction strength in nature." *Annual Review of Ecology, Evolution, and Systematics* 36:419–444.

Worthington, A. H. 1989. "Adaptations for avian frugivory—Assimilation efficiency and gut transit-time of *Manacus vitellinus* and *Pipra mentalis.*" *Oecologia* 80 (3):381–389. doi: 10.1007/bf00379040.

Wright, J. J. 2012. "Adaptive significance of venom glands in the tadpole madtom *Noturus gyrinus* (Siluriformes: Ictaluridae)." *Journal of Experimental Biology* 215 (11):1816–1823.

Yodzis, P. 1984. "How rare is omnivory?" *Ecology* 65 (1):321–323.

Yosef, R., and N. Yosef. 2010. "Cooperative hunting in brown-necked raven (*Corvus rufficollis*) on Egyptian Mastigure (*Uromastyx aegyptius*)." *Journal of Ethology* 28 (2):385–388. doi: 10.1007/s10164-009-0191-7.

Zalasiewicz, J., C. N. Waters, M. Williams, A. D. Barnosky, A. Cearreta, P. Crutzen, E. Ellis et al. 2015. "When did the Anthropocene begin? A mid-twentieth century boundary level is stratigraphically optimal." *Quaternary International* 383:196–203. doi: 10.1016/j.quaint.2014.11.045.

Zanette, L. Y., A. F. White, M. C. Allen, and M. Clinchy. 2011. "Perceived predation risk reduces the number of offspring songbirds produce per year." *Science* 334 (6061):1398–1401.

Zaret, T. M. 1980. *Predation and Freshwater Communities.* New Haven, CT: Yale University Press.

Zavaleta, E. S., R. J. Hobbs, and H. A. Mooney. 2001. "Viewing invasive species removal in a whole-ecosystem context." *Trends in Ecology & Evolution* 16 (8):454–459.

Zeigler, J. M., and G. W. Whitledge. 2010. "Otolith trace element and stable isotopic compositions differentiate fishes from the Middle Mississippi River, its tributaries, and floodplain lakes." *Hydrobiologia* 661 (1):289–302. doi: 10.1007/s10750-010-0538-7.

Zeigler, J. M., and G. W. Whitledge. 2010. "Assessment of otolith chemistry for identifying source environment of fishes in the lower Illinois River, Illinois." *Hydrobiologia* 638 (1):109–119. doi: 10.1007/s10750-009-0033-1.

Zeppelin, T. K., D. S. Johnson, C. E. Kuhn, S. J. Iverson, and R. R. Ream. 2015. "Stable isotope models predict foraging habitat of northern fur seals (*Callorhinus ursinus*) in Alaska." *Plos One* 10 (6). doi: 10.1371/journal.pone.0127615.

Zhang, S., D. Midthune, P. M. Guenther, S. M. Krebs-Smith, V. Kipnis, K. W. Dodd, D. W. Buckman, J. A. Tooze, L. Freedman, and R. J. Carroll. 2011. "A new multivariate measurement error model with zero-inflated dietary data, and its application to dietary assessment." *Annals of Applied Statistics* 5 (2B):1456–1487. doi: 10.1214/10-aoas446.

Zuk, M., and G. R. Kolluru. 1998. "Exploitation of sexual signals by predators and parasitoids." *Quarterly Review of Biology* 73 (4):415–438.

Index

Page numbers followed by f, t, and, b indicate figures, tables, and boxes, respectively.

Milton Keynes UK
Ingram Content Group UK Ltd.
UKHW020322111024
449327UK00040B/1847